Automated Stratigraphic Correlation

FURTHER TITLES IN THIS SERIES

1. A.J. Boucot
   EVOLUTION AND EXTINCTION RATE CONTROLS

2. W.A. Berggren and J.A. van Couvering
   THE LATE NEOGENE — BIOSTRATIGRAPHY, GEOCHRONOLOGY AND PALEOCLIMATOLOGY OF THE LAST 15 MILLION YEARS IN MARINE AND CONTINENTAL SEQUENCES

3. L.J. Salop
   PRECAMBRIAN OF THE NORTHERN HEMISPHERE

4. J.L. Wray
   CALCAREOUS ALGAE

5. A. Hallam (Editor)
   PATTERNS OF EVOLUTION, AS ILLUSTRATED BY THE FOSSIL RECORD

6. F.M. Swain (Editor)
   STRATIGRAPHIC MICROPALEONTOLOGY OF ATLANTIC BASIN AND BORDERLANDS

7. W.C. Mahaney (Editor)
   QUATERNARY DATING METHODS

8. D. Janóssy
   PLEISTOCENE VERTEBRATE FAUNAS OF HUNGARY

9. Ch. Pomerol and I. Premoli-Silva (Editors)
   TERMINAL EOCENE EVENTS

10. J.C. Briggs
    BIOGEOGRAPHY AND PLATE TECTONICS

11. T. Hanai, N. Ikeya and K. Ishizaki (Editors)
    EVOLUTIONARY BIOLOGY OF OSTRACODA. ITS FUNDAMENTALS AND APPLICATIONS

12. V.A. Zubakov and I.I. Borzenkova
    GLOBAL PALAEOCLIMATE OF THE LATE CENOZOIC

Developments in Palaeontology and Stratigraphy, 13

Automated Stratigraphic Correlation

F.P. Agterberg

Mathematical Applications in Geology Section, Geological Survey of Canada, 601 Booth Street, Ottawa, Ont., K1A 0E8, Canada

ELSEVIER
Amsterdam — New York — Oxford — Tokyo 1990

ELSEVIER SCIENCE PUBLISHERS B.V.
Sara Burgerhartstraat 25
P.O. Box 211, 1000 AE Amsterdam, The Netherlands

Distributors for the United States and Canada:

ELSEVIER SCIENCE PUBLISHING COMPANY INC.
655, Avenue of the Americas
New York, NY 10010, U.S.A.

ISBN 0-444-88253-7

This book is printed on acid-free paper.

Printed in The Netherlands

FOREWORD

Geological correlation of strata plays a key role in sedimentary basin analysis. Such correlation, particularly when scaled in linear time, requires that a series of unique points for non-recurrent events like occurrences of fossils must first be determined, common to the sedimentary record as observed at different sites. An important contention of geological correlation is that once such events, probably grouped in biozones, have been properly determined and defined, these units can indeed be used for correlation. This statement, which might seem to be trivial, is made here because existing stratigraphic codes show how to construct stratigraphic units but they do not define how to correlate them. The actual correlation generally takes place in the subjective domain of regional experts on a particular basin or time period. Procedures for correlation or stratigraphic equivalence depend on subjective evaluation of the unique relation of each individual site record to the derived and accepted standard. It follows that correlation as practiced in geology cannot be readily verified without a detailed, and probably exhaustive review of all the underlying facts. Traditionally there is no method of formulating the uncertainty in fixation of individual records to the standard. Hence biostratigraphy often is more considered an art rather than a science. The problem of using subjective judgement only is not so much that it leads to right or wrong stratigraphy, but that a single solution is proposed. It should be attempted to establish reasonable criteria for successful correlation by providing insight into the actual uncertainty in correlation, either in millions of years or in depth in meters.

This book is an important review on 25 years of progress in computer-based stratigraphic correlation of fossil data. The best methods should combine sound mathematical logic with sound stratigraphic reasoning, and allow the user to retain full control over input and results. The author of this study is at the forefront of research and development in quantitative stratigraphy, particularly with respect to methods that apply to fossil distributions as frequently found in exploration wells in frontier basins. The ten chapters systematically explore the foundations and objective applications of quantitative biostratigraphy. This will bring us a step closer to a more automated procedure of correlation, applicable in a wide range of sedimentary basin analyses.

F.M. Gradstein, Chairman,
Committee on Quantitative Stratigraphy,
Dartmouth, Nova Scotia, January 1990

PREFACE

The purpose of this book is to provide an introduction to recent developments in automated stratigraphic correlation using computer programs for ranking and scaling of stratigraphic events. It is intended for advanced geology students, research workers and teachers with a background in stratigraphy and an interest in using computer-based techniques for problem-solving. The mathematical background provided is sufficient to justify the methods that are used but the equations are relatively few and concentrated in specific sections (mainly in Chapters 3, 6 and 8) and may be skipped by readers who are not mathematically inclined. Occasionally, use is made of elementary statistical techniques (t-test, chi-squared test or analysis of variance) on which additional explanations can be found in one of the numerous excellent introductory textbooks on probability and statistics in existence.

After data inventory for a region or time period, the stratigrapher first proceeds to establish a regional zonation which later can be used for correlation. Age calibration is a requirement for constructing this zonation as well as for the process of stratigraphic correlation. The computer can play an integral rôle in these procedures. In this book, the emphasis is on worked-out examples of application of ranking, scaling and correlation of stratigraphic events using relatively small datasets, for illustration of the intermediate steps made within the computer between input and output. It should be clear to the reader that automated stratigraphic correlation is not a simple automatic process such as alphabetic sorting. The stratigrapher has to integrate vast amounts of information which cannot possibly be stored in large databanks. Every piece of evidence or link between different pieces of evidence or hypotheses has its own sources of uncertainty associated with it. Using a computer for problem-solving may violate uncertainties that cannot be quantified. Computer input, therefore, always should be evaluated critically by expert stratigraphers and paleontologists.

In total there are ten chapters. The purpose of the first two chapters is to introduce the probabilistic method for automated stratigraphic correlation and to discuss principles of quantitative stratigraphy. Applications of mathematical statistics and computer science not specifically dealing with ranking and scaling but of interest to stratigraphers and paleontologists are presented in Chapter 3. Coding and file management of stratigraphic information (Chapter 4) provides the input required for ranking and scaling of biostratigraphic events by means of the RASC method treated in the next two chapters. A number of topics including rank correlation, precision of the scaled optimum sequence, normality testing and the modified RASC method are presented separately (in Chapters 7 and 8) as extensions and refinements of the RASC method. The chapter on event-depth curves and multi-well comparison (Chapter 9) contains examples of regional applications with automated correlation between stratigraphic sections. Finally, in Chapter 10, much of the material on methods presented in earlier chapters is summarized in a general description of the micro-RASC system of computer programs for ranking, scaling and regional correlation of stratigraphic events.

I am indebted to many individuals and organizations for support. Foremost among these is Felix Gradstein of the Atlantic Geoscience Centre of the Geological Survey of Canada who started me thinking about automated biostratigraphic correlation in 1978. From 1979 to 1986, I had the privilege of being the Leader of Project 148 (Quantitative Stratigraphic Correlation Techniques) of the International Geological Correlation Programme co-sponsored by Unesco and the International Union of Geological Sciences. This project and later the Committee on Quantitative Stratigraphy of the International Commission on Stratigraphy provided the framework for regular discussions with most colleagues active in method development for quantitative stratigraphy. I have used suggestions of many of these colleagues, especially P.O. Baumgartner (Université de Lausanne, Switzerland), G.F. Bonham-Carter (Geological Survey of Canada, Ottawa), J.C. Brower (Syracuse University, Syracuse, New York, U.S.A.), J.M. Cubitt (Poroperm, Chester, U.K.), E. Davaud (Université de Genève, Switzerland), P.H. Doeven (Petro-Canada, Calgary, Canada), C.W. Drooger (University of Utrecht, the Netherlands), L. Edwards (U.S.G.S., Reston, Virginia, U.S.A.), C.M. Griffiths (University of Trondheim, Norway), J. Guex (Université de Lausanne, Switzerland), C.W. Harper, Jr. (University of Oklahoma, Norman, U.S.A.), W.W. Hay (University of Colorado, Boulder, Colorado, U.S.A.), I. Lerche (University of South Carolina, Columbia, S.C., U.S.A.), D.F. Merriam (Wichita State University, Wichita, Kansas, U.S.A.), M. Rubel (Academy of Sciences, Estonian SSR, Tallinn, U.S.S.R.), W. Schwarzacher (Queen's University, Belfast, U.K.), B. Stam (Shell Syria, Damascus), J.E. Van Hinte (Free University, Amsterdam, the Netherlands) and M. Williamson (Shell Canada, Calgary, Canada).

Thanks are due to these individuals for their critical remarks during development of the ranking and scaling techniques to be discussed. I am grateful for assistance by computer programmers at the Geological Survey of Canada especially to Ning Lew, Louis Nel and Jacqueline Oliver, and to Dan Byron, Marc D'Iorio, and Kazim Nazli as my students at the Ottawa-Carleton Geoscience Centre.

For this book I have made extensive use of material in publications authored or co-authored by me during the past 10 years. On eight occasions, I was one of the lecturers of the one-week Quantitative Stratigraphy Short Course given under the auspices of IGCP Project 148 and the Committee on Quantitative Stratigraphy in Canada (2×), Brazil, China, Holland, India, U.K. and U.S.A. Mostly attended by stratigraphers and quantitative geoscientists from oil companies, this course provided a stimulating environment for jointly exploring and testing ideas on how to use computers intelligently. Those familiar with the earlier work will find many extensions of the RASC method made during the past three years especially in the fields of coding the original stratigraphic information, comparison with other methods and statistical evaluation. For example, it was well known that ranges on average range charts constructed by means of RASC tend to be shorter than those resulting from most other methods. The new modified RASC method yields range charts with wider ranges connecting entries to exits for taxa in those stratigraphic sections where these taxa were observed at their lowest and highest positions relative to all other taxa considered.

The Geological Survey of Canada has allowed me to work on this book project which involved extensive support including drafting and photography. The project would not have been possible without the invaluable help in word-processing received from Janet Gilliland, Shirley Kostiew, Guylaine Leger and Diane Winsor. Martin Tanke of Elsevier has provided guidance and encouragement. Last but not least I thank my wife Codien for her help and understanding.

F.P. Agterberg,
Ottawa, January 1990

CONTENTS

CHAPTER 1

PROBABILISTIC METHOD FOR AUTOMATED STRATIGRAPHIC CORRELATION

1.1 Introduction

From 1976 to 1986 about 150 scientists in 25 countries collaborated under the auspices of the International Geological Correlation Programme in Project 148: Evaluation and Development of Quantitative Stratigraphic Correlation Techniques. More recently similar work has been performed within the context of the Committee for Quantitative Stratigraphy of the International Commission on Stratigraphy. Although individual paleontologists and stratigraphers had used quantitative methods before, the collaboration in IGCP-148 led to new mathematical methods of stratigraphic correlation, mainly in biostratigraphy but also in chronostratigraphy and lithostratigraphy. These methods are reviewed in this book with emphasis on those developed by the author and his colleagues in Canada.

Sequencing methods deal with the relative order of stratigraphic events such as the highest occurrences of fossil taxa as observed in many sections. Intervals between successive events in an ordered sequence can be estimated (scaling) and the results expressed in linear time if a subgroup of the stratigraphic events can be dated. Such methods have been used extensively, e.g. to construct biozonations for Jurassic and younger sediments along the NW Atlantic margin (Gradstein et al., 1985) and, recently, to develop a new deep water benthic foraminiferal zonation for the Cenozoic strata of the Central and Viking Grabens, North Sea (Gradstein et al., 1988; Agterberg and Gradstein, 1988). Several regional hiatuses of 2 to 5 million years (Ma) in duration, stand out and match changes in sea level. The same methods have been employed for automated isochron contouring with error bars in depth or time units in Cenozoic and Cretaceous basins, off eastern Canada. Such information may be used for automated basin history analysis.

Time-successive assemblages of fossils also can be established by using multivariate methods on co-occurrences of events or with Guex's (1987) method of Unitary Associations in conjunction with graph theory on the overlap of stratigraphic ranges. Other methods for stratigraphic correlation to be reviewed in this book include Shaw's (1964) composite standard method and various uses of cubic spline functions for smoothing and interpolation. Attractions of quantitative stratigraphy are the use of rigorous methodology which highlights many properties of the data, the ability to handle large and complex data bases in an objective manner, and statistical evaluation of the uncertainty in the results. Generally, little conceptual orientation is required in order to use these methods and thereby gain more information from a particular dataset.

1.2 IGCP Project 148

The IGCP Project "Evaluation and Development of Quantitative Stratigraphic Correlation Techniques" was initiated in 1976 for the purpose of developing computer-based mathematical theory and analysis of geological information which can be applied to obtain automated correlation techniques in stratigraphy. These techniques are especially important in analysis of hydrocarbons and coal bearing basins. The project was terminated in 1986 and final results were described in Agterberg and Gradstein (1988). The rapid growth of data in stratigraphy has led to an increased demand for quantification of the data for machine-handling and graphic display. Quantitative stratigraphy is useful in this because it helps to organize the data in novel ways. Specific problems can be solved by establishing regional standards of ordered stratigraphic events and performing correlations on the basis of these standards preferably with estimates of uncertainty. Comprehensive descriptions and computer programmes have been prepared for different techniques which were applied to the same datasets in order to evaluate their respective advantages and drawbacks. The purpose of these evaluations is to select those techniques which are relatively simple and easily understood, achieve maximum resolution also in comparison with traditional methods of stratigraphic correlation, and can be implemented on computers of different types including microcomputers.

Studies in the fields of biostratigraphy, lithostratigraphy (especially well logs) and sedimentology make successful use of the quantitative modelling approach. Statistical and other numerical techniques can be used for erection of biozonations, correlation of zones and events, classification and matching of lithofacies in well logs or sections, lithofacies pattern recognition, and modelling of geological processes relative to the numerical time scale. The IGCP-148 participants were conducting research mainly in the fields of biostratigraphy and lithostratigraphy. Special attention was given to the performance of computer-based quantitative techniques in comparison with the results obtained by conventional qualitative stratigraphic correlation methods.

During the first years of existence (1976 to 1981), the emphasis within IGCP-148 was on method development. The statistical problems encountered when attempting to describe quantitative methods of stratigraphic correlation in a cohesive manner are far more complex and difficult to solve than one might expect. Some of the studies made under the auspices of IGCP-148 would not have been possible without recent advances in the theory of mathematical statistics, especially graph theory for order relationships between stratigraphic events or co-occurrences of fossil species, and spline-curve fitting theory for age-depth relationships with error analysis.

Later the primary activity in IGCP-148 shifted from method development to application, for solving specific stratigraphic problems using large data bases for regions in North America, Europe and India. Deep Sea Drilling Project data sets in the Atlantic and Pacific Oceans were also analyzed. Except for subprojects on the Silurian in the Baltic region and the Cambrian in Texas, the participants have been working mostly on Cenozoic, Cretaceous and Jurassic stratigraphy.

Research on the following major problems was mostly completed:

. Creation and definition of a mathematical theory of stratigraphic relationships.

. Establishment of standards and codes for the biostratigraphic, lithological and environmental information attainable from well logs, cores, and surface sections.

. Development of a mathematical theory for stratigraphic correlation.

. Development of practical methods of biostratigraphic correlation concentrating on quantification of assemblage zones, sequencing methods, set theoretical approaches, morphometric chronoclines and multivariate methodology.

Development of practical methods of correlation concentrating on methods of spectral analysis (frequency domain), methods of stretching and zonation (time domain), methods of stratigraphic interpolation and multivariate statistical analysis.

Over 200 publications emanating from IGCP-148, including computer programs, have been listed in Geological Correlation and the IGCP Catalogues. This includes collections of papers in books and special issues of scientific journals (Cubitt, Editor, 1978; Gill and Merriam, Editors, 1979; Cubitt and Reyment, Editors, 1982; Agterberg, Editor, 1984; Gradstein et al., 1985; Agterberg and Rao, Editors, 1988; Oleynikov and Rubel, Editors, 1989). After 1986, the international co-operation achieved was continued under the auspices of the Committee on Quantitative stratigraphy of the International Commission on Stratigraphy which recently has provided an indexed list of 637 publications on quantitative biostratigraphy (Thomas et al., 1988). For other recent papers see Agterberg and Bonham-Carter (1990, Part III: Quantitative Stratigraphy).

A comprehensive review of quantitative biostratigraphy for the period 1830-1980 already had been published by Brower (1981). Tipper (1988) reviewed 400 articles in the general field of quantitative stratigraphic correlation providing an annotated bibliography. Both Brower (1981) and Tipper (1988) noted that the development of mathematical techniques has tended to outstrip their acceptance by practicing stratigraphers. It is true that sophisticated techniques not only require more mathematical background from the user but, if not used knowledgeably, could lead to unrealistic or erroneous results more readily than simple methods. On the other hand, techniques that are easy to understand may be too simplistic for application in the real world. The best methods should provide new insights by combining mathematical logic with sound stratigraphic reasoning and allowing the user to retain full control over input and output.

In the International Stratigraphic Guide of the Subcommission on Stratigraphic Classification of the International Commission on Stratigraphy (Hedberg, Editor, 1976) a clear distinction is made between

(1) Lithostratigraphy in which strata are organized into mappable units based on their lithologic character;

(2) Biostratigraphy with correlative units based on fossil content of strata; and

(3) Chronostratigraphy with superimposed units based on the relative age relations of the strata.

In this book, as in IGCP Project 148, emphasis is on biostratigraphy, a field in which relatively few quantitative methods were available 12 years ago.

In order to explore the relation between qualitative and numerical methods, this book is started with a review of principles and definitions in stratigraphy in this chapter and the next one, emphasizing the biosphere record.

1.3 Quantitative biostratigraphy

Numerical methods in biostratigraphy make use of the quantified fossil record in sedimentary rock sections for precise recording and correlation of extinct biological events in space and time. They can be grouped into six basic categories:

(1) Sampling and delineation of environments with fossils that occur in patches (instead of displaying random spatial distributions);

(2) Automated microfossil recognition;

(3) Analysis of evolutionary sequences;

(4) Measurement of the attributes of index fossils;

(5) Determination of the most likely (scaled) sequence of biostratigraphic events as recorded in different stratigraphic sections; and

(6) Analysis of assemblage zones and concurrent range zones.

Emphasis in this book is on subjects (1), (5) and (6). This includes the construction of range charts depicting periods of existence for different fossil taxa in comparison with one another.

There are few basic studies that shed light on the actual distribution of fossils in rocks from a statistical point of view. For a review and applications to modern benthic Foraminifera and Late Cretaceous molluscs, see Buzas et al. (1982). The geological factors affecting the chance of event detection generally remain unknown and cannot be modelled prior to extensive sampling and stratigraphic analysis itself. On the other hand, it is widely known from repeated observations that for many groups of organisms, the majority of taxa is found at relatively few sampling sites and with few specimens. Figure 1.1 shows the cumulative number of highest or lowest occurrences of taxa in well or outcrop sections in different areas of a large number of taxa of Mesozoic radiolarians, Cenozoic dinoflagellates, Cenozoic Foraminifera and Cretaceous nannofossils. The radiolarian and nannofossil data use lowest and highest occurrences; the dinoflagellates and foraminifers highest occurrences only.

The graphs of Figure 1.1 show that the number of lowest or highest occurrences of taxa found in at least 1, 2, 3, ..., n sites, decreases steadily. In other words, the majority of species (events) occur at few sites and few species (events) are ubiquitous. It is noted that the sections used for the examples vary in density and spacing and the shapes of the curves in Figure 1.1 are influenced by methods of sampling. In Figure 1.1, dinoflagellate events are most localized and nannofossils least. The use of first and last occurrences increases traceability of taxa as shown for the radiolarians and nannofossils. Obviously, quantitative stratigraphic methods may want to cull the data so as to avoid use of species for which the number of events is limited and enhances "noise". Thresholds in, for example, ranking and scaling (RASC) are set such that no use is made of events that occur in less than k_c sections; k_c is set by the user. Rare events of value for age calibration can be re-introduced later, during final analysis.

Several computer-based methods are available for determining the most likely sequence of biostratigraphic events recorded in different stratigraphic sections and for the construction of quantitative range charts. The resulting zonations can be of either the average or conservative types. In general, average zonations will underestimate the position of the highest occurrence of a range zone at a given place while they overestimate its base. On the other hand, the concept of an average is tied to that of a probability distribution. This allows bases and tops to be fitted with confidence limits (see later). Conservative zonations are produced by sequencing methods designed to give the stratigraphically

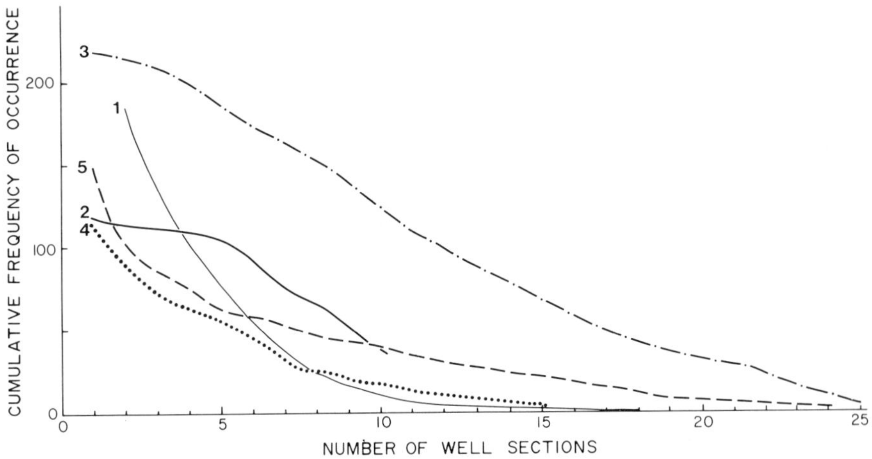

Fig. 1.1 Cumulative frequency distributions of stratigraphic first and last occurrences of microfossils in Mesozoic and Cenozoic strata: *1* = number of dinoflagellates occurring in 2, 3, ... wells; data for 249 last occurrences of Cenozoic dinoflagellates in 19 wells, northwestern Atlantic margin; *2* = data for 119 first and last occurrences of late Cretaceous nannofossils in 10 wells, northwestern Atlantic margin; *3* = data for 220 first and last occurrences of Mesozoic radiolarians at 76 sites, Mediterranean and Atlantic realms; *4* = data for 116 last occurrences of Mesozoic foraminifers in 16 wells, northwestern Atlantic margin; *5* = data for 147 last occurrences of Cenozoic foraminifers in 29 wells, central North Sea (from Agterberg and Gradstein, 1988).

highest possible estimate of the top of a range zone and the stratigraphically lowest estimate of the base of a range zone. Their drawback is that they are sensitive to anomalous situations arising when, locally, fossils were moved upwards or downwards in a stratigraphic section due to mixing of sediments later in geological time or because of contamination. When a fossil was poorly preserved, misidentification may also be a reason that its range of occurrence in a section is under- or overestimated. Assemblage zones, concurrent range zones and other types of zones are easily derived from dissecting the sequence of all events. Assemblage zones can also be determined by means of multivariate statistical methods such as cluster analysis. In the latter methods, the order of successive events in time is not used but zonations are obtained from co-occurrences of different species in the samples.

A new approach (Unitary Associations method; see later) developed during the past 12 years by J. Guex and E. Davaud in Switzerland uses graph theory to establish the order relationships of events formed by overlap of stratigraphic ranges. The final associations are mathematically successive assemblages of fossil ranges which are equivalent to the Oppel zones of traditional biostratigraphy (Guex, 1987). Baumgartner (1984) employed the Unitary Associations method to propose a comprehensive

Tethyan radiolarian zonation with 14 zones in 43 Middle Jurassic — Early Cretaceous sections. All zones are defined and identified in the sections. Several zones would not have been detected without the quantitative method employed for this study mainly because of patchiness of the fossil record.

Special properties of the paleontological record form the basis of biostratigraphy. These properties include first appearance datum (entry), range, peak occurrence, and last appearance datum (exit) of fossil taxa. Paleontological correlation for geological studies depends on comparing similar fossil occurrences in or between regions by means of a paleontological zonation. The observed order of paleontological events is generally different from place to place. In correlating wells drilled for oil, occurrences of the same event in different wells normally are connected by straight lines in stratigraphic profiles or fence diagrams. If there is a reversal in order for two events in two wells, these lines will cross. The cross-over frequency for pairs of events, therefore, provides a measure of inconsistency.

During the late 1950s and early 1960s, Shaw (1964) had developed a simple semi-objective method (Composite Standard method) of the conservative type for dealing with inconsistencies. First and last appearances of paleontological events in two sections are plotted against each other. Next a line is fitted by using the method of least squares and used for combining the two sections (line of correlation). The updated positions of first or last appearances are those that are respectively lower or higher in either of the two sections. A new section is plotted against the combination of the first few sections. The procedure of adding other sections is repeated until the "composite standard" is obtained that reflects the maximum ranges of taxa. Shaw's (1964) methodology was to a large extent based on original work by earlier quantitative paleontologists, notably Brinkmann (1929) who introduced basic concepts of statistical biostratigraphy.

Shaw's approach continues to be widely used. There is similarity between it and the methods advocated in this book. The RASC approach first gives a composite standard and lines of correlation are constructed later. Computer-based variants of Shaw's method include those developed by Edwards (1984; 1989) and Gradstein and Fearon (1990). Edwards' method is computer-based in that the stratigrapher combines sections and subjectively fits lines while displaying intermediate results on the screen of a computer terminal. The method of Gradstein and Fearon is microcomputer-based and employs De Boor's (1978) cubic splines for curve-fitting. In both methods intermediate results can be modified until a satisfactory composite standard is obtained at the end of a session.

So-called probabilistic methods which produce average ranges view biostratigraphic sequences as random deviations from a true solution. The solution faces four sources of uncertainty:

(1) The uncertainty due to the fact that the optimum, or "true", sequence of fossil events has not been established. Under the influence of Hay's (1972) paper, ranking of events in time to arrive at their stratigraphic order is often referred to as "Probabilistic Stratigraphy". Binomial theory was used to evaluate superpositional relations between events for statistical significance. However, as Agterberg and Nel (1982a,b) have pointed out, there are no simple models to rank stratigraphic events according to a numerical probability. The problem is that order in time should be based both on direct and on indirect estimates. For example, in Hay's binomial theory the fact that event A occurs above B in several sections ranks the same as that A in some sections occurs above events C, D, E, F and G, and that in some other sections C, D, E, F and G occur above B. Both situations lead to the conclusion that A occurs above B, although there is no simple way to express this in terms of numerical probability and more advanced mathematical methods for multiple comparison have to be used.

(2) The uncertainty due to the fact that the intervals between fossil events along a relative time scale are not known (spacing or scaling problem). In conventional biostratigraphy extensive use is made of distances in time between events or (non) overlap of ranges to produce assemblage zones. In the simple, graphical technique of the composite standard as developed by Shaw (1964), distance between two or more successive events is a function of the relative dispersion of each event in the sections considered; first occurrence levels are minimized and last occurrence levels are maximized, but no direct standard errors are available for the composite positions.

(3) The uncertainty due to the fact that the geographic distribution of an event is not known. Drooger (1974) refers to this as traceability. As pointed out earlier, few taxa are ubiquitous and most species are rare.

Consequently, recovery is strongly affected by the vagaries of lateral change in facies. Nevertheless, given enough sampling points and counts, interpolations may be used to predict the potential presence of each species.

(4) The error in the determination of biostratigraphic events at the scale of a well, or outcrop section. This is basically a sampling error which calls for an understanding and mathematical expression of errors in field and laboratory techniques.

In order to arrive at an optimum zonation and to attach confidence limits to correlations, considerable quantitative insight into these four sources of uncertainty is required.

For the purpose of coping with numerous inconsistencies in a database, containing many benthonic Foraminifera in wells along the Canadian Atlantic margin (see Section 4.7), a computer program for the ranking and scaling of events (RASC program) was developed by the author in collaboration with F.M. Gradstein and co-workers in Canada which produces three types of biostratigraphical answers:

(a) The optimum (or average) sequence of stratigraphic events along a relative time scale.

(b) The clustering in relative time, of these events, based on the cross-over frequencies of the events, weighted for the number of occurrences, using the optimum sequence of (a) as input. This results in a scaled optimum sequence with variable distance interval between each pair of successive events along the RASC scale.

(c) The stratigraphic and statistical normality (or comparison of order relationships) of the events in individual sections compared with the scaled optimum sequence.

In large-scale applications, the RASC computer program has produced range charts and assemblage zonations which superseded micropaleontological resolution previously available. For example, D'Iorio (1986) used this method for integration of large Cenozoic foraminiferal and dinoflagellate datasets from wells drilled on the Grand Banks and Labrador Shelf, northwestern Atlantic Margin. In comparison with optimum sequences for Foraminifera and dinoflagellates taken separately, an increase in stratigraphic resolution of the regional biozones and a minor reordering of successive events resulted from this process of integration (see Section 9.12). Although a dataset for a single fossil group is enlarged when microfossils from other groups are added, the gain in statistical precision because of larger sample sizes may be counteracted by the introduction of new sources of bias related to differences in environmental control and completeness of information, between the different fossil groups.

1.4 Quantitative chronostratigraphy

An approach in which biostratigraphy, paleoecology, lithostratigraphy, and geochronology are combined with one another is called burial history (cf. Stam et al., 1987) or geohistory analysis (Van Hinte, 1978; also see Lerche, 1990). It deals with subsidence and sedimentation in time. Data from wells or sections are organized linearly with the rates of subsidence, sedimentation and thermal maturation of organic matter, expressed in years, thousands of years, or larger time units. Special emphasis is placed on a method for decompaction of subsurface sedimentary units, using sonic logs or porosity data.

The prerequisite of this approach is a good calibration of fossil zonations with respect to the geochronologic scale. The determination of trends is the primary objective and individual errors in calibration are less important. This is because the trends can be generalized and used for extrapolation, whereas errors in calibration produce localized "noise" which should be eliminated if possible.

Information on rates of sedimentation, change in paleo-waterdepth, unconformities, and other factors can be integrated in time with sediment thickness data and paleo-waterdepth plots (cf. Doveton, 1986). Refinements include corrections for compaction and loading which provide information on seafloor or basement subsidence, evaporite movements, undercompaction phenomena and exact timing of important changes in geological history. The linear time perspective significantly clarifies geological history and therefore exploration geology. This is primarily so because it allows "dynamic" reconstruction of sedimentary basin history, e.g. the time of maturation and migration of hydrocarbons in a region may be postulated in linear time.

"Explorationists" also can establish a numeric chronostratigraphy for well sections and calculate estimates for the extent in time of the missing section at unconformities (cf. Van Hinte, 1978; Mohan, 1985). Consequently, a new kind of cross-section can be constructed that shows isochrons imaging chronostratigraphic depositional patterns just like the seismic record does. As their geochronologic resolution normally will be higher than that of seismic sections, isochron cross-sections are most useful in the calibration and the interpretation of the seismic record.

As a follow-up to the RASC (ranking and scaling) program, a computer-based method of quantitative correlation was proposed, which uses a numerical geologic time scale resulting from RASC. The computer program is called CASC (Correlation And SCaling in time). Both mainframe and microcomputer versions of CASC have been developed. The mainframe version (Agterberg et al., 1985) provides two types of displays. Initially, an event-depth curve is constructed for each stratigraphic section or well considered. Later the results for different sections are correlated.

Figure 1.2 shows a CASC multi-well comparison for five offshore wells on the Labrador Shelf. Briefly, the method runs as follows. A separate set of biostratigraphic events (exits of microfossils only) was observed in each well. By using the RASC computer program, a scaled optimum sequence was obtained for a group of 21 wells. The RASC distances of 54 events each occurring in 7 or more wells were transformed into ages in millions of years using a subgroup of 23 Cenozoic foraminiferal events for which literature-based ages were available. This allowed the construction of event-depth curves for individual wells. A probable age can be computed for any point along the depth-scale of a well, together with an error bar expressing the uncertainty of this estimate.

Three types of error bars are shown in Figure 1.2. A local error bar is estimated separately for each individual well. It is two standard deviations wide and has the probable isochron location at its center. Use is made of the assumption that the rate of sedimentation is linear in the vicinity of each isochron computed. Consideration of nonlinear sedimentation rates results in the asymmetrical modified local error bar of Figure 1.2B. Like the local error bar a global error bar (Fig. 1.2C) is symmetric but it is based on estimates of uncertainty in age which are computed from the uncertainty in distance of the 54 foraminiferal events in the scaled optimum sequence based on all (21) wells.

In a large-scale application, Williamson (1987) used the Ranking and Scaling method to erect eleven biozones for the Hibernia oil field region, Grand Banks, Canada (also see Chapter 9). Using the CASC method for a regional time-scale interpretation of the zonation and isochron correlation, Williamson proposed a subsurface correlation framework that to a considerable extent matches the results of subsurface seismic sequence analysis and provides chronostratigraphic correlation. He pointed out that these computer programs put many of the concepts and philosophies that have been used for many years by biostratigraphers on a statistical basis, and as such, prospective users of the techniques would require little

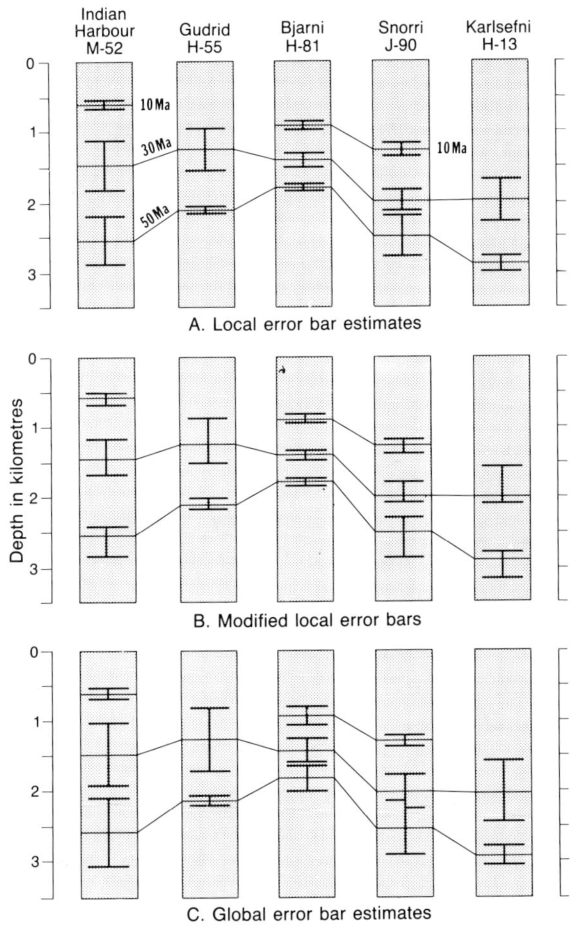

Fig. 1.2 Example of CASC multi-well comparison with three types of error bar. The probable positions of the time-lines were obtained from event-depth curves fitted to the biostratigraphic information of individual wells. For further explanation see text.

conceptual orientation in order to use these methods and thereby gain more information from a particular data set.

1.5 Quantitative lithostratigraphy

Lithostratigraphic correlation can be defined as the correct identification of lithological boundaries in different locations. When the correlated points are connected, they reproduce the shape of the rock body (lithosome). This type of correlation is not probabilistic and, in the stratigraphic sense, it is not even measurable. By establishing quantitative methods, a probability measure of whether a proposed correlation is right or wrong may be found. The similarity between two sections is a measurable quantity. If two portions in the sections are identical, this can be called a match and the number of matches is used as a measure of the similarity. An example of a simple matching technique for estimating the similarity between two successions of lithologies is to divide the number of matches by the total number of comparisons made. This technique called "cross-association" is explained in detail by Davis (1986, pp. 234-239). Elaborating on these concepts, Vrbik (1985) obtained statistical properties of the number of runs of matches between two random stratigraphic sections. Olea (1988) has developed an interactive computer system for lithostratigraphic correlation of wireline logs.

A fundamental prerequisite for such quantitative approach is the meaningful numerical coding of lithologies. In addition, most quantitative modelling studies require interpolation between equal intervals. This can be accomplished by linear interpolation between irregularly spaced points along sections or by using more sophisticated tools such as the cubic spline function. Smoothing factors in spline interpolation can be determined by interactively using a computer terminal, or by employing statistical methods such as cross-validation (see Section 9.5).

Because of differences in the rate of sedimentation, stretching or shrinking of sections is normally required before lithostratigraphic correlation is possible (cf. Mann and Dowell, 1978; Shaw, 1978; Kwon and Rudman, 1979; Kemp, 1982). An example of a new technique is the slotting method for pairwise comparison of sections (cf. Gordon, 1982). Suppose that two sections with observed lithological parameters, A_1, A_2, ..., A_n and B_1, B_2, ..., B_n are to be slotted. One series, e.g. A_1, A_2, B_1, A_3, B_2, A_4, A_5, ..., can be created in which the successive data points show a minimum of dissimilarity. This method works best with continuous lithological variables as obtained in well logging (Gordon and Reyment, 1979). Clark (1989) has developed a randomization test for comparison of ordered sequences obtained by slotting or other matching techniques.

In addition to differences in rate of sedimentation, hiatuses can present a problem in lithostratigraphic correlation. Smith and Waterman (1980) introduced a stratigraphic correlation algorithm designed to deal with the gap problem. This technique was originally used in studies of evolution of genetic sequences in molecular biology (Waterman et al., 1976). Their approach is also closely related to "time-warping" in speech recognition (Sankoff and Kruskal, Editors, 1983). An essential property of these methods is the ability to include gaps in correlations. A single stratigraphic unit can be made a gap (not matched) and several adjacent units can be treated as a single gap. The single-gap method was programmed by Howell (1983). In its most general form (Waterman and Raymond, 1987), one or several adjacent strata in a column can be matched with one or several strata in a second column and deletions within one of these multiple matches also are possible. The latter new algorithms include a method of minimum distance and a method of maximum similarity. Within this context, a similarity algorithm is given to locate and correlate the best matching segments or intervals from each lithostratigraphic column considered.

1.6 Recent developments in stratigraphy

Radiometric methods provide estimates of age in millions of years. However, any radiometric method is subject to a measurement error which is usually much greater than the uncertainties associated with the relative ordering of events using methods of stratigraphic correlation (e.g. biostratigraphic or magnetopolarity methods). Relatively imprecise isotope determinations can be combined to produce more precise estimates of the age of stage and chronozone boundaries (cf. Section 3.9).

Recently, the International Commission on Stratigraphy has published a global stratigraphic chart with geochronometric and magnetostratigraphic calibration (Cowie and Bassett, 1989) incorporating information of numerous subcommissions, working groups and committees. A considerable amount of uncertainty remains associated with some stage boundaries mainly because different radiometric methods

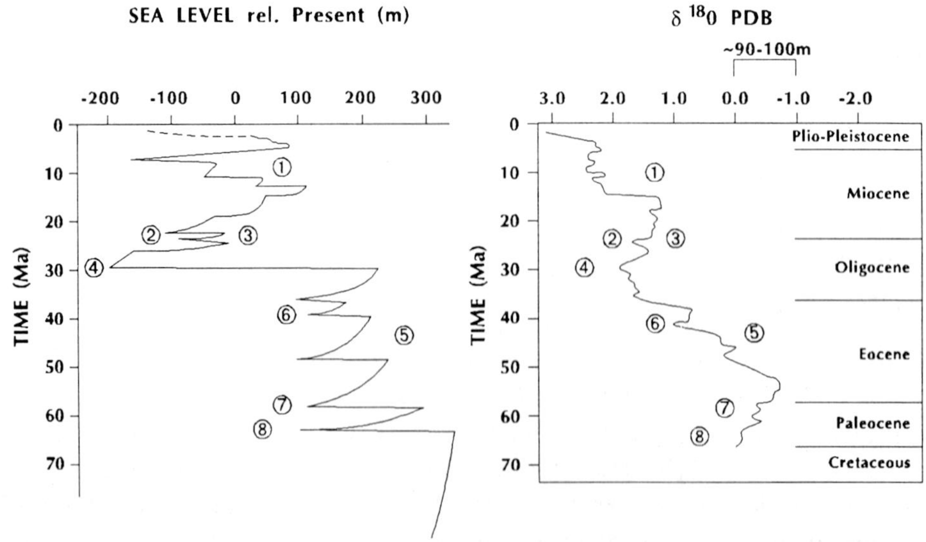

Fig. 1.3 Comparison of the magnitudes of sea level events of the Tertiary as inferred by Vail et al. (1977) from seismic stratigraphy, and the composite benthic $\delta^{18}O$ record according to Miller and Fairbanks (1985). The encircled numbers refer to particular rises and falls examined by Williams et al. (1988). Also see Table 1.1.

may yield results that are significantly different. For example, Odin (1982) estimated the age of the Jurassic-Cretaceous boundary at 130 ± 3 Ma but Harland et al. (1982) obtained 144 ± 5 Ma. These 95 percent confidence intervals do not overlap indicating unresolved problems of methodology. This subject will be discussed in more detail in Section 3.12.

Menning (1989) has provided a synopsis of 30 complete and partial geochronological time scales for the Phanerozoic published over a 70-year period to 1986. It is remarkable how close the most recent time scales are to the first scale of Barrell (1917). For example, Barrell's estimate of the Jurassic-Cretaceous boundary was 135 Ma which is identical to the age estimate for this boundary in the above-mentioned 1989 global stratigraphic chart. On the other hand, many geologists prefer the 144 Ma estimate of Harland et al. (1982) and Kent and Gradstein (1985) for the age of the Jurassic-Cretaceous boundary (cf. Section 3.12).

Seismic stratigraphy and isotope chronostratigraphy (Williams et al., 1988) are providing new tools for the stratigrapher. For example, Figure 1.3 is a comparison of the magnitude of particular sea level events of the Tertiary as inferred from seismic stratigraphy (Vail et al., 1977) and the composite benthic $\delta^{18}O$ record (Miller and Fairbanks, 1985). The two patterns exhibit a similar long-term trend. Table 1.1 (after Williams et al., 1988) compares magnitudes of 8 Tertiary sea level events (rises or falls) based on the two methods. These are 3rd order events. In almost all instances, the inferred sea-level change using sequence boundary patterns yielded larger estimated changes than the $\delta^{18}O$ signal. The overall agreement is not good at this level of detail but both these types of methodology are new and subject to continuous improvement. For a recent review of this topic and other approaches of chemical stratigraphy to time-scale resolution, see Williams (1990).

Quantitative dynamic stratigraphy (cf. Cross, Editor, 1990) is the application of mathematical procedures to the analysis of geodynamic, stratigraphic, sedimentogic and hydraulic attributes of sedimentary basins. These are viewed as features produced by the interactions of dynamic processes operating on physical configurations of the Earth at specific times and places. A typical model of this type may represent currents of water in sedimentary basins that alternately erode, transport and deposit sediments. These processes can be represented by means of differential equations that are solved repeatedly with numerical parameters which control their rate. Philosophies and strategies of model building in this field are discussed by Lerche (1990).

TABLE 1.1

Comparison of the magnitude of particular sea level rises and falls based on seismically defined unconformities with the $\delta^{18}O$ record (after Williams et al., 1988, Table II, p. 112).

Event	Type	Timing(Ma)	Agreement	Seismic(m)	$\delta^{18}O$(m)
1	fall	15.5-6.6	poor	~300	<50
2	fall	24	good	<50	<50
3	rise	30-15.5	poor	>300	<100
4	fall	30	poor	>400	<50
5	fall	52-37	poor	<100	~250
6	fall	40	good	~100	~100
7	fall	59	poor	<150	<50
8	fall	62.5	poor	~200	<50

CHAPTER 2

PRINCIPLES OF QUANTITATIVE STRATIGRAPHY

2.1 Introduction

The original meaning of stratigraphy is "description of layers" and like most earth science disciplines it is essentially a natural philosophy. This implies that stratigraphy is rooted in a body of organized, historically-accumulated observations, governed by a series of widely accepted principles and rules. The two physical principles of this philosophy are:

1) geological time is irreversible because it is directed along the arrow of time; and

2) sedimentary layers are laid down sequentially, one after another and become younger upwards if left undisturbed (law of Steno; cf. Nowlan, 1986).

Over the last 200 or more years the science of stratigraphy has developed into several major categories of effort and knowledge.

Lithostratigraphy is concerned with the classification, description and lateral tracing or matching of rock units, characterized mainly by their physical properties like sediment-type, degree of fossilization and alteration, texture, and color. Modern techniques for classification also make use of properties like seismic velocity (seismostratigraphy), or emission and propagation of a host of physical signals in boreholes (log analysis). The principal problem that besets classification and tracing or matching (whether automated or not) is that lithological characteristics are non-unique and repeat themselves in geological time. As a result, there is a fundamental difference between the quantitative treatment of single sections and quantitative approaches to lithostratigraphic tracing based on multiple comparison of sections. Since the principal unit of lithostratigraphy is the formation, which is a so-called mappable unit of distinctive lithology, it is more appropriate to use tracing as a proof of original continuity of strata, rather than correlation, which should be reconstructed from biostratigraphy or magnetostratigraphy. Correlation requires that a series of unique points for non-recurrent events must first be determined, common to the stratigraphic record as observed at different sites. An excellent introduction to this field of study is by Schwarzacher (1985a,b).

The properties of the paleontological or fossil record form the basis of biostratigraphy, which generally is called upon to determine the unique points of correlation, mentioned earlier. In the stratigraphic record the paleontologist recognizes fossil taxa and from the continuous change of taxa through time stratigraphic events are reconstructed. A taxon is defined as a stable unit consisting of all individuals (fossils) considered to be morphologically sufficiently alike to be given the same (Linnean) name. For stratigraphic purposes, a taxon (species, or unit of different rank) is recognized by a qualified paleontologist, whether based on single specimens or "populations". Commonly, categories intermediate between such taxa are not used.

Biostratigraphic events are defined by the presence of a taxon in its time context, as derived from its position in a rock sequence. For stratigraphic purposes relatively few events per taxon are considered only, such as the first occurrence (appearance, entry), the last occurrence (disappearance, exit), and possibly the most common or peak occurrence between an entry and an exit. These events are the result of the evolution of life on Earth. They differ from physical events in that they are unique, non-recurrent, and that their order is irreversible. As a result, the three-fold division of geological time into (1) prior to, (2) during, and (3) after the existence of a taxon, is not ambiguous and provides a basic tool for stratigraphic correlation. It is implied that each taxon was potentially present at all points in time between its entry and exit. Absences within its range are either environmental or preservational. This principle for constructing ranges also was discussed by Cheetham and Deboo (1963). Subsequent authors (cf. Brower, 1981; Tipper, 1988) referred to it as the "range-through" method.

2.2 Zones in biostratigraphy

The principal unit of "measurement" in biostratigraphy is the zone. A zone is a body of strata commonly characterized by the presence of certain fossil taxa. The most common types of zones are (after Hedberg, ed., 1976): (1) *assemblage zone* ----- a group of strata characterized by a distinctive

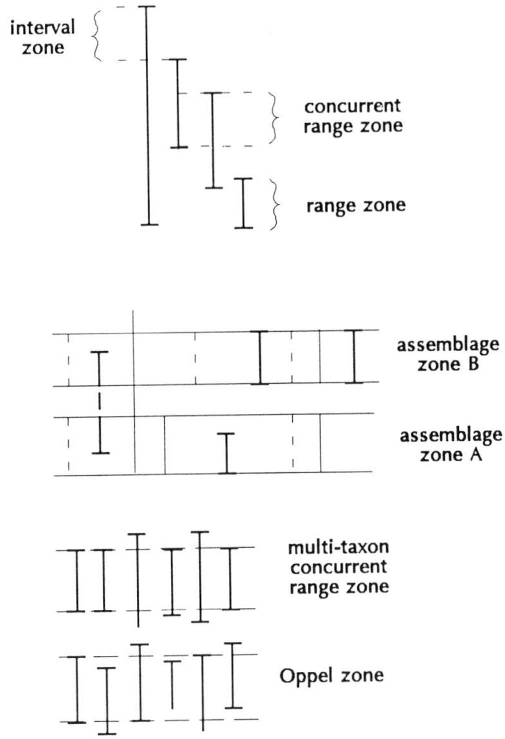

Fig. 2.1 Types of zones commonly used for biostratigraphic correlation (simplified from Hedberg, Editor, 1976). See text for further explanation.

assemblage of fossil taxa; (2) *range zone* ----- a group of strata corresponding to the stratigraphic range of a selected taxon in a fossil assemblage; (3) *concurrent range zone* ----- the overlapping part of the range zones of two or more selected taxa. The use of two or more taxa whose range zones overlap reinforces correlation; (4) *phylo-zone* ----- a body of strata containing a segment of a morphological-evolutionary lineage for a taxon, defined between the predecessor and the successor. The taxon is part of a lineage with morphologically well defined increments assumably in stratigraphic order; and (5) *interval zone* ----- the stratigraphic interval between two successive biostratigraphic events. In general, zones based on drill cutting samples are interval zones.

Several types of zones are schematically represented in Figure 2.1. Assemblage zones, multi-taxon concurrent range zones and Oppel zones are based on many taxa. The taxa in assemblage zones may have lived together or were accumulated together under similar conditions.

Assemblages may recur in a stratigraphic sequence and then can be useful as indicators of environments. They may represent a given geological age, although they are not controlled by the end points of ranges of taxa. In general, evolutionary changes have been sufficient to make assemblages of one age distinctive from those of another age. Multi-taxon concurrent range zones and Oppel zones both are based on the endpoints of ranges of taxa. According to Hedberg (Editor, 1976), the concept of the Oppel Zone largely embodies the concept of the concurrent-range zone but relaxes its strict interpretation sufficiently to allow supplementary use of biostratigraphic criteria other than range-concurrence that are believed to be useful for demonstrating time equivalence. Thus the Oppel zone is more subjective, more loosely defined and more easily applied than the concurrent range zone.

The techniques to be described in this book are automated so that large databases can be treated by computer-based statistical techniques using stratigraphic principles. In several of the automated techniques to be described, biozonations and correlations will be based on average end points of many local ranges. Figure 2.2 illustrates the concept of an average interval zone. Highest occurrences for two taxa (A and B) were determined in nine sections (1-9). In most (7 out of 9) sections, the taxon A exits above B. In two sections (numbered 3 and 9 in Fig. 2.2), B exits above A. A variety of methods can be used to estimate the average exit of taxon A which occurs above the average exit of taxon B. Together these average end points define an average interval zone.

Average interval zones can be combined with one another in order to construct regional biozonations. Suppose that the eight exits in the

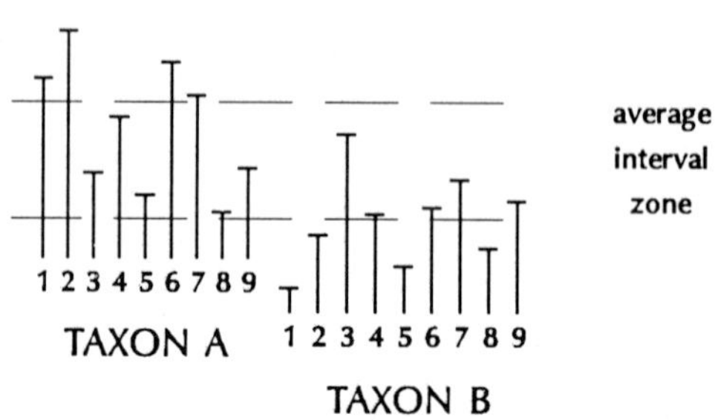

Fig. 2.2 RASC zonations are based on average stratigraphic events. The average interval zone between the exits of taxa A and B begins before the highest occurrence of B in section 3 and ends before the highest occurrence of A in section 2.

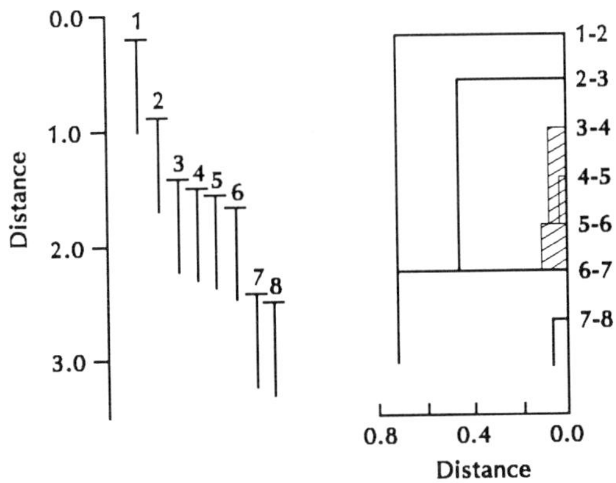

Fig. 2.3 Construction of dendrogram for scaled highest occurrences of eight taxa. Intervals between successive (average) exits are plotted along the distance scale of the dendrogram. Events which are close together along the distance scale on the left (such as exits 3 to 6) form clusters which can be shaded in the dendrogram. Clusters separated by longer distances can be useful as (RASC) zones in a regional biozonation. Because average exits are used, events belonging to the same cluster are characterized by more frequent cross-overs of tie-lines between sections.

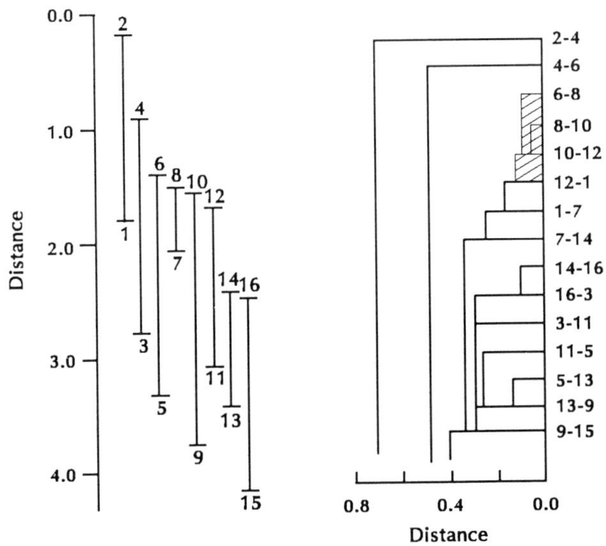

Fig. 2.4 Same as Fig. 2.3 using lowest and highest occurrences to construct the dendrogram.

example of Figure 2.3 are averages. The seven intervals between them were plotted along the distance scale to the right and a dendrogram was obtained by constructing perpendicular lines moving downward from the points that represent the average interval zones. Each perpendicular line ends when it meets the co-ordinate of an average interval zone. The resulting dendrogram shows clusters for average exits that are close together along the original distance scale. These clusters can be useful for biostratigraphic correlation. An example of this technique using lowest occurrences in addition to highest occurrences is shown in Figure 2.4.

Zonations emphasize the temporal and spatial restriction of morphologically distinct fossil taxa, arranged in zones. Good zonations have zonal units with well-defined upper and lower limits, are easily recognizable in many sections, correlate well and have been compared to other regional or extra-regional zonations.

Correlation is one of the most widespread, abstract undertakings of the mind and refers to causal linkage of present or past processes and events. Such events can be inorganic, organic or abstract. Geological correlation generally expresses the hypothesis that a mutual relation exists between stratigraphic units. In a more narrow sense it means that samples (or imaginary samples) from two separate rock sections occupy the same level in the known sequence of stratigraphic events. Without correlation, successions of strata or events in time derived in a specific area would not contribute to our understanding of earth history elsewhere (McLaren, 1978).

Suppose that the stratigraphic distribution of hundreds of taxa has been sampled in dozens of wells or outcrop sections. Following a detailed analysis, a range chart is proposed that synthesizes the information on all ranges to arrive at total (maximum) ranges for each taxa. The range chart is segmented, using co-existences of taxa and discrete taxon events, in order to establish time-successive intervals. Each interval is called a zone. When only last occurrences of fossils are known, such a chart portrays a succession of events or partial ranges.

The critical and least understood step in the practice of correlation is to actually tie the zones (back) to the individual sections. This may be a difficult undertaking when the individual stratigraphic record shows frequent inconsistencies due to sampling problems, reworking, unfilled ranges because of facies changes, and other factors.

Ideally, the individual fossil record as observed in each rock section should be compared to a regional standard prior to actual correlation. Insight should be gained in the likelihood that observed events occur where the standard (zonation) suggests that they should be found. In practice, the paleontologist will make a judgement on the outliers, or events to be rejected or moved up or down in a section. Next, the paleontologist will in each rock section define the successive zones in such a manner that a minimum number of (key) taxa for each of the zones fall outside the suggested zonal limits. Mismatch of the zones and the individual record is explained as noise or strictly local correlation character of the zones. Obviously, this is ideal terrain for a quantitative approach where more than one solution can be proposed depending on thresholds selected and where error bars may show uncertainty of correlation and zonal limits.

Partially under the influence of a paleomagnetic reversal scale, which promises virtually isochronous correlations for horizons in which a paleomagnetic event has been unambiguously determined, efforts have been made to establish detailed sequences of evolutionary fossil data. This effort has been particularly successful in the siliceous and calcareous marine plankton record of the last 150 m.y., as preserved in Deep Sea Drilling Program sites. In theory this allows for more or less reliable point correlation in time, but in practice, independent corroboration using the correlation of as many types of events as possible remains desirable.

In this vein, it is important to establish the separation by necessity of the reference framework of fossil taxa and rocks from abstract geological time. Biostratigraphy, the global or regional record of paleontological events or zones and their limits, used to correlate rock sequences, is the common link between lithostratigraphy and chronostratigraphy. Commonly it is assumed that correlation lines correspond to time lines, but this remains a hypothesis (Drooger, 1974). To equate biostratigraphy with chronostratigraphy and a priori substitute biozone for chronozone is misleading. Although biostratigraphically perfect correlation can be strongly diachronous, it may nevertheless be of value in sedimentary basin analysis. The assumption of contemporaneity has to be verified through other means, particularly by comparison to correlations using a particular zone elsewhere and through superposition of multiple correlative units.

Chronostratigraphy, which has led to the development of the commonly used scale of geological stages, is essentially relative. As a measure of relative age in geological history, reference is made to the standard chronostratigraphic scheme made up of successive stages like Cenomanian, Turonian, Coniacian in the Cretaceous system. The stage unit is a well-delimited body of rocks of an assigned and historically agreed upon relative age, younger than typical rocks of the next older stage, and older than typical rocks of the next younger stage.

The accurate portrayal of geological history demands that relative and subjective scales be modified into a numerical, linear scale. The conversion of a relative to a so-called absolute scale, measured in units of linear time like one million years is embodied in geochronology. Numerous well-identified stratigraphic samples with accurate radiometric age determinations are needed to calibrate the bio-magnetostratigraphic scales in linear time.

2.3 Quantitative versus qualitative stratigraphy

In stratigraphy, there has been a considerable amount of discussion regarding whether or not a probabilistic approach should be used. Harper (1981) has stressed the need for a quantitative and statistical approach for inferring succession of fossils in time. He has argued that most, if not all, stratigraphic paleontologists make subjective assessments of the probabilities of competing hypotheses regarding the ranges of taxa in time. According to Harper (1981, p. 445), these assessments can and should be backed up by quantitative methods and statistical tests. Others (e.g. Jeletzky, 1965) have pointed out that quantitative methods either explicitly or implicitly bring in new assumptions which could be too restrictive. The greatest drawback of some types of quantitative methods is that unequal things may be treated equally. Jeletzky (1985, p. 138) based zonal schemes on index fossils replacing or completely ignoring a great many other, facies-bound or long-ranging fossils often comprising the bulk of the faunas concerned. A naive statistical approach based on counts of all fossils would have led to inferior results.

It seems obvious that statistical methods are most useful in subfields of paleontology which are rich in sampling points and taxa, especially if use is made of standardized sampling methods and if valid conclusions should be drawn by the elimination of "noise" for decision-making (e.g. from micropaleontological information in oil exploration). The following quotations from Schindewolf (1950, p. 79-80) as translated by Jeletzky (1965, p. 139) for relation between quantitative "faunal" and qualitative "species zone" methods remain valid to-day as a summary for the relation between quantitative and qualitative methods:

"It would seem to me that there is no need to make a choice here, that is, the two methods are not usually exclusive but complementary. It is indeed not at all possible to draw a sharp boundary between them. In order to achieve a greater precision in chronology, we use sometimes (in the case of species zones), second or third series of species in addition to our principal evolutionary series of species. We compare, furthermore, the time ranges of individual species with one another and so succeed in recognition of a number of subzones. In such instances, one already considers a certain percentage of the total fauna. This naturally constitutes a transition to the faunal method. In practice, the latter method also does not ever utilize the sum total of forms available but only a selection therefrom. The long-ranging, chronologically useless representatives of a fauna, which usually form its percentage wise predominant element, are in this case quietly denied any consideration."

"A community of organisms is a complex thing, the components of which are characterized by very different behavior. Some of the individual forms (taxa) are extremely dependent on facies. They only bloom under quite definite, narrowly limited conditions of life. If these conditions are altered, they become extinct locally in some instances. In other instances, they emigrate and reappear sometimes, at least in the instances of long-ranging species in considerably younger horizons, the conditions of deposition of which have satisfied their specific bionomic requirements. Other organisms are less facies-dependent. However, their sensitivety varies so that the individual forms concerned (taxa), in turn, behave very differently whenever the conditions of life undergo changes. The changes of facies are therefore apt to result in faunal discordances and strong variations in the composition of the faunas concerned."

Amongst quantitative stratigraphers, there has been discussion about whether one should adopt a probabilistic or a non-probabilistic (axiomatic, wholly deductive, or deterministic) approach. Harper (1981, p. 442) has argued that a non-probabilistic approach may lead to relative age hypotheses which should not be proposed because they are neither falsifiable nor verifiable. As a starting point for discussion, Harper made the following three assumptions:

1. The principle of superposition applies at any given sample site. Owing to facies changes, the principle is best restricted, where possible, to individual sites where superpositional order can actually be seen in outcrop, or where it is obvious as in a borehole in a structurally simple area.

2. The range of a taxon at any given sample site has not been extended upward by reworking (Jones, 1958; Wilson, 1964) or downward by stratigraphic leaks (Jones, 1958; Foster, 1966). (In exploration

micropaleontology, one also has to avoid downward extension due to cave-ins in wells.)

3. If two taxa occur together in a given narrow sample horizon (bed), then their temporal ranges overlap in geological time (Edwards, 1978, p. 248).

Harper (1981, p. 443) remarked that assumptions 1 and 2 are essential to a non-probabilistic approach. Assumption 3 is expendable if co-occurrences by themselves are not used to infer overlap. According to Harper, there are 13 basic relative age hypotheses for any pair of taxa A and B (Fig. 2.5). Hypotheses numbered 10A-B and 11A-B which assess that the two taxa are sequential in time, may be falsified but not verified using the three assumptions (1-3). Hypotheses 1-9 taken individually can neither be verified nor falsified. No single one of them can be verified since any conceivable available data will be consistent with the other eight. Harper (1981) concluded that a non-probabilistic approach of this type is not fruitful. On the other hand, a probabilistic approach working

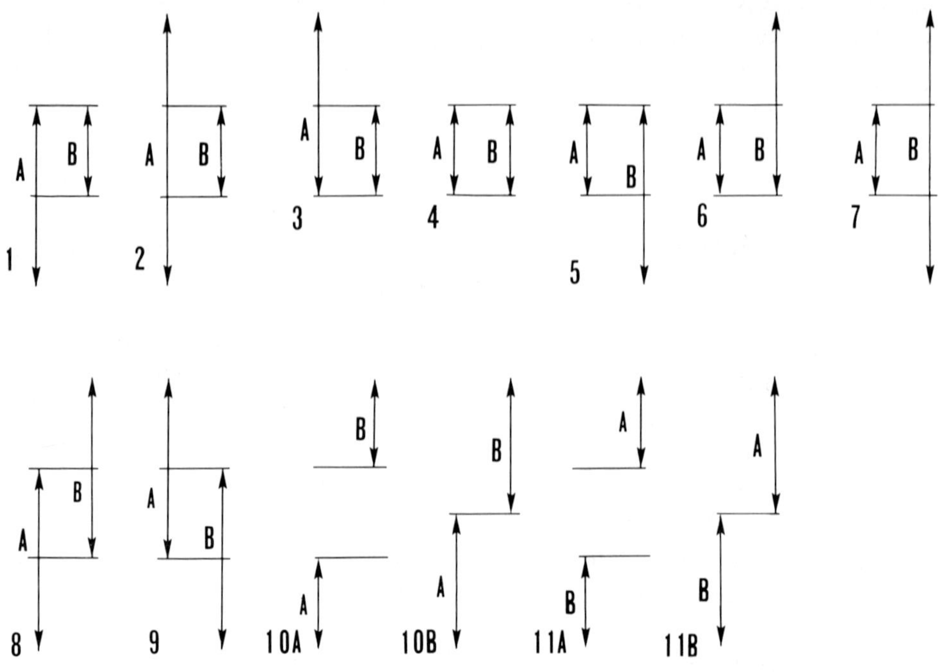

Fig. 2.5 Possible relative age hypotheses for two taxa A and B according to Harper (1981). Vertical line segments with arrows indicate ranges of taxa in time. Two hypotheses (10 and 11) are further divided on the basis of presence or absence of a time gap between ranges of the two taxa.

with preferred sequences rather than all individual sequences allows significance tests that are based on a comparison between "sample" means and hypothetical "population" means.

Fossils, taxa and events

From the previous discussions it is clear that in biostratigraphy relatively little use is made of possible variables such as frequency of individual fossils belonging to a specific taxon; e.g. measured per sample or per unit area of outcrop. To a large extent, the various types of biostratigraphic zones are defined on presences and absences of taxa rather than abundance data. The paleontologist looking for fossils in the field commonly attempts to recognize as many different taxa as possible. The ranges of these taxa are of special interest. The paleontologist usually tries to find the stratigraphically lowest as well as the highest occurrence of each taxon within a section (local range) or region. In general, it is more efficient to recognize among the hundreds or thousands of fossils the presence of one or more fossils belonging to a specific taxon, rather than to attempt to classify and count all individual fossils. It will be discussed in Chapter 3 that microfossil abundance data can be useful for correlation in biostratigraphy. However, very large samples and much effort may be required to obtain fossil abundance data which are relatively precise. It is more effective to establish the presence or absence of a taxon, because, in general, more information is provided by presence-absence data of many taxa than by precise abundance data for relatively few taxa.

Nevertheless, the presence of a taxon in a bed is determined by its abundance in this bed. This abundance reflects the chances that the taxon occurred at a given place, became fossilized, was found and correctly identified, which in themselves reflect hit-or-miss processes. It will be seen that when quantitative correlation of the presence-absence data for taxa in different stratigraphic sections is attempted, this effort is commonly hampered by existence of numerous inconsistencies which must be resolved before meaningful correlation is possible.

The quantitative analysis of abundance data can be useful in specific subfields of paleontology such as palynology. For example, Christopher (1978) successfully performed pairwise comparison of time series for quantitative palynologic correlation of Upper Cretaceous sections from the Atlantic coastal plain.

2.4 Local versus regional ranges of taxa

Each fossil taxon has a lowest and a highest occurrence in the local range for a continuous outcrop section or a single well, as well as in the regional composite range for a number of stratigraphic sections. A regionally-based range chart is more useful for stratigraphic correlation than the local ranges showing superpositional relations that often are mutually inconsistent. The positions of highest occurrences for a regional range chart commonly are underestimated, and those of lowest occurrences overestimated when distances to observed ends are measured from the base of each stratigraphic section upward and averaged between sections. This problem will be discussed at length in the next section. Suppose, however, that this type of bias can be neglected and that it has been possible to measure the local ranges for a number of taxa in a number of sections. Then combining sections with one another to construct a single range chart may give misleading results for a number of other reasons. The problem was illustrated by Davaud (1982) as follows.

Figure 2.6 is a theoretical example showing distribution in space and time of 7 different taxa and their true chronological succession. Obviously, the local ranges in the four sections A-D differ from the true regional succession of the biological events. Differential preservation of the taxa during fossilization may create further differences between local and regional ranges. So do the processes of sedimentation, compaction, and other processes. Figure 2.7 illustrates possible influence of differential sedimentation on the ranges for a single species. Disregarding other factors, a combination of the living range factor (Fig. 2.6) and the differential sedimentation factor (Fig. 2.7) resulted in the sedimentary record of Figure 2.8. Obviously, the local ranges of Figure 2.8 do not provide good estimates of the local ranges in Figure 2.6. Neither can a composite range chart based on Figure 2.8 provide an approximation to the chronological succession of "biological" events in Figure 2.6.

Fortunately, it generally is possible in practice to design experiments in order to check whether or not the factors illustrated in Figures 2.6 to 2.8 have significant effects. For example, differences in living range can be evaluated by performing separate data analyses on subsets of a regional

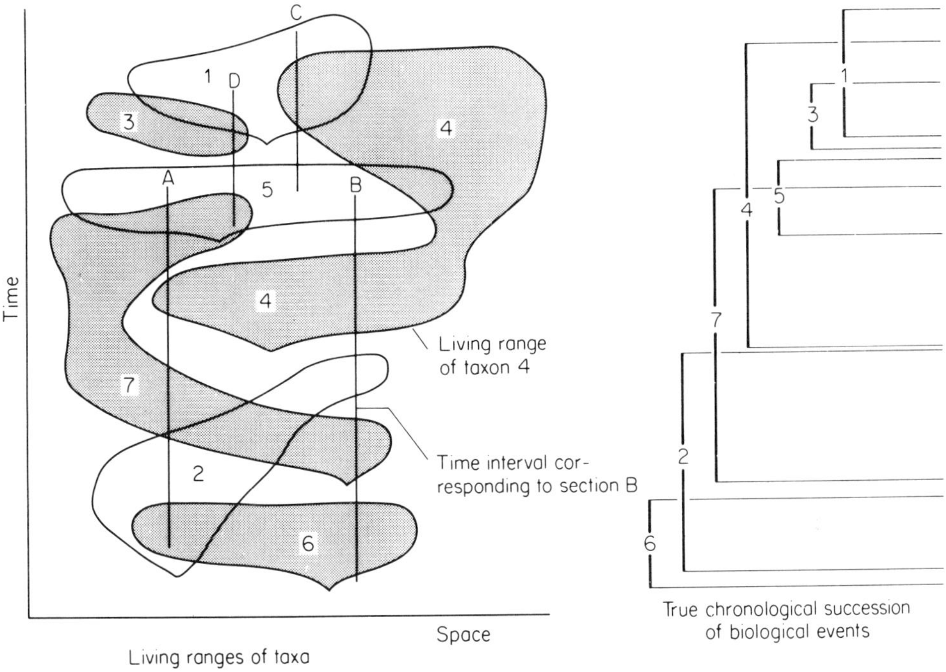

Fig. 2.6 Theoretical example of Davaud (1982) showing distribution in space and time of seven different taxa with true chronological succession.

database (cf. Section 4.7). These subsets which correspond to geographical subregions would yield different results if there were large shifts in the living ranges of the taxa. It also may be possible to evaluate this factor by means of multivariate analysis using the geographical locations of the stratigraphic sections as variables (cf. Section 2.4). The influence of differences in rates of sedimentation between stratigraphic sections can be evaluated if sufficient information is available to establish the sediment accumulation histories for individual sections using the numerical geological time scale (see Chapter 9).

2.5 Estimation of the highest and lowest occurrences of taxa

Figure 2.9 illustrates the relationship between fossil finds, ends of observed local range and "true" ends of the local range of a taxon. In recent years, several methods have been developed for estimating the "true" highest and lowest occurrences of a taxon (Jasko, 1984; Springer and Lilje, 1988; Strauss and Sadler, 1989). This type of estimation is only possible if simplifying assumptions are made, e.g. constant facies with

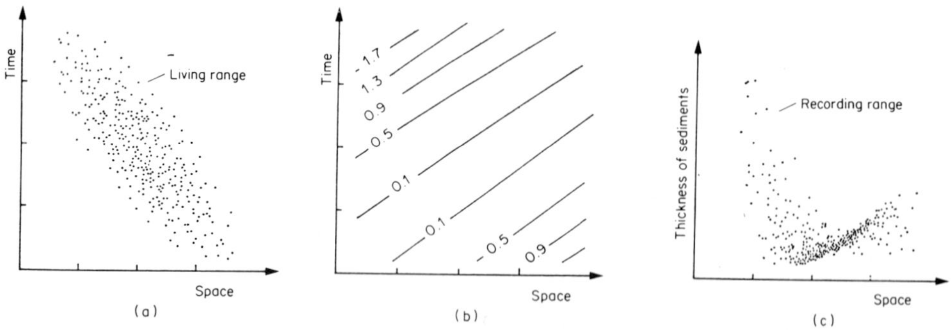

Fig. 2.7 Diagrams to illustrate how biological events are recorded in sediments (after Davaud, 1982). Diagram (a) shows time-space domain for a particular species. Population density is reflected by points density. Diagram (b) illustrates that during same period of time and in same geographic area, the sedimentation rate changed. When the sedimentation rate is applied to points of diagram (a) and integrated over time, the points are moved to new positions in the sedimentary record as shown in diagram (c). If the probability of detection is proportional to density of points in the sedimentary record, the end point of the chronological range of a species could be underestimated, especially if sedimentation rate was high at time of biological disappearance of the species.

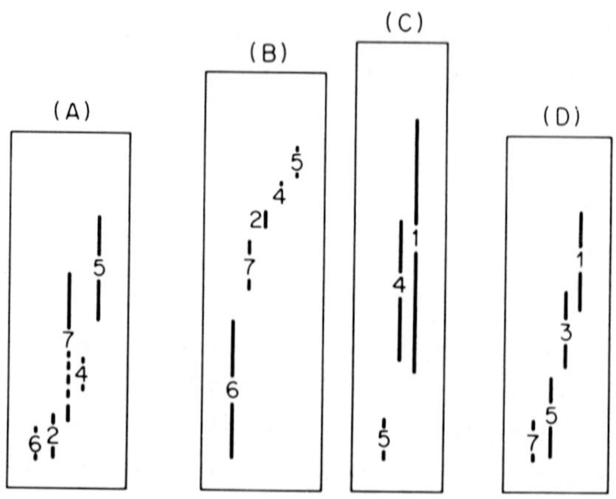

Fig. 2.8 Sedimentary record of biological events in four stratigraphic sections corresponding to the theoretical example of Fig. 2.6. Distortion due to differential role of sedimentation was similar to the one shown in Fig. 2.7 (b).

constant average rate of sedimentation. Figure 2.10 (from Strauss and Sadler, 1989) shows local ammonite ranges in late Cretaceous strata of Seymour Island, Antarctic Peninsula. The observed local ranges and finds are from Macellari (1986). The highest occurrences were obtained by

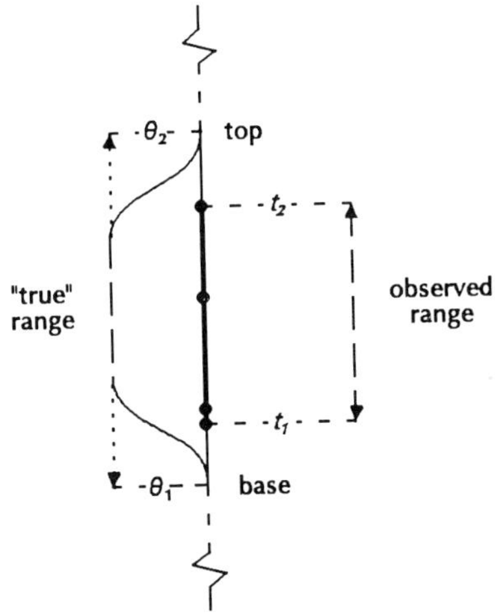

Fig. 2.9 Relationship between observed range extending from time t_1 to t_2 and "true" range extending from time θ_1 to θ_2. Strauss and Sadler (1989) assumed that the probability of finding a fossil is constant across its true range. If a species was less abundant at its time of appearance or disappearance, as illustrated by the density curve in the diagram, it becomes more difficult to estimate the true range even if facies and sedimentation remained constant.

Strauss and Sadler as unbiased point estimators and their upper range extension to 95 percent confidence interval. These authors used the Dirichlet distribution which results from a Poisson process for uniform sedimentation. It was assumed that each fossil existed for an unknown period of time. The chances of finding it remained equal during this period. The density curve for highest finds has a tail that extends in the stratigraphically downward direction under these conditions.

Jasko (1984) used a different model to estimate precision of the observed lowest occurrence of a taxon. He assumed that initially the population of a taxon increases its size exponentially as established e.g. for bacterial colonies in the laboratory. The average number of specimens per unit volume would follow a Poisson distribution. The combination of these two distributions leads to a new (compound Poisson) frequency distribution permitting estimation of the average range (r) and its standard deviation (d) for a given number of specimens (see Table 2.1). In practice, it may be possible to determine the local range from the observations (see Table 2.2) and to set it equal to the average range. The corresponding standard deviation then expresses the uncertainty in the position of the lowest occurrence. In the example of Table 2.2, the compound Poisson distribution provides a good fit from 2700 ft downward.

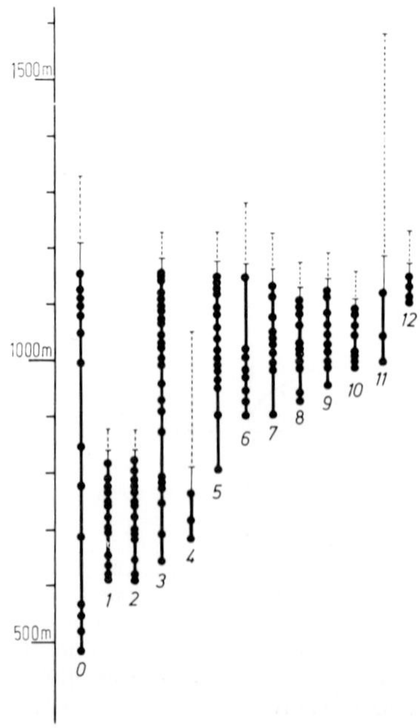

Fig. 2.10 Ammonite ranges in late Cretaceous strata of Seymour Island, Antarctic Peninsula. Observed local ranges (heavy vertical lines) and actual finds (solid circles) after Macellari (1986, Fig. 5). Extrapolated end-points of ranges according to Strauss and Sadler (1989, Fig. 1). Light vertical lines represent upper range extensions to unbiased point estimators. Dashed vertical lines are upper range extensions to 95 percent confidence intervals. Numbers assigned to taxa are as follows: *0* = *Diplomoceras lambi; 1* = *Maorites seymourianus; 2* = *Kitchinites darwini; 3* = *Grossouvrites gemmatus; 4* = *Maorites weddelliensis; 5* = *M. densicostatus* morphotype-alpha; *6* = *Kitchinites laurae; 7* = *Anagaudryceras seymouriense; 8* = *Maorites densicostatus* morphotype-gamma; *9* = *Pachydiscus riccardi; 10* = *Maorites densicostatus* morphotype-beta; *11* = *Pseudophyllites loryi; 12* = *Pachydiscus ultimus*.

This is indicated by the close correspondence between observed frequencies and expected frequencies based on the statistical model. In total, 25 microfossil forms were observed for the bottom 3 classes in Table 2.2. The ratio of standard deviation to range is 0.348 if $n=25$. Because the lowest occurrence was observed in a sample at 3446 ft., the local range is 3446-2700=746 ft. The standard deviation for the lowest occurrence is estimated to be $0.348\times746=260$ ft. If the position of the lowest occurrence would be normally distributed (i.e. satisfying the Gaussian curve model), there would be a 95% probability that the true lowest occurrence is below $3446+1.645\times260=3874$ ft.

TABLE 2.1

Averages (r), standard deviation (d) and their ratio ($V = d/r$) as functions of sample size (n) as obtained by means of computer simulation experiments (after Jasko, 1984).

n	r	d	V	n	r	d	V
1	.000	.985	-----	16	3.111	1.259	.405
2	.864	1.093	1.265	17	3.203	1.259	.393
3	1.355	1.128	.832	18	3.231	1.247	.386
4	1.663	1.163	.699	19	3.285	1.263	.385
5	1.910	1.191	.623	20	3.323	1.273	.383
6	2.112	1.188	.562	21	3.370	1.267	.376
7	2.263	1.199	.530	22	3.432	1.288	.375
8	2.412	1.209	.501	23	3.514	1.270	.361
9	2.541	1.206	.475	24	3.534	1.277	.361
10	2.638	1.227	.465	25	3.586	1.249	.348
11	2.737	1.247	.456	26	3.563	1.276	.358
12	2.817	1.237	.439	27	3.648	1.287	.353
13	2.893	1.250	.432	28	3.692	1.272	.345
14	2.971	1.250	.421	29	3.698	1.269	.345
15	3.052	1.254	.411	30	3.777	1.292	.342

Possible models for the shape of the frequency distribution for positions of highest and lowest occurrences will be discussed in the next section. It is noted here that Strauss and Sadler's model for highest occurrences implies that this distribution is not symmetrical. Theoretically, in their model, the last find has a distribution with a longer tail in the stratigraphically downward direction. Instead of this, the distribution of Strauss and Sadler's estimated end of the range has a long narrow tail that extends upwards, especially for fossils with relative few finds such as *Maorites weddelliensis* (4) and *Pseudophyllites loryi* (11) in Fig. 2.10. Jasko's model for lowest occurrences (Table 2.2) implies an asymmetrical frequency distribution with its long narrow tail extending downward. The estimated lowest occurrence is skewed in the same direction. Thus the 95% confidence limit of 3874 ft for the lowest occurrence estimated in the preceding paragraph is probably incorrect because it was based on the symmetric Gaussian distribution model. If Jasko's model is correct, the 95% confidence limit has a depth value greater than 3874 ft.

A third model for sampling bias resulting in artificial range truncation was developed by Signor and Lipps (1982). These authors deal with the phenomenon that taxa begin to disappear from the fossil record before mass extinctions actually take place. Figure 2.11 illustrates this idea. The line in Figure 2.11A represents an abrupt change in the diversity of various taxa coinciding with mass extinction (e.g. at the

TABLE 2.2

Jasko's (1984) example of frequency (= number of specimens) of a microfossil species in a borehole section. Lowest occurrence in sample at 3446 ft.

Depth interval in ft	Actual frequency	Expected frequency
2100 - 2400	41	40.1
2400 - 2700	26	23.6
2700 - 3000	11	13.9
3000 - 3300	9	8.2
3300 - 3600	5	4.8

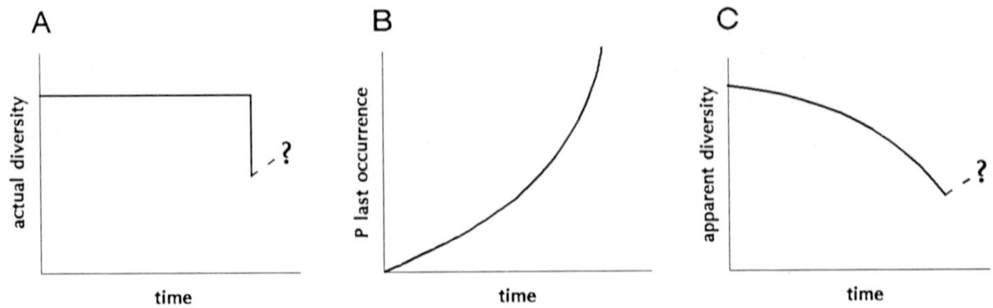

Fig. 2.11 Model of Signor and Lipps (1982) for alteration of diversity patterns by artificial range truncation. In Fig. 2.11A, diversity is suddenly reduced by a catastrophic extinction event. Imposing the artificial range truncation model illustrated in Fig. 2.11B on the pattern of Fig. 2.11A produces the apparent gradual decline in diversity of Fig. 2.11C.

Cretaceous-Tertiary boundary). Figure 2.11B plots an arbitrary probability curve giving the probabilities of different degrees of range truncation. This produces the apparent diversity curve shown in Figure 2.11C. Note that the slope of the hypothetical curve in Figure 2.6B continues to increase until the time of the mass extinction. Different sedimentary sections would be characterized by different curves. For example, if the curve of Figure 2.11B is representative for nearshore marine and terrestrial sections, the deep sea plankton record would have a curve whose slope increases less initially and becomes steeper near the time of the mass extinction (Signor and Lipps, 1982, p. 294). Thus the apparent diversity curve for oceanic microplankton is closer to actual diversity than e.g. the curve for dinosaurs below the Cretaceous-Tertiary boundary (cf. Russell, 1975, 1977; Van Valen and Sloan, 1977).

2.6 The frequency distributions of highest and lowest occurrences of taxa

Figure 2.12 shows a hypothetical relationship between relative abundance, observed highest occurrence and relative time for two taxa. Agterberg and Nel (1982b) introduced this example to illustrate that the abundance of a taxon may have changed through time. The range of the frequency curve of its observed highest occurrence is narrower than the range of the abundance curve although these two curves end at the same value along the time axis. Especially if a systematic sampling procedure is carried out such as obtaining cuttings at a regular interval (e.g. 30 ft or 10 m) along a well in exploratory drilling, the highest occurrences of two taxa with overlapping frequency curves may be observed to be coeval. The fact that two taxa have observed highest occurrences in the same sample does not necessarily mean that they disappeared at the same time. Rare taxa such as taxon B in Figure 2.12 are likely to have wider ranges for their highest occurrences.

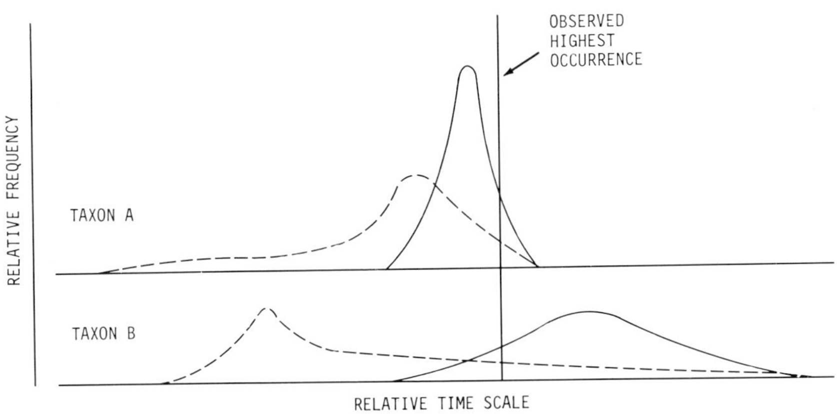

Fig. 2.12 Schematic diagram representing frequency distributions for relative abundance (broken lines) and location of observed highest occurrence (solid lines) for two taxa. Vertical line illustrates that observed highest occurrences of two taxa can be coeval even when the frequency distributions of these two taxa are different.

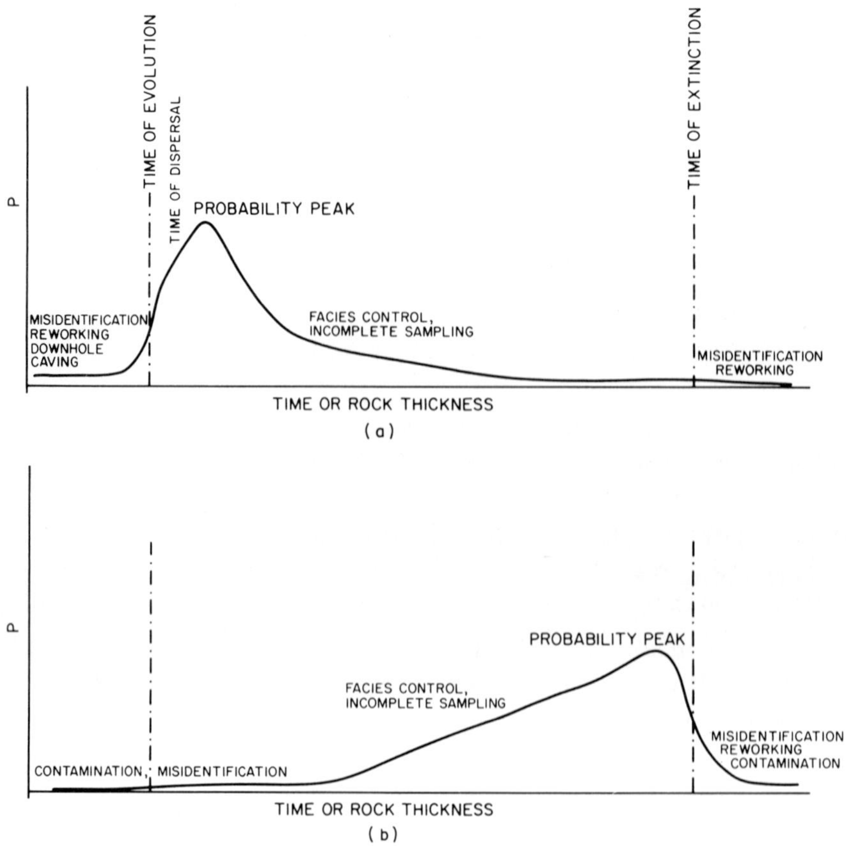

Fig. 2.13 Edwards' (1982a) model to display probability of observing lowest - or highest-occurrence event relative to "true" time of evolution or extinction in outcrop or core material for (a) first occurrence event; and (b) last occurrence event. According to Edwards (1982), details for curves will vary for every individual taxon, and gross shapes of curves will vary with kind of organism (e.g. rapidity of dispersal, facies control) and nature of sample material (core, outcrop, cuttings).

Figure 2.12 shows symmetrical, "normal" curves for the observed highest occurrences. It can be assumed that, in reality, these curves are not symmetric but skewed. Figure 2.13 (from Edwards, 1982a) is an attempt at displaying asymmetric curves for lowest and highest occurrences along with the main factors controlling the shapes. It is noted however, that Edwards' assumption on the nature of the skewness differs from that implied by Jasko's model, in which the tail of observed lowest occurrences extends in the stratigraphically downward direction (In Edwards' model it extends upward). In the model of Strauss and Sadler, the tail for highest occurrences points downward which is in agreement with Edwards' assumption. Likewise, the model of Signor and Lipps (Fig. 2.11B) is in agreement with that of Edwards because the slope of their curve continues to increase in the stratigraphically upward direction.

Figure 2.14 from Baumgartner (1986) also supports the model of Edwards (Fig. 2.13). It is illustrated in this diagram why a composite range based on many sections generally is relatively short ($=i_{AB}$) when it is based on mean positions of the frequency distributions for highest and lowest occurrences. In the Unitary Associations method, stratigraphic correlation is based on the three zones in the column on the right of Figure 2.14. The range of taxon A extends higher than the interval e_A and that of Taxon B occurs below e_B. The latter two intervals are based on the symmetrical Gaussian curves. A curve of this type has the property that 68 percent of the observations deviate less than one standard deviation from its mean. If e_A and e_B would be extended to points located two standard deviations from their mean, the probabilistic range chart becomes approximately equal to the zonation resulting from the Unitary Associations method. These wider probabilistic ranges would contain approximately 95 percent of the observations.

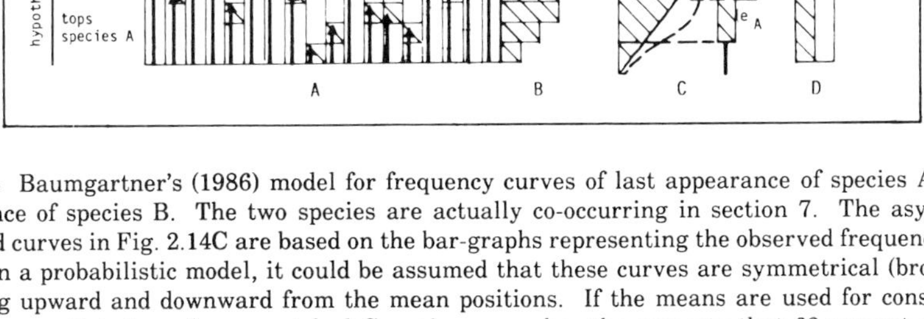

Fig. 2.14 Baumgartner's (1986) model for frequency curves of last appearance of species A and first appearance of species B. The two species are actually co-occurring in section 7. The asymmetrical smoothed curves in Fig. 2.14C are based on the bar-graphs representing the observed frequencies of Fig. 2.14B. In a probabilistic model, it could be assumed that these curves are symmetrical (broken lines) extending upward and downward from the mean positions. If the means are used for constructing a range, the result is i_{AB}. A symmetrical Gaussian curve has the property that 68 percent of the area undder the curve is contained between its inflection points located at the mean plus or minus one standard deviation. These intervals are shown as e_A and e_B. The Unitary Associations method would result in the overlapping ranges for species A and B shown in Fig. 2.14D. The latter result would also be obtained by using the Gaussian curves and assuming that e_A and e_B would extend two instead of one standard deviations on either side of the mean.

Edwards (1982b) has pointed out that if both highest and lowest occurrences of taxa are used, there is a possibility that in some methods of ranking, the highest occurrence of a taxon would end up below its lowest occurrence. Possible and impossible arrangements for the events resulting from 2 taxa are shown in Figure 2.15. Note that all impossible arrangements have in common that either A (lowest occurrence of first species) occurs above B (highest occurrence of first species) or that C occurs below D for the second species. If in a statistical method all events were to be treated independently, the final ranking might contain impossible arrangements. A problem of this type can be avoided, e.g. by recognizing during the coding of the stratigraphic events or within the computer program for statistical analysis, that the lowest occurrence is below the highest occurrence for each taxon in theory and practice.

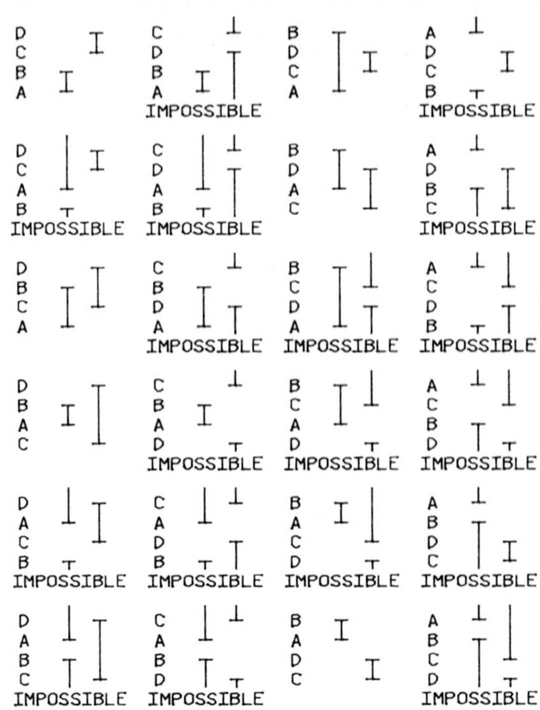

Fig. 2.15 The 24 arrangements of 4 events, where A and B are first and last occurrences of one species, and events C and D are first and last occurrences of a second species. Only 6 of these arrangements are possible (from Edwards, 1982b). Quantitative stratigraphers should always look for impossible arrangements in computer output and modify their algorithm if required.

Several possible frequency distribution models for highest and lowest occurrences are shown in Figures 2.16 and 2.17. The spike (A) represents abrupt disappearance of a taxon in Figure 2.16 and its immediate widespread appearance in Figure 2.17. Because the spike is symmetrical, the frequency curve also must be symmetrical when it is narrow (possibly B in Figs. 2.16 and 2.17). Wider frequency curves have different values for their mode (1), median (2) and mean (3), respectively. Curves for which the order of the mode, median and mean is 123 are positively skew in the direction of time. Those with order 321 are negatively skew.

Symmetrical curves have coinciding mode, median and mode. As shown in the captions of Figures 2.16 and 2.17, all models discussed so far correspond to one of the 12 possibilities. It can be assumed that, with the possible exceptions of A and C in Figures 2.16 and 2.17, all these frequency curves exist in the fossil record. In practice, it is almost always impossible to precisely measure the shapes of the frequency distributions of the highest and lowest occurrences of a taxon because one would need large numbers of sections that are calibrated precisely according to time-lines.

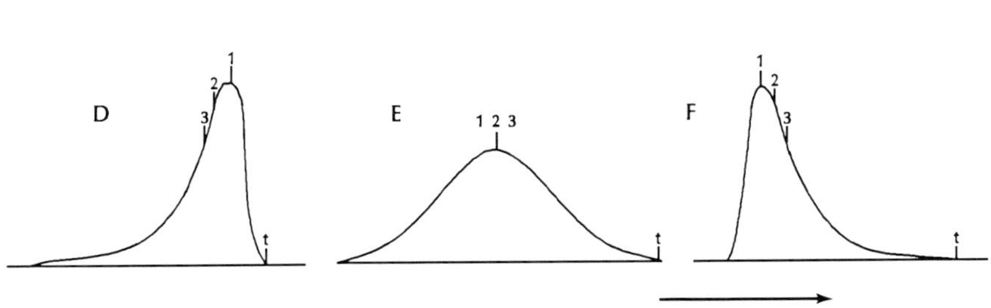

Fig. 2.16 Six possible shapes for the frequency distribution of the observed last occurrence of a taxon. the top (t) is the truly last occurrence. The numbers 1, 2 and 3 represent mode, median and mean, respectively. These three statistics coincide for a symmetrical curve. Most paleontologists assume that Fig. 2.16D is the most widespread shape. Arrow points in direction of time.

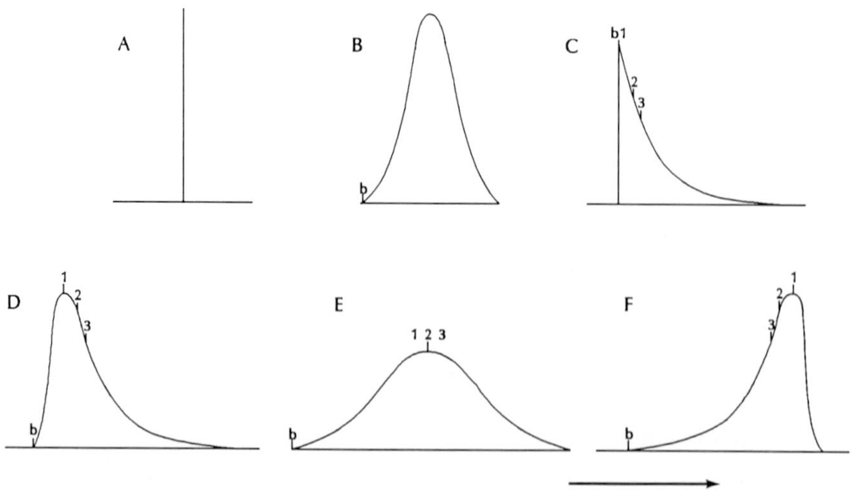

Fig. 2.17 Six possible shapes for the frequency distribution of the observed first occurrence of a taxon. The base (b) is the truly first occurrence. The numbers 1, 2 and 3 represent mode, median and mean, respectively. Opinions are divided as to which shape (D or F) is most widespread.

The subject of shapes of frequency distributions of highest and lowest occurrences largely remains in the realm of speculation, as is indicated by the fact that no concensus has been reached in literature. It seems that, in the absence of outliers due to reworking and other disturbing factors, the majority of paleontologists assume the shape of Figure 2.16D for the frequency distribution of the tops and that of Figure 2.17F for the bases. Both distributions have their longest tail in the stratigraphically downward direction. Figure 2.17F as the preferred model for first appearance data is contrary to the models of most quantitative stratigraphers (see before). However, as pointed out by Shaw (1964, p. 94), many paleontologists assume that there is a period (Shaw's "hemera") in the history of any species before it reaches its acme (Shaw's "epibole") in terms of numbers of individuals. Such a model is most likely to result in the shape of Figure 2.17F. Later in this book (see Chapter 9), a method will be discussed for actually measuring the skewness of the frequency distributions of bases and tops. However, the number of applications of this method remains too small to decide which models are most widespread.

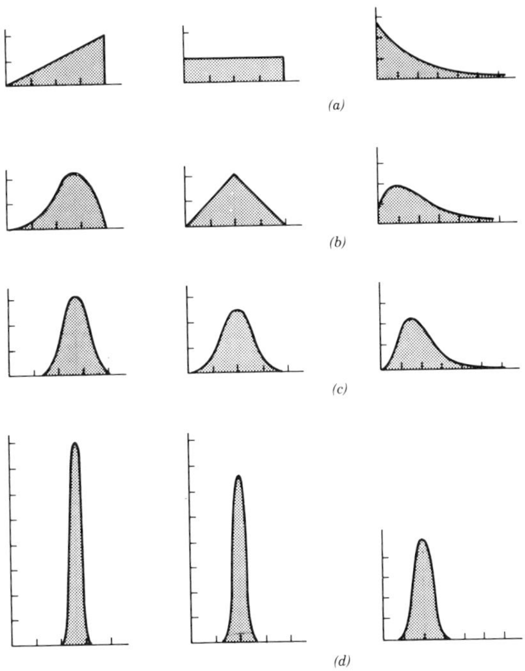

Fig. 2.18 Examples of the effect of averaging illustrate the central limit theorem of mathematical statistics. No matter what shape the frequency distribution of the original observations (a), taking the average of two (b), four (c) or 25 (d) observations not only decreases the variance but brings the curve closer to the normal (or Gaussian) limit (after Lapin, 1982; and Davis, 1986).

In the RASC method of ranking and scaling, the initial objective is to estimate the mean value (3 in Figs. 2.16 and 2.17) of the highest and lowest occurrences as precisely as possible. Biozonations as well as stratigraphic correlations are based on these mean values. The advantage of this procedure is that the mean can be precisely estimated regardless of the shapes of the frequency distributions of the events. This relative independence of shape is due to the central limit theorem of mathematical statistics (see Fig. 2.18) which states that addition or averaging of n independent random variables gives new random variables that become normally distributed when n increases. In the scaling part of RASC, distances between successive mean event locations are estimated by averaging many indirect distance estimates. Each of the latter estimates is a value originating from a frequency distribution that itself is an average of the frequency distributions for three separate stratigraphic events. Although the shapes of the original distributions may not be normal, the resulting frequency distributions based on sets of three events

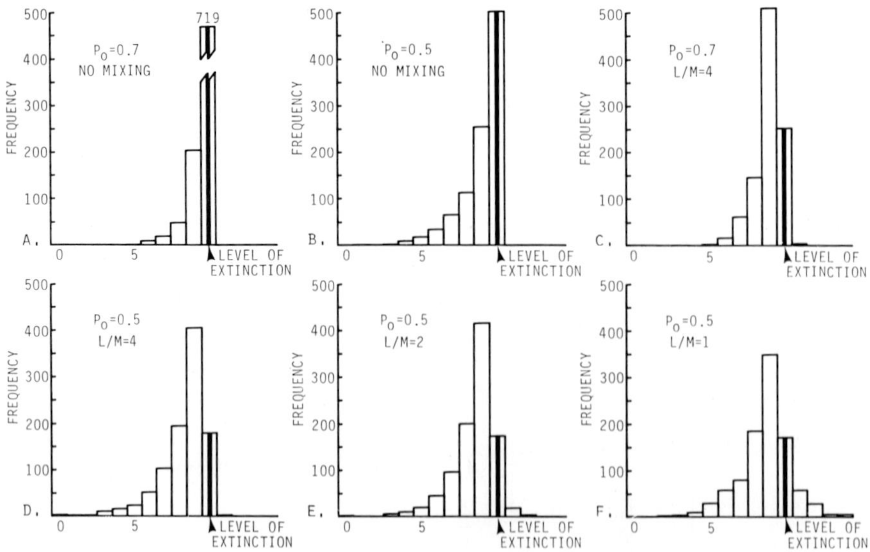

Fig. 2.19 Frequency histograms for finding a taxon within its range before and after mixing (from Edwards, 1982b). See text for further explanation.

are probably approximately normal. Further averaging of many indirect estimates yields mean event locations along the RASC scale that can be very precise.

Ranges based on mean positions are shorter than ranges resulting from attempts to estimate the locations of the true tops and bases (t and b) in Figures 2.16 and 2.17. Such maximal ranges attempt to represent the periods of time that taxa existed in a region. Estimation of the true end points is more difficult than estimating the mean event locations for several reasons: (1) statistically, the largest or smallest value in a sample of n values drawn from a population has a standard deviation which is greater than that of the mean of all values; and (2) the influence of "outside" values not belonging to the statistical population on the average range is much smaller than their influence on the maximal range. This is because maximal ranges would be based on values due to outside factors such as misidentification, contamination, downhole caving or reworking (cf. Fig. 2.13) unless these factors can be identified with certainty so that all outside values can be eliminated.

It is possible that the shape of the frequency distribution is changed because of one or more outside factors. Berger and Heath (1968) proposed a model for postdepositional mixing which was used by Edwards (1982) in computer simulation experiments. Figures 2.19 shows results for two initial distributions (A) and (B) after variable amounts of mixing (to degrees 1, 2 and 3). Degree 1 ($L/M = 4$) mixing led to a downward shift of the modes as shown in the resulting frequency curves (C) and (D). The effect of increased mixing to degrees 2 ($L/M = 2$) and 3 ($L/M = 1$) is shown in (E) and (F) for the second initial distribution only. Edwards (1982b) used the formula $P = P_0 \exp(-L/M)$ of Berger and Heath (1968) where P_0 and P represent the probability of finding the taxon within its range before and after mixing, respectively; L is the sample interval, and M is the thickness of the zone of mixing. The tail on the right (in direction of time) is increasing in length and the end product after mixing becomes nearly symmetrical in Figure 2.19F.

CHAPTER 3

APPLICATIONS OF MATHEMATICAL STATISTICS AND COMPUTER SCIENCE TO ZONATION, CORRELATION AND AGE INTERPOLATION

3.1 Introduction

This chapter contains background information for various applications of mathematical statistics and computer science. It can be skipped by readers who are not primarily interested in mathematically-based theory. Concepts and methods to be discussed include: (1) probabilities, Bernoulli trials and the binomial model; (2) graph theory; (3) multivariate analysis; (4) method of maximum likelihood; and (5) smoothing splines. Most of these techniques are illustrated by means of geological examples of interest in paleontology and stratigraphy although the emphasis in this chapter is on mathematical background. Not all mathematical discussions are contained in this chapter. Other techniques will be introduced in separate sections within later chapters as needed.

Modern mathematics and the theory of probability and statistics are formally based on set theory. There have been several interesting attempts to formulate conventional stratigraphy in strict logico-mathematical terms (Dienes, 1974; 1982; Dienes and Mann, 1977; Carimati et al., 1982). The language of set theory, although a necessity in pure mathematics, is not of immediate practical usefulness in stratigraphy which has a well-developed language of its own. Although superpositional relations between stratigraphic events can be precisely formulated in terms of sets, the nomenclature of set theory is unpalatable to most stratigraphers as pointed out by Tipper (1989, p. 480).

The mathematical techniques introduced in this chapter are required for statistical applications and for use in computer-based graphs and graphics. Although these techniques are widely applied in other fields of science, and may be elementary to those trained in mathematical statistics, they have been used hardly at all in stratigraphy. The purpose of this chapter is not only to review statistical methods that have been applied in stratigraphy, but also to show that other methods (e.g. maximum likelihood method) can be used to refine existing methodologies.

3.2 Binomial test for randomness

The binomial test for randomness will be briefly discussed (cf. Hay, 1972; Southam et al., 1975; Blank and Ellis, 1982). If the sequence of a pair of biostratigraphic events is random, the probability of one event preceding the other is $p = 1/2$. Each observed superpositional relation is thought to be the outcome of a Bernoulli trial. Suppose that two events (A and B) both occur in N sections. Then the probability that A occurs above B k times satisfies

$$P(k) = {}_N C_k 2^{-N} \tag{3.1}$$

with the binomial coefficient being

$$_N C_k = N! \left[k!\,(N-k)! \right]^{-1} \tag{3.2}$$

For example, if $N = 5$, then $P(0) = P(5) = 1/32$; $P(1) = P(4) = 5/32$; and $P(2) = P(3) = 10/32$. These probabilities add to one. It is also possible to write $P(0 \text{ or } 5) = 1/16$, $P(1 \text{ or } 4) = 5/16$ and $P(2 \text{ or } 3) = 10/16$. In practice, the observation that A occurs k times above B generally cannot be distinguished from B occurring k times above A when the hypothesis $p = \mathrm{E}(K/N) = 1/2$ is being tested. In this expression, E(...) denotes expected value. K denotes the binomial random variable with observed frequencies k (=0, 1, 2, ..., N). The test hypothesis obviously cannot be rejected if K/N becomes equal to 1/2, a situation which may be observed when N is even. For $k > N/2$, the probability

$$P_c(k) = 2 \sum_{r=k}^{N} {}_N C_k 2^{-N} \tag{3.3}$$

may be computed where the subscript c denotes that this probability is cumulative. For the preceding example, $P_c(5) = 1/16$, $P_c(4) = 1/16 + 5/16 = 6/16$, and $P_c(3) = 6/16 + 10/16 = 1$. This probability was tabulated by Hay (1972, Table 1 on p. 264). Next a level of significance (e.g. $\alpha = 0.05$) can be selected. Then the hypothesis $p = 1/2$ will be rejected only if $P_c(k) < \alpha$.

The binomial test is useful when only two events are being compared to each other. If many events are to be considered simultaneously while most values of N are small, this approach is less useful. For example, in Figure 4.2 of Chapter 4 (see later), event Δ occurs 4 times above event $<$. According to the binomial test $P_c(4) = 1/8 = 0.125$ for $N = 4$. This exceeds $\alpha = 0.05$ and the hypothesis that events 1 and 10 are coeval ($p = 1/2$) therefore may not be rejected. Strictly speaking, it would have to be accepted . On the other hand, event Δ is separated from event $<$ by 4 intermediate levels with other events in 3 of the 4 sections considered. This would suggests that event Δ probably occurs above event $<$.

A multivariate statistical approach would be needed to test whether or not two events are coeval when observations on many other events also are available. Later, an approach (scaling method) will be developed which permits the use of significance tests in which all events can be considered simultaneously.

3.3 Binomial distribution model for microfossil abundance data

This section deals with statistical analysis of microfossil abundance data. The microfossil record of the Portugese Oxfordian black shales (Stam, 1986; Agterberg et al., 1990) will be used for example. In this case history study it will be investigated whether, and to what extent, foraminiferal abundance data can be used for detailed biostratigraphic correlation in two sections of the black shale in the Montejunto area of central Portugal.

In general, most biostratigraphic correlation is based on biozonations derived from range charts using highest and lowest occurrences of species. For example, in exploratory drilling a sequence of samples along a well in the stratigraphically downward direction is systematically checked for first occurrences of new species. The probability of rejecting a species in a single sample depends primarily on its abundance. As a measure, relative abundance (to be written as p) of a species in a population of microfossils is commonly used. Together with sample size (N), p specifies the probability of the binomial distribution with general equation:

$$P(K=k) = P(k) = {}_N C_k \, p^k (1-p)^{N-k} \quad (k = 0, 1, ..., N) \tag{3.4}$$

which represents the probability that k microfossils of the taxon with relative abundance p will be found in a sample of N microfossils. Note that for $p = 1\text{-}p = 0.5$, this probability reduces to the one used in the binomial test for randomness (Eq. 3.1). If p is very small, the binomial probability can be approximated by the probability of the Poisson distribution.

$$P(k) = e^{-\lambda} \lambda^k / k! \quad (k = 0, 1, ..., N) \tag{3.5}$$

which is determined by a single parameter λ. The Poisson distribution can be derived from the binomial distribution by keeping $\lambda = Np$ constant and letting N tend to infinity while p tends to zero.

The expected (or mean) value for a binomial distribution is $E(K) = Np$ and for a Poisson distribution: $E(K) = \lambda$. The variance $\sigma^2(K)$ of the binomial distribution is $Np(1\text{-}p)$ while the variance of the Poisson distribution satisfies $\sigma^2(K) = E(K) = \lambda$.

Figure 3.1 (after Dennison and Hay, 1967) shows probability of failure to detect a given species for different values of p as a function of sample size ($=N$). For example, in a sample of $N=200$ microfossils, a species with $p=1$ percent has probability of about 15 percent of not being detected. This implies that the chances that one or more individuals belonging to the species will be found are good. Unless its relative abundance is small, the first occurrence of a species in a sequence of samples can be established relatively quickly and precisely.

It is noted that the two scales in Figure 3.1 are logarithmic and that the lines are approximately straight unless p is relatively large. This is because the equation for zero probability of the Poisson distribution, which provides a good approximation when p is small, plots as a straight line on logarithmic graph paper. If 10 is used as the base of the logarithms, the equation of each line in Figure 3.1 is simply $\log_{10} N = \log_{10} \lambda - \log_{10} p$ with $P = P(K=0) = \exp(-\lambda)$ as follows from Equation (3.5).

The binomial distribution model on which Figure 3.1 is based also can be used to estimate confidence intervals for any specific proportion value (p). Unfortunately, it turns out that large samples would be needed to estimate, with precision, the relative abundances of many different species. In general, proportions estimated from actual samples are

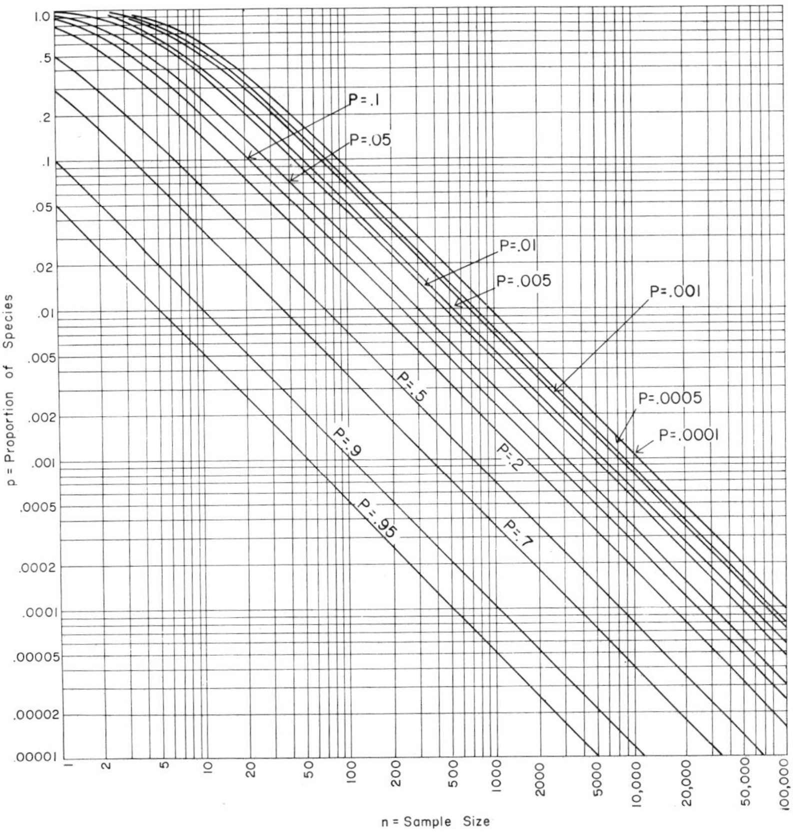

Fig. 3.1 Size of random sample (n) needed to detect a species occurring with proportional abundance (p) in population with probability of failure to detect its presence fixed at P (after Dennison and Hay, 1967).

uncertain. Moreover, the use of the binomial distribution model is based on the assumption that the underlying population is a homogeneous random mixture. This condition may hold true only locally, at the precise place where a sample was actually taken. The proportions of the species may change parallel and, in general more rapidly, perpendicular to bedding. It is hard to establish such changes because of the uncertainty in the estimated values.

For these reasons, it is hazardous to use measured proportion values for biostratigraphic correlation although it will be shown in the following case history study that some species (e.g. *Epistomina mosquensis*) can be useful for this purpose. The precision of proportion values also has been studied in detail by palynologists. Maher (1972) has published nomograms for computing 0.95 confidence limits of pollen data. A related topic is to study the precision of microfossil concentration measurements by employing samples spiked with marker grains (Maher, 1981; White, 1990).

Geological background

Both syn-rift fault tectonics and changes in eustatic sealevel influenced Jurassic carbonate through clastics marine sedimentation in the Montejunto Basin, Portugal (cf. Stam, 1986; Agterberg et al., 1990).

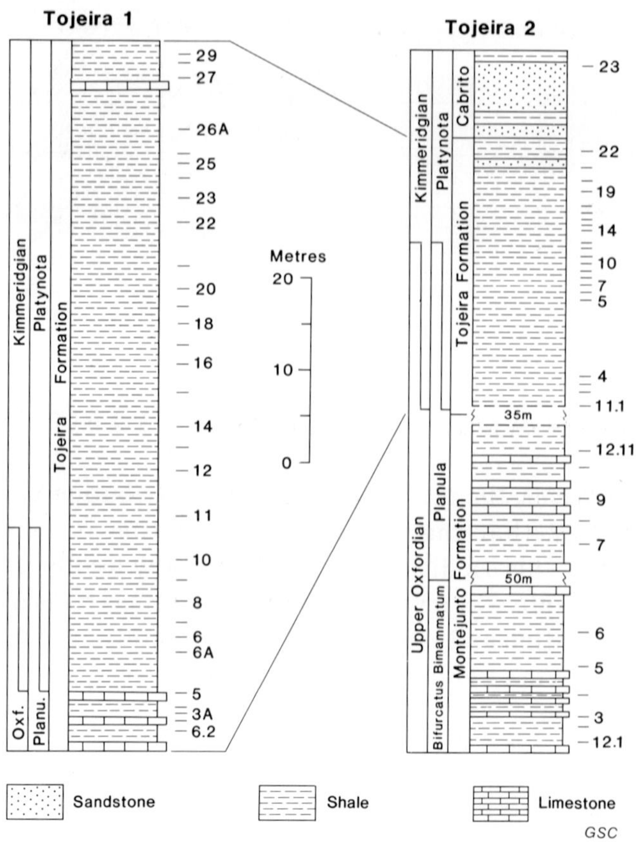

Fig. 3.2 **Left side**: Tojeira 1 section with sample members 6.2-6.29 (after Stam, 1986); ammonite zones (*Planula* and *Platynota* Zones) of Mouterde et al. (1973) also are shown. This section is immediately overlain by the poorly exposed sandy Cabrito Formation. **Right side**: Tojeira 2 section with sample numbers 12.1-12.11 and 11.1-11.23 (after Stam, 1986).

Bathonian through Callovian carbonate bank and shelf apparently became emergent in latest Callovian time due to widespread uplift or sealevel fall. Renewed transgression in Middle Oxfordian led to bituminous algal and micritic to oolithic limestones of the Cabacos Formation, changing upward into thick-bedded micritic brachiopod biostromes of the Montejunto Formation. Rapid deepening in latest Oxfordian to early Kimmeridgian time, when conditions became more humid, led to sedimentation of dark grey shales of the Tojeira Formation, followed upward by massive terrigenous-clastic fill (Cabrito and Abadia Formations).

In Oxfordian time (approximately 150 Ma ago), at the onset of the late Jurassic, a transition from one sedimentary mega-sequence into another one took place. For example, in the North Sea Basin, the Lusitanian Basin and the southern margin of Tethys ocean, now occupying the belt between the central Himalayans and Tibet, the Oxfordian saw the sudden onset of black shale deposition lasting up to 15 Ma or more. Climate must have become more humid; the black shale facies was probably also related to regional basinal deepening, in the absence of major relief rejuvenation that would induce terrigenous clastic supply. In places, the shales constitute major hydrocarbon source rock.

Location of Tojeira sections; summary of Stam's quantitative results

The Lusitanian Basin originated in the late Triassic — early Jurassic as a result of movements along Hercynian basement faults including the prominent Nazare strike slip fault. Several cross-sections in the Montejunto area were sampled by Stam (1986) for quantitative analysis of Middle and Late Jurassic Foraminifera in Portugal and its implications for the Grand Banks of Newfoundland. The so-called Tojeira 1 section with sample numbers 6.2-6.29 (after Stam, 1986) is shown in Figure 3.2 (left side). It is continuously exposed and occurs about 2km southeast of the Tojeira 2 section (Figure 3.2, right side) with Stam's sample numbers 12.1-12.11 and 11.1-11.23. The Tojeira 2 section is not continuously exposed; two missing parts are estimated to be equivalent to 35m and 50m in the stratigraphic direction, respectively.

Tojeira shales contain a rich and diversified (over 45 taxa) planktonic and benthonic foraminiferal fauna, including *Epistomina mosquensis*, *E. uhligi*, *E. volgensis*, *Pseudolamarckina rjasanensis*, *Lenticulina quenstedti*, and *Globuligerina oxfordiana*. Stam determined from 21 to 43 species per sample in Tojeira 1; between 301 and 916 benthos was counted per sample; proportions were estimated for 14 species. The plankton/benthos (P/B) ratio also was determined for each sample. Correlation coefficients for relative abundance estimates of the benthonic Foraminifera are close to zero but several of these coefficients were shown by Stam (1986) to be significantly greater or less than zero. R- and Q-mode factor analysis and cluster analysis gave separate assemblages of mutually associated species. For example, the group with *E. mosquensis, P. rjasanensis, O. strumosum* and agglutinants prefers the deep-water Tojeira shales to the underlying shallow-water Montejunto Formation. Similar results were obtained by Stam for the Tojeira 2 section.

Additional sampling and Nazli's autocorrelation analysis

Gradstein and Agterberg (1982) had worked previously with highest occurrences of Foraminifera in offshore wells drilled on the Labrador Shelf and Grand Banks. The samples were cuttings obtained during exploratory drilling by oil companies. Such samples are small, taken over large intervals and subject to down-hole contamination so that only highest occurrences (not lowest occurrences) of Foraminifera can be determined. These problems associated with exploratory drilling can be avoided on land if continuous outcrop sampling is possible. According to paleogeographic reconstructions (see Stam, 1986), the Lusitanian and Grand Banks Basins were close to one another during the Jurassic and had comparable sedimentary, tectonic and faunal history. On land continuous outcrop sampling can be undertaken in the Lusitanian Basin only.

After preliminary statistical autocorrelation analysis of Stam's data, new samples from the two Tojeira sections were collected during the summer of 1986. F.M. Gradstein identified the foraminiferal taxa. Only relatively few samples were taken at exactly the same places where Stam had sampled before. Figure 3.3 shows typically poor correlations between proportions estimated from Stam's and Gradstein's counts for species in samples taken at the same spots. These scattergrams reflect random (binomial) counting errors, local spatial variability of the (unknown) mean proportion values, as well as possible determination errors. In another sampling experiment, five samples were taken laterally at 5m interval from the same stratigraphic horizon at the base of Tojeira 1. Estimated

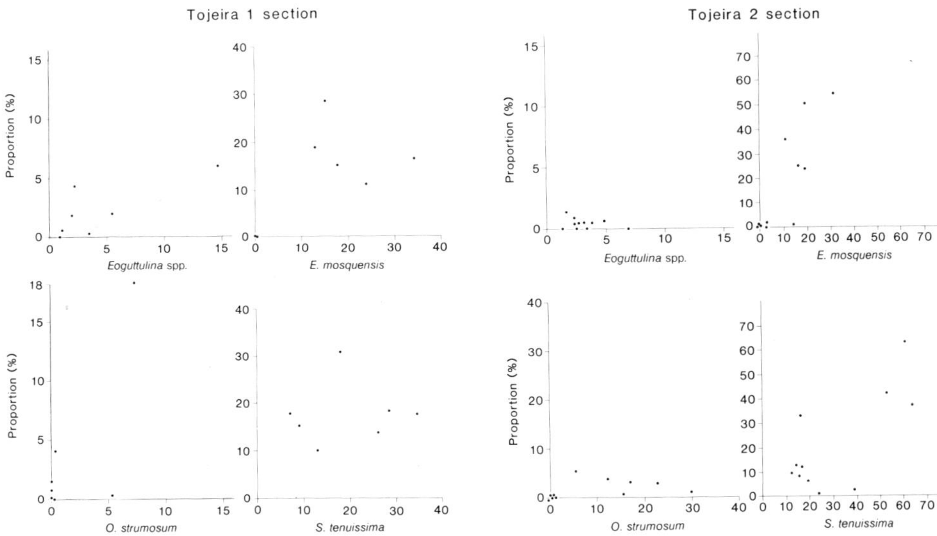

Fig. 3.3 **Left side**: Proportions of four benthonic Foraminifera for seven replicate samples from same sites in Tojeira 1 section based on determinations by Stam (horizontal axis) and Gradstein (vertical axis). **Right side**: ditto for eleven replicate samples in Tojeira 2 section. See text for discussion of lack of agreement.

proportion values as well as total benthos counted for these 5 samples were shown in Agterberg et al. (1990, Table 1). The measured proportions are markedly different, again illustrating the uncertainty commonly associated with microfossil abundance data.

As a first step for an M.Sc. project, Nazli (1988) subjected Stam's data for 14 benthonic species in 31 samples from Tojeira 1 to the ARIMA (Auto Regressive Integrated Moving Average) procedure of the Statistical Analysis System (SAS) as implemented on the IBM mainframe computer at the University of Ottawa in 1986. SAS (Statistical Analysis System) is a statistical software package with separate versions for mainframes and personal computers (available from SAS Institute Inc., Box 8000, Cary, NC, U.S.A.). The ARIMA method was originally developed by Box and Jenkins (1976). The first part of SAS ARIMA output for *E. mosquensis* is shown in Figure 3.4. In autocorrelation, successive values along a time series are correlated with one another for different lags (= intervals along the series). Normally in applications of ARIMA, the values are equally spaced along the time axis. The decompacted sedimentation rate during deposition of the Tojeira Formation was about 5cm per 1000 years. Although the shale is homogeneous in composition, it cannot be taken for granted that sampling it at equal intervals would yield a series with points

```
                                SAS
                          ARIMA PROCEDURE
                    Tojeira 1: E. mosquensis
                        AUTOCORRELATIONS

LAG COVARIANCE CORRELATION -1 9 8 7 6 5 4 3 2 1 0 1 2 3 4 5 6 7 8 9 1
  0    160.079    1.00000  |                    |********************|
  1    79.9485    0.49943  |                .   |**********          |
  2    85.2347    0.53245  |               .    |***********         |
  3    58.3794    0.36469  |              .     |*******   .         |
  4    32.2471    0.20145  |              .     |****       .        |
  5    27.9955    0.17489  |              .     |***        .        |
  6    14.9058    0.09312  |              .     |**         .        |
  7    25.9934    0.16238  |              .     |***        .        |
  8    23.4033    0.14620  |              .     |***        .        |
  9    19.8307    0.12388  |             .      |**          .       |
 10    12.4919    0.07804  |             .      |**          .       |
                        . MARKS TWO STANDARD ERRORS

                                                                  GSC
```

Fig. 3.4 Partial output of SAS ARIMA procedure for *E. mosquensis* proportions in Stam's 31 samples from Tojeira 1 (for complete print-out, see Nazli, 1988, Fig. 4-12, p. 98). ARIMA maximum likelihood estimation gave three statistically significant coefficients for first order autocorrelation coupled with two-term moving average. This result is compatible with assumption of signal-plus-noise model in Figure 3.5.

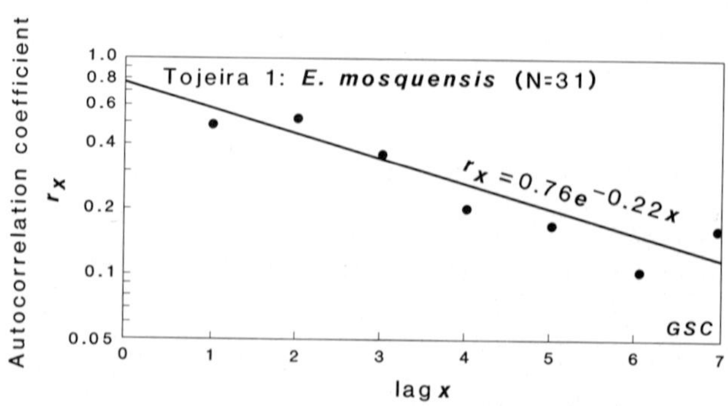

Fig. 3.5 Estimated autocorrelation coefficients of Figure 3.4 plotted along logarithmic scale and approximated by exponential function.

that are equally spaced in time. The 31 samples used for Figure 3.4 are approximately equally spaced in the stratigraphic direction (see Fig. 3.2, left side). The resulting autocorrelation pattern for *E. mosquensis* is approximately exponential. In Figure 3.4, the first few estimated autocorrelation coefficients (lags 1 and 2) are greater than zero with a probability of over 95 percent as indicated by the confidence limits (for two standard deviations) in the plot on the right-hand side of Figure 3.4. The approximately exponential nature of the pattern is brought out more clearly in Figure 3.5 where a logarithmic scale is used for the vertical axis, so that an exponential function with equation $r_x = c.\exp(-ax)$ plots as a straight line. Nazli (1988) has applied other statistical tests including spectral analysis available as SAS procedures to the microfossil abundance data. He established that most autocorrelation patterns can be interpreted as white noise (random variability) with the following exceptions: In Tojeira 1, *Eoguttulina* sp., *E. mosquensis* and *Ophthalmidium strumosum* exhibit non-random patterns with approximately exponential autocorrelation functions. *E. mosquensis* and *O. strumosum* show similar non-random patterns in Tojeira 2 where exponential patterns were also established for *Spirillina tenuissima* and agglutinants. For these seven sequences, straight lines were constructed on semi-logarithmic plots as exemplified in Figure 3.5 for *E. mosquensis* in Tojeira 1. For the three species in Tojeira 1, the analysis was repeated for a combined series of 41 samples by adding the samples taken in 1986 at ten new sample sites.

Each straight line was interpreted as representative of a signal-plus-noise model (cf. Jenkins and Watts, 1968; Agterberg, 1974). The standard deviation (s_N) of the noise component for local random variability then can be estimated from the intercept (c) of the straight line with the vertical axis. For example, in Figure 3.5, $c=0.76$. This is the proportion of variance accounted for by the signal. It leaves a proportion of ($1-c=$) 0.24 for the noise component. The variance of the 31 values was 0.0160079 (cf. Fig. 3.4). Multiplication of this value by 0.24 and taking the square root yields $s_N = 6.2$ percent. One would expect this standard deviation to be at least as large as the standard deviation (s_B) arising from the binomial counting process. The value s_B can be estimated from the average proportion ($=\bar{p}$) and average number ($=\bar{n}$) of counts per sample. For example, $\bar{n}=443$ for Stam's 31 Tojeira 1 samples; the corresponding average proportion value for *E. mosquensis* is $\bar{p}=22.5$ percent. From the binomial variance for proportions with equation $s^2_B=\bar{p}\ (1-\bar{p})\ /\ \bar{n}$, it then follows that $s_B=1.98$ percent. Because for the ratio, $s_B/s_N=0.32$, this result would mean that 32 percent of the measured random variability for *E. mosquensis* in Tojeira 1 (Stam's 31 samples only) is due to counting errors whereas the remaining 68 percent can be ascribed to local random variability in the rock. This result is shown in Table 3.1 together with similar statistics for the other species with approximately exponential autocorrelation functions in the Tojeira sections.

Discussion

Binomial theory has been widely used in paleontology and stratigraphy for estimating the precision of relative abundance with (cf. Shaw, 1964; Dennison and Hay, 1967). A graph (Fig. 3.1) can be used to rapidly estimate the probability of not detecting a species if it is present. Several other graphical methods of calculating sums of binomial probabilities have been developed. For a summary, see Johnson and Kotz (1969). The latter publication also contains various approximations for the binomial, and references to tables containing values of individual probabilities and sums of probabilities.

TABLE 3.1

Comparison of standard deviations (in percent) due to counting (s_B) and total local random variability (s_N) for species with average proportion $\bar{p}$ (in percent) and approximately exponential autocorrelation function (after Agterberg et al., 1990).

	$\bar{p}$	c	s_N	s_B	s_B / s_N
Tojeira 1 (31 samples; $\bar{n}=443$)					
(a) *Eoguttulina* spp.	2.77	0.76	2.2	0.78	0.36
(b) *E. mosquensis*	22.47	0.76	6.2	1.98	0.32
(c) *O. strumosum*	1.93	0.50	1.7	0.59	0.37
Tojeira 2 (30 samples; $\bar{n}=250$)					
(a) *E. mosquensis*	13.84	0.88	3.8	2.19	0.57
(b) *S. tenuissima*	25.75	0.90	5.5	2.76	0.50
(c) *O. strumosum*	11.25	0.91	2.8	2.00	0.71
(d) Agglutinants	10.42	0.58	3.2	1.93	0.61
Tojeira 1 (41 samples; $\bar{n}=408$)					
(a) *Eoguttulina* spp.	2.20	0.48	2.9	0.71	0.25
(b) *E. mosquensis*	23.76	0.52	8.4	2.11	0.25
(c) *O. strumosum*	2.39	0.60	1.8	0.76	0.41

It should be kept in mind that binomial theory only can provide approximate estimates of precision of relative abundance estimates. The main reason for this is that, as when red balls are drawn at random from a vase with balls of many colors, binomial theory applies to random mixtures. In practice, the random variability model only may account for part of total spatial variability. In this section, a more general model was applied with $X_i = S_i + N_i$; $N_i = N_{Li} + N_{Bi}$. It is assumed that at each sample location (i) an observed proportion value (X_i) is the sum of a signal (S_i) and a noise (N_i) component. The signal is "random" with constant autocorrelation function as generally is assumed in statistical time-series analysis and mining geostatistics. (However, a deterministic trend or drift component also could exist in and might need special consideration). By systematically comparing relative abundance values for samples taken at different distances from one another (mainly perpendicular but also parallel to bedding), it is possible to estimate separate variances of signal and noise. In the practical example (Tojeira sections, Portugese Oxfordian black shales), the existence of "signal" could be established for only 2 of 14 species in both sections although 3 other taxa showed systematic change in abundance through time in one of the sections only. The "noise" component can be imagined as resulting from local random variability that arises when samples are taken very close to one another but not exactly at the same locations. This noise is the sum of the binomial counting error (N_{Bi}) and a local noise component without counting error (N_{Li}). Theoretically the latter component is independent of sample size. In Table 3.1 it is shown that for the 3 taxa with "signal" in Tojeira 1, the sampling error (s_B) is about one third of the standard deviation (s_N) of total noise. The ratio s_B/s_N is close to 0.6 for the 4 taxa with "signal" in Tojeira 2. Later (in Section 3.6) it will be shown for *E. mosquensis* that the signal can be extracted by eliminating the total noise component.

The purpose of the material presented in this section was not only to show how binomial theory can be applied to estimated microfossil proportion data but also to indicate that probabilities and standard deviations estimated by means of this theory may be valid only for random mixtures of microfossils derived from the samples as taken in the field. In this respect, microfossil abundance data resemble, for example, assay values in mining for which special geostatistical techniques have been developed (see e.g. David, 1977).

3.4 Multiple pairwise comparison

Hudson and Agterberg (1982) listed several trinomial models by means of which three probabilities p_1, p_2 and p_3 (for occurrence of A_1, A_2 or A_3) can be estimated using all possible pairwise comparisons of two stratigraphic events. Here A_1 denotes the situation that an event E_i occurs above another event E_j in a section, A_2 is for E_j above E_i, and A_3 for the situation that E_i and E_j are coeval. These models include Glenn and David's (1960) model, and Davidson's (1970) model (also see Section 6.10).

Davidson's model was successfully applied by Edwards and Beaver (1978) and later by Hudson and Agterberg (1982) to several data sets. Drawbacks, pointed out in the latter publication, were that this method, because of many iterations required, becomes time-consuming even for digital computers when the number of events exceeds 40. Also, the model is not able to handle the situation that many events in the upper parts of a large stratigraphic column occur with certainty above many events in its lower parts. Agterberg (1984) showed that a modification of Glenn and David's model is not subject to these constraints and can be used in situations where Davidson's model is definitely not applicable.

Glenn and David's model is an extension of the so-called Thurstone-Mosteller model (cf. Mosteller, 1951) which uses Gaussian curves for the distribution of positions of events along a linear scale as is done in the RASC model. The original Thurstone-Mosteller model does not permit ties. (In stratigraphy ties are coeval events.) As a first step for calculating average distances between events along this linear scale, the observed cross-over frequencies are converted to Z-values according to the transformation $\Phi^{-1}(P) = Z$. This is the inverse of $P = \Phi(Z)$ where Φ denotes the fractile (cumulative frequency) of a normal distribution in standard form. Mosteller (1951) has shown that, under certain conditions, the best position of an event along the scale is obtained by averaging all Z-values for pairwise comparisons of this event to all other events. The resulting position is "best" in a least squares sense. If the RASC model would be used in a situation that none of the frequencies P_{ij} are missing or equal to one, then the unweighted method (simple averaging of Z-values regardless of sample sizes) would yield results nearly identical to those of the Thurstone-Mosteller model. Modifications were made in the RASC model to avoid missing values and frequencies equal to one or zero. These modifications can also be applied to Glenn and David's model. This trinomial model successfully estimated the probability that two events are coeval in several applications (see Section 6.10).

In the RASC model, observed ties are not ignored but each tie of two events E_i and E_j is scored as a 50 percent probability that E_i occurs above E_j and a 50 percent probability that E_j occurs above E_i. Observed scores S_o can be compared with estimated frequencies $S_e = P_s \times R$ in which the estimated probabilities P_s (for E_i occurring above E_j) satisfy $P_s = \Phi(d_s)$; d_s may be estimated by means of the weighted scaling option of the RASC computer program in which variations of sample size R are considered. The agreement between observed and estimated scores was excellent for Cenozoic Foraminifera on the Labrador Shelf – Grand Banks (see Section 6.10, for details). The chi-squared test for goodness of fit was used for making this comparison. This shows that the scaling method of RASC permits the use of significance tests for comparing pairs of events with one another on the basis of probabilities estimated from the order relationship of all events considered simultaneously.

3.5 Applications of graph theory

Several authors including Guex (1977), Smith and Fewtrell (1979) and Agterberg and Nel (1982b) have used graphs for representing relationships between biostratigraphic events . The applications in this section will be to co-occurrences and superpositional relationships of fossil taxa. Graph theory is a branch of applied mathematics in which properties of graphs are established and used to solve specific problems. Roberts (1976, 1978) has provided an excellent introduction to the topic (also see Berge, 1973; and Carré, 1979).

Guex (1987) has made an important contribution to quantitative stratigraphy by adopting a graph theoretical approach. The Guex approach differs from the probabilistic one underlying the methods discussed in this book in that co-occurrences of fossils are used as the basic building stones for constructing "Unitary Associations" of fossils which can be used for correlation. Guex and Davaud (1984, p. 71) stated that "observed co-occurrences between species must be accepted as true unless the contrary is demonstrated. No deterministic analysis of the problem can be performed otherwise". Later in this volume, results obtained by the RASC computer program will be compared with results obtained by the Unitary Associations method for several examples. The purpose of this

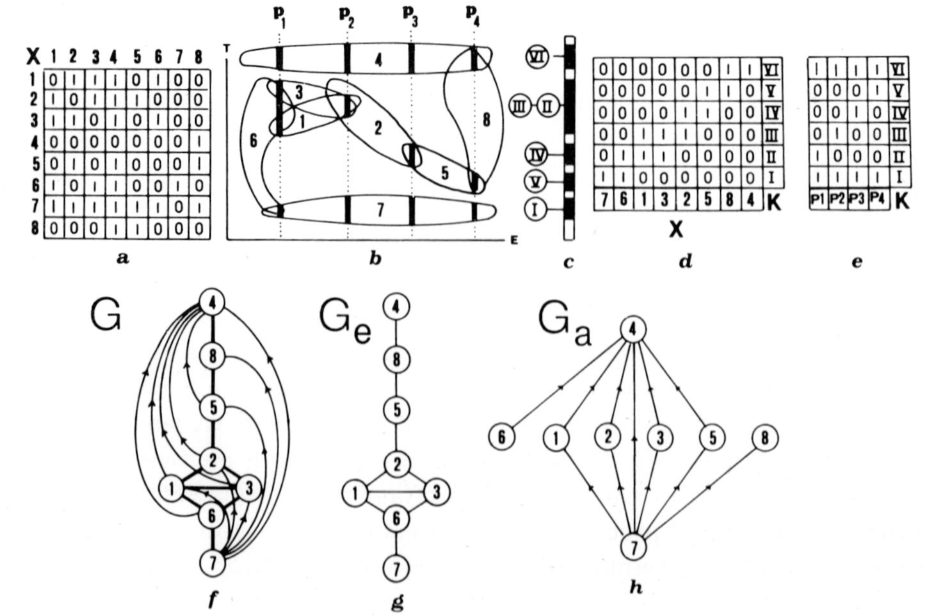

Fig. 3.6 Example of concepts of graph theory applied in biostratigraphy (after Guex, 1980). (a) Adjacency matrix containing same information as Fig. 3.6f for sections in Fig. 3.6b; (b) space-time relationship of 8 species numbered 1 to 8; heavy black vertical lines represent stratigraphic sections with observations on domains of existence (closed regions) of the eight species; T = time, E = space; (c) relative chronological position of the intervals I to VI for maximal cliques representing "Unitary Associations" derived from Figs. 3.6d and 3.6g; (d) matrix relating maximal cliques (K) of Fig. 3.6g to the eight species (X); (e) maximal cliques (K) identified in four sections (p_1-p_2) of Fig. 3.6b; (f) biostratigraphical graph G representing co-occurrences and superpositional relationships between the 8 species as observed in the four sections; (g) undirected graph G_e representing co-occurrences of Fig. 3.6f only; (h) directed graph G_a with arcs for superpositional relationships. The original purpose of this diagram was to illustrate, for a simple example, that construction of an interval graph (see Fig. 3.7) normally does not result in a chronological ordering. Only "reproducible Unitary Associations" are chronologically ordered as shown in Fig. 3.6e (Guex, 1980).

section is to introduce the additional concepts of graph theory needed for this. Figure 3.6 (from Guex, 1980) will be used for illustration.

Graphs consist of vertices and arcs or edges. An arc is an edge with an arrow indicating the direction for an ordered pair of vertices. Hypothetical space-time domains of eight fossil species are shown in Figure 3.6. Observations were made in four stratigraphic sections (heavy black lines in Fig. 3.6b). All observed relationships of co-occurrence or superposition are shown in the graph G of Fig. 3.6f which can be decomposed into an undirected graph (Fig. 3.6g, G_e with edges only) and a directed graph (Fig. 3.6h, G_a with arcs only). The same information is contained in the so-called adjacency matrix of Figure 3.6a. Each of the fossils has a row and a column in Figure 3.6a. If two species are observed to co-occur, this is shown by a pair of ones in the adjacency matrix (e.g. 1 and 2). An ordered pair (e.g. 4 and 1) is coded by means of a one in the column for 4 (and row for 1 above the diagonal of zeros in Fig. 3.6a) and a zero in the row for 4 and column for 1 (below the diagonal). If a fossil is observed above another fossil in one or more sections and below it elsewhere, this pair of fossils will be scored as a pair of ones in the adjacency matrix.

An undirected graph G_e is called complete if it contains all possible edges. A complete subgraph of an undirected graph is called a clique. A clique is maximal if it is not contained in a larger clique. Figure 3.6g has six maximal cliques labelled I to VI in Figures 3.6c-e. For example, the subgraph (4, 8) is complete in Figure 3.6g. It is referred to as maximal clique VI with two consecutive ones in the matrix of Figure 3.6d. Another example of a maximal clique is III (for fossils 1, 2 and 3) with three consecutive ones in Figure 3.6d. In the example of Figure 3.6, the maximal cliques are "Unitary Associations" which can be recognized in individual sections without ambiguity (see Fig. 3.6e) and used for

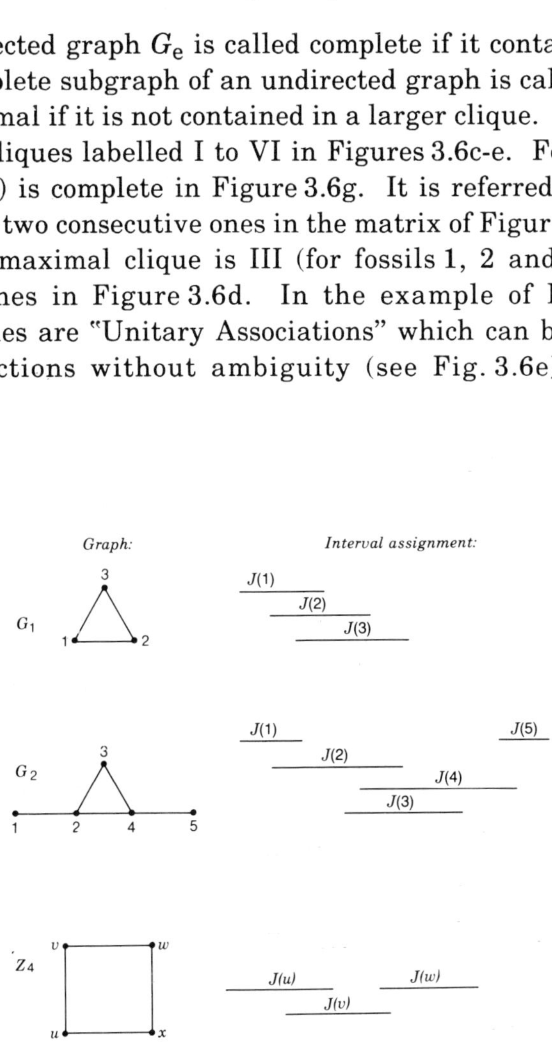

Fig. 3.7 G_1 and G_2 are examples of interval assignments $J(i)$, $i = 1, 2, \ldots$ for undirected graphs. An interval assignment for Z_4 with vertices u, v, w and x does not exist (after Roberts, 1976).

correlation. In general, the situation is more complex than that shown in the example of Figure 3.6 and additional concepts and methods of graph theory are needed.

In general, a set of intervals on the real line can be represented by means of a so-called interval graph. Only graphs with a interval assignment (Fig. 3.7 from Roberts, 1976) are interval graphs. The interval $J(i)$ of a vertex i of an interval graph overlaps at least in part with the intervals of vertices to which i is connected by an edge.

The special graph Z_4 (Fig. 3.7c) is not an interval graph because it is not possible to assign intervals to it. The vertices of Z_4 are labelled u, v, w and x in Figure 3.7c. According to the preceding definition of an interval assignment, the intervals $J(u)$ and $J(v)$ would have to overlap because u and v are connected by an edge. $J(v)$ extends to the right of $J(u)$ in Figure 3.7c because it cannot completely lie within $J(u)$ (otherwise, $J(w)$ could not be overlapping $J(v)$ without overlapping $J(u)$ as required). According to the relationships drawn in Z_4, $J(w)$ overlaps $J(v)$ but not $J(u)$ and must be depicted in the interval assignment as shown. It is not possible now to draw the interval for $J(x)$ which should overlap with $J(w)$ and $J(u)$ but not $J(v)$. This completes the proof that Z_4 does not have an interval assignment and is not an interval graph.

A graph G_e with vertices V and edges E can be written as $G_e = (V, E)$. A graph $H_e = (W, F)$ is a subgraph of $G_e = (V, E)$ if W is a subset of V and F a subset of E. H_e is called a generated subgraph if F consists of all edges from E joining vertices in W. It can be seen that if G_e is an interval graph, then every generated subgraph (but not every subgraph) must also be an interval graph.

Any graph G_e representing associations of fossil species should be an interval graph because pairs of fossils coexisted during specific time intervals with or without overlap. The question of when a graph is an interval graph can be answered in several ways. Fulkerson and Gross (1965) have proved the theorem that a graph G_e is an interval graph if and only if there is a ranking of the maximal cliques of G_e which is consecutive. A ranking $K_1, K_2, ..., K_p$ of the maximal cliques of G_e is called consecutive if whenever a vertex u is in K_i and K_j for $i < j$, then for all $i < r < j$, u is in K_r. It is easy to see that the maximal cliques of G_e in Figure 3 6 are consecutive. Consequently, G_e of Figure 3.6 is an interval graph.

Gilmore and Hoffman (1964) proved the following theorem: A graph G_e is an interval graph if and only if it satisfies the following conditions: (a) Z_4 is not a generated subgraph of G_e, and (b) G_e^c is transitively orientable. G_e^c is the complementary graph of G_e. It has the same vertices as G_e but edges only between those vertices which are not connected by edges in G_e. If G_e is an interval graph, G_e^c has edges connecting vertices representing nonoverlapping intervals only. Suppose that arrows are assigned to these edges thus changing them into arcs either pointing in the direction for "before" or "after". It is easy to see that, if G_e is an interval graph, these arrows all point either in the forward or in the backward direction of the real line. Conversely, if G_e^c has the preceding property, then G_e (without Z_4's) is an interval graph according to the theorem of Gilmore and Hoffman. The formal definition of a transitively oriented graph G_a is that, if (travelling in the directions of the arrows) a vertex v can be reached from another vertex u, and a vertex w from v, then w can be reached from u.

A graph G representing stratigraphic relationships (e.g. Fig. 3.6f) generally is a mixture of an undirected graph G_e and a directed graph G_a. From the preceding two theorems, it can be seen that the complement of G_e for the example (Fig. 3.6g) is transitively orientable. The directed graph G_a (Fig. 3.6h) for observed superpositional relationships is a subgraph of the oriented complement of G_e. In a situation that the relationships between all possible pairs of fossils are fully known, the biostratigraphic graph G would be the union of G_e and its oriented complement. If G_e is an interval graph, G cannot contain any if a number of "forbidden" generated subgraphs. For example, the Guex's cycle C_3 is a frequent forbidden structure with 3 vertices (u, v, and w) showing u before v, v before w and w before u. This is comparable with the 3-event cycle for stratigraphic events to be introduced in Chapter 5 on ranking (e.g. cycle ABC in Fig. 5.7). In a biostratigraphical graph G, C_3 is not a possible generated subgraph because it would mean that G_e^c is not transitively orientable and G_e is not an interval graph.

C_3 constitutes the most frequently encountered forbidden structure in biostratigraphical graphs G. C_3's are likely to occur in the strong component of G if it exists. The strong component of a graph is defined as the generated subgraph which is strongly connected and has the maximum number of vertices. A directed graph is called strongly connected if for every pair of it vertices u and v, v is reachable from u and u from v. Guex and Davaud (1984) introduced a special coefficient $s = c/r$ for each arc (e.g. u to v) where c represents number of times this arc occurs in a C_3 within the strong component and r is the total number of times the arc occurs in the strong component. If the coefficient s of an arc is high, this may indicate reworking or contamination. If reworking is suspected, v is omitted in beds where it was observed to occur above u. For contamination, u would be removed from below v.

Guex and Davaud (1984) have developed further rules for interactive or automated elimination of other forbidden structures from G. For example, Z_4 is removed by assuming "virtual" co-occurrence for either a pair of two or all four of the fossils involved. Two fossil species are said to co-occur virtually if their co-occurrence was not observed but inferred. After elimination of all inconsistencies, the biostratigraphic graph G yields an interval graph G_e of which the maximal cliques can be determined. These are the Initial Unitary Associations (I.U.A.'s). They are called "initial" because Guex and Davaud (1984) added the following method for combining some of the I.U.A.'s with one another in order to form the U.A.'s. The I.U.A.'s are identified in sections as previously illustrated for the Unitary Associations in Figure 3.6e. A complete I.U.A. may not be observed in a section. However a given I.U.A. is fully characterized by anyone of its unique species or pairs of species. I.U.A.'s characterized by "virtual" (inferred, not observed) co-occurrences of fossils only cannot be identified in sections. Guex and Davaud (1984) then proceeded by constructing the directed graph G_k of superpositional relations between the I.U.A.'s as identified in the sections. The construction of G_k with the I.U.A.'s as vertices is identical to the extraction of G_a for the original biostratigraphical graph G. Next they find the I.U.A.'s with the longest path in G_k. In general, a vertex in a directed graph G_a is connected to another vertex by means of a "path" if the arrows on the arcs between these two vertices point in the same direction. Each I.U.A. not on the longest path is combined with the I.U.A. on the path with which it has an interval in common. This gathering process yields the final Unitary Associations (U.A.'s) which are identified in the sections as the I.U.A.'s were before. If the new I.U.A.-U.A. method is applied to the example of Figure 3.6, the Initial Unitary Associations II and III would be combined with one another.

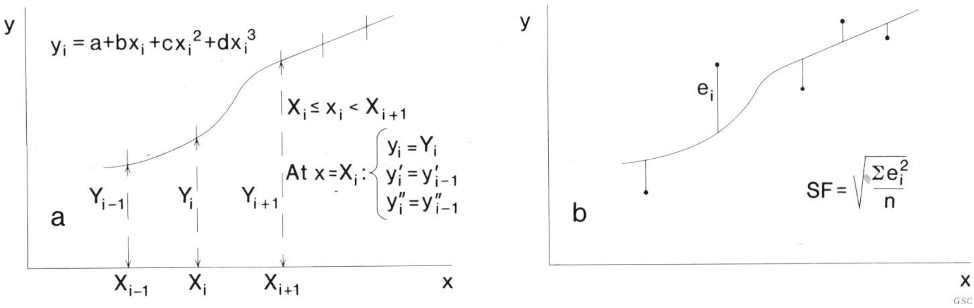

Fig. 3.8 Schematic diagrams of cubic interpolation spline and cubic smoothing spline. The cubic polynomials between successive knots have continuous first and second derivatives at the knots. The smoothing factor (SF) is zero for interpolation splines. Here as well as in later applications, the abscissae of the knots coincide with those of the data points.

3.6 Use of cubic smoothing splines for removing "noise" from microfossil abundance data

Two benthonic species (*E. mosquensis* and *O. strumosum*) show exponential autocorrelations in the Tojeira 1 and 2 sections introduced in Section 3.3 and are good candidates for attempts to filter out the noise in order to retain systematic patterns of change of abundance in the stratigraphic direction which may be useful for biostratigraphic correlation. *E. mosquensis* was selected for further work because it is relatively abundant throughout the entire shale section of Tojeira 1 and 2 whereas *O. strumosum* is nonexistent or rare in the lower half of the Tojeira Formation.

Various statistical methods are available for elimination of noise from data. These include curve-fitting using polynomial or Fourier series, geostatistical "Kriging", signal extraction as in statistical theory of communication, and the construction of smoothing splines. A variant of the latter technique will be used here because it is particularly well suited for coping with the problem of irregular sampling intervals in one dimension.

Figure 3.8 illustrates the concepts of interpolation and smoothing spline functions. Although splines of higher and lower orders can be constructed, the third-order or cubic spline seems to be optimum for irregularly spaced sampling intervals (see later). Spline functions have a long history of use for interpolation; e.g. in numerical integration. Their use for smoothing is a relatively recent development which commenced in the late 1960s after the discovery of smoothing splines by Schoenberg (1964) and Reinsch (1967, 1971). Whittaker (1923) had proposed an early variant.

The interpolation spline curve passes through all (n) observed values. Along the curve, there are a number of knots where various derivatives of the spline function are forced to be continuous. In the example of Figure 3.8, the knots coincide with the data points. A separate cubic polynomial with 4 coefficients is computed for each interval between successive data points. These cubics must have continuous first and second derivatives. After setting the second derivative equal to zero at the first and last data points, the continuity constraints yield so many conditions, that all ($4n$-4) coefficients can be computed. Smoothing splines have the same properties as interpolation splines except that they do not pass through the data points. Instead of this, they deviate from the observed values by an amount that can be regulated by means of the smoothing factor (SF) representing the average mean squared deviation.

For each specific value of SF, which can be set in advance, or estimated by cross-validation (see Section 10.4), a single smoothing spline is obtained. In his recent book on spline smoothing and non-parametric regression, Eubank (1988, e.g., p. 153) discusses that unequally spaced data points may give poor results for smoothing splines. De Boor (1978) pointed this out for interpolation splines. In order to avoid poor results obtained by following cubic smoothing splines to biostratigraphic data for constructing age-depth curves, Agterberg et al. (1985) proposed the simple "indirect" method to be discussed in more detail in Section 9.3. The age data in this approach have relatively large errors while the depths are irregularly spaced. First, a cubic spline is fitted to the ages using relative depths (levels) at a regular interval instead of the actual, irregularly spaced depth measurements. For this purpose the actual depth levels are equally spaced with interval distance set equal to unity.

A separate spline is fitted to the depth measurements along a depth scale, but expressing them as a monotonically increasing function of level. In practice this second curve is nearly an interpolation spline. Combination of the two curves, accompanied by further smoothing if required, yields the final cubic spline for the age-depth relationship. This

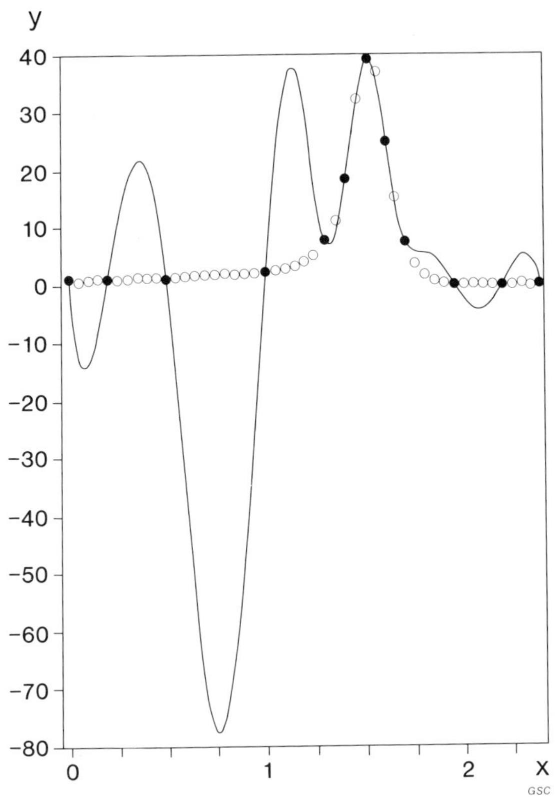

Fig. 3.9 De Boor (1978, Fig. 8.1, p. 224) simulated irregular spacing along x-axis by selecting 12 points (solid circles) from set 49 regularly spaced measurements of a variable (y) as a function of another variable (x). The optimum fifth order interpolation spline (with 7 knots) provides poor fit except around the peak.

result is not subject to unrealistic oscillations as may arise in data gaps if a spline-curve is directly fitted to the data. In the next section, the indirect method will be applied to microfossil abundance data. These data show increases as well as decreases in the stratigraphic direction; oscillations due to irregular spacing in the stratigraphic direction arise even more frequently than in age-depth curve applications for which the spline-curves must be monotonically increasing with age and depth.

The following experiment with interpolation splines illustrates how the problem of unrealistic oscillations can be avoided, using the indirect method. It should be kept in mind that the problem of oscillations in data gaps becomes even more serious if the data are subject to "noise" as in applications to microfossil abundances. Figure 3.9 is from De Boor (1978,

p. 224). In total, 49 observations were available for a property of titanium (y) as a function of temperature (x). These data points have regular spacing along the x-axis. Irregular spacing was simulated by De Boor by selecting $n=12$ data points which are closer together on the peak than in the valleys. De Boor used this example to illustrate that poor results may be obtained even if use is made of a method of optimal spline interpolation in which best locations are computed for (n-k) knots of a k-th order spline. For the example of Figure 3.9, $k=5$ so that 7 knots were used. Although these seven knots have optimal locations along the x-axis, the result is obviously poor, because the shape of the relatively narrow peak is reflected in nonrealistic oscillations in between the more widely spaced data points in the valleys. De Boor (1978, p. 225) pointed out that using a lower-order spline would help to obtain a better approximation. In subsequent applications, use is made of cubic splines only ($k=3$). Figure 3.10A shows the cubic interpolation spline for the 12 irregularly spaced points of Figure 3.9 using knots coinciding with data points. Contrary to the 5th order spline with 7 knots, the new result provides a good approximation. Deletion of 3 more points from the valleys (Fig. 3.10B) begins to give the relatively poor cubic interpolation spline of Figure 3.10C which has unrealistic oscillations in the valleys because all intermediate data points were deleted.

Figure 3.10 also shows results obtained by applying the indirect method in the situation that led to the worst cubic-spline result for the previous example (7 data points, Fig. 3.10C). Figure 3.10D is the cubic interpolation spline for regularly spaced "levels". Figure 3.10E is a monotonically increasing cubic smoothing spline with a small positive value of SF for the relation between x and level. Figure 3.10F is the combination of the curves of Figures 3.10D and E. The approximation to the original pattern for 49 values (Fig. 3.9) is only relatively poor in the valleys where no data were used for control. Unrealistic oscillations were avoided by the use of the three-step indirect method of Figure 3.10(D-F).

3.7 Biostratigraphic correlation between Tojeira 1 and 2 sections in central Portugal using *E. mosquensis* abundance data

Figures 3.11A and B show sequences of samples (combined Stam and Nazli data) for the Tojeira 1 and 2 sections. Distances in the stratigraphic direction are given in meters measuring downward from Stam's

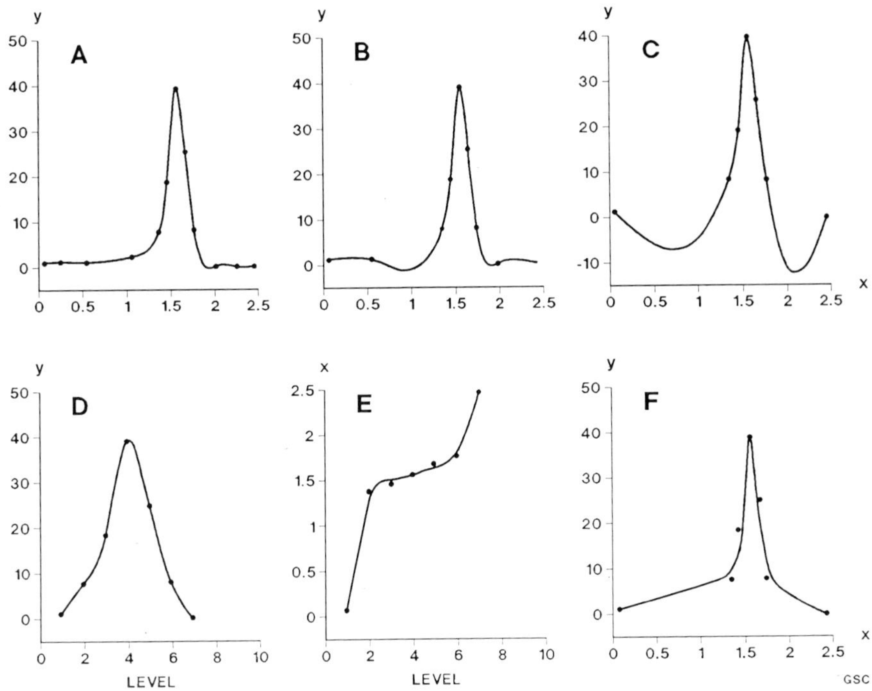

Fig. 3.10 **Top part**: Cubic interpolation splines with knots at data points fitted to irregularly spaced data. (A) Use of same 12 points as in Fig. 3.9 gives good result; (B) deletion of 3 points in the valleys still gives fair interpolation spline although local minima at both sides of the peak are not supported by original data set of 49 measurements; (C) deletion of 2 more points in the valleys results in poor cubic interpolation spline. **Bottom part**: Indirect method of cubic spline-fitting. (D) The six intervals along the x-axis between data points were made equal before calculation of cubic interpolation spline; (E) non-decreasing cubic spline with small positive value of smoothing factor (SF = 0.038) was fitted to interval as function of "levels"; (F) curves of (D) and (E) were combined with one another and re-expressed as cubic spline function which does not show the unrealistic fluctuations of the cubic interpolation spline of Fig. 3.10C.

stratigraphically highest sample (No. 6.29) in Tojeira 1. This sample was taken just below the base of the overlying Cabrito Formation. The stratigraphically highest sample in Tojeira 2 (No. 11.19) occurs about 6m below this base. It is noted that 3 samples taken by Stam in Tojeira 2 above No. 11.19 (cf. Fig. 3.2, right side) contained too few Foraminifera for abundance data to be determined.

The data for *E. mosquensis* plotted in Figure 3.11, were tabulated in Agterberg et al. (1990, Table 3). As shown by Nazli (1988), Tojeira microfossil abundances are normalized when the probit transformation is applied. (The probit transformation consists of converting a proportion to

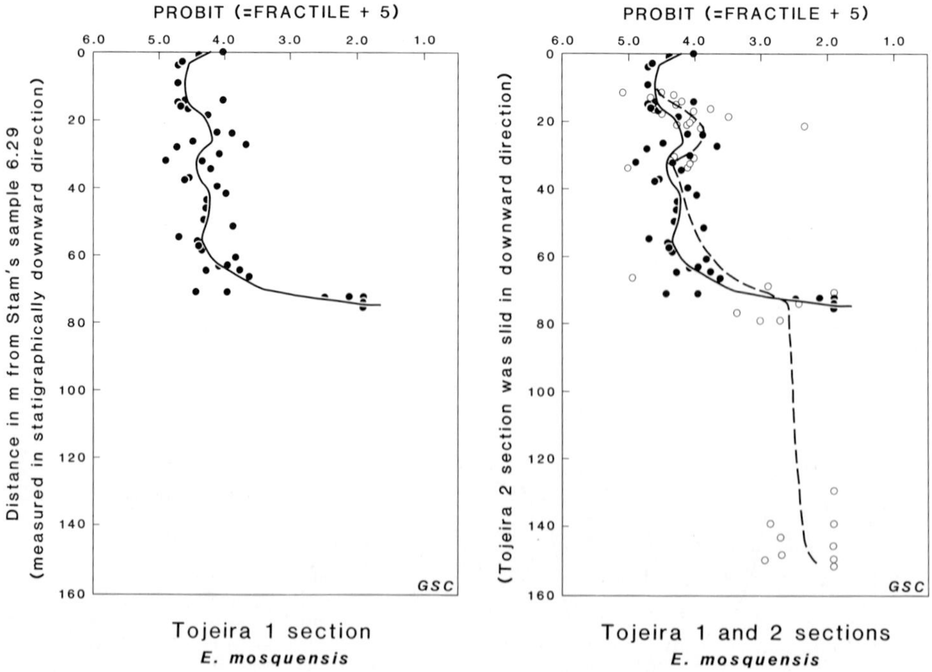

Fig. 3.11 **Left side**: Indirect method of cubic spline-fitting illustrated in Fig. 3.10 (D-F) applied to probits of *E. mosquensis* abundance data for Tojeira 1 section. **Right side**: Same with observations and spline-curve for Tojeira 2 section superimposed. Patterns were slid with respect to one another until a reasonably good fit was achieved. Zero distance (at sample 6.29 in Tojeira 1) falls just below base of overlying Cabrito Formation (cf. Fig. 3.2). Correlation between the two sections is poorest along the 35m data gap in Tojeira 2.

its fractile of the normal distribution in standard form and adding 5 to the result). The purpose of the latter expression is to reduce the relative influence of both relatively high and low values. Such "normalization" is desirable because smoothing splines are fitted by using the method of least squares in which the influence of each deviation from the curve increases according to the square of its magnitude. The smoothing factor (SF) should not be mainly determined by relatively few values only.

Results for the indirect method applied to *E. mosquensis* in Tojeira 1 and 2 are shown in Figures 3.11A and B, respectively. The two spline-curves were slid with respect to one another until a "best" fit was found (see Fig. 3.11B). A 10m downward movement of the Tojeira 2 sequence, which places the base of the overlying Cabrito Formation in nearly the same stratigraphic position in both sections, produces the best correlation.

It is noted that there is a 35m data gap in the Tojeira 2 section so that the local maximum and minimum located within the equivalent of this gap in Tojeira 1 could exist in Tojeira 2 as well. For Tojeira 1, sampling was restricted to the shales of the Tojeira Formation whereas samples for the underlying Montejunto Formation in which *E. mosquensis* is absent or rare were also obtained and used for Tojeira 2. In real distance, the two sections are about 2km apart. It may be concluded from the pattern of Figure 3.11B that it is likely that both Tojeira 1 and 2 share essentially the same relative changes in abundance of *E. mosquensis* during deposition of the approximately 70m of late Jurassic shale in this part of the Lusitanian Basin.

Stam's (1986) plots for the P/B (plankton/benthos) ratio in the Tojeira sections suggested that there may exist several oscillations with peaks where benthos and plankton are nearly equally abundant separated by valleys with little or no plankton. Precise correlation of these peaks and valleys is not possible because of "noise" which even became more prominent when P/B ratios for Nazli's samples were added. Agterberg et al. (1989) showed results obtained by the indirect method of spline fitting applied to the transformed data for P/B ratio in the two sections. Locations of samples were shown with respect to Stam's sample 6.29 in both sections (Tojeira 2 was slid 10m downward as in Fig. 3.11B). Although, on the average, more plankton was deposited in the area of Tojeira 2, the spline-curves display patterns that can be interpreted as similar. In total, there were probably four peaks in the P/B ratio indicating successive periods of planktonic bloom during deposition of the upper Jurassic shale. This result collaborates the one described for the *E. mosquensis* abundance data (see Fig. 3.11).

Not only abundance data can be used for correlation. Reyment (1980) has reviewed basic techniques combining statistics and time series analysis applied to morphometrics of evolutionary sequences. Ecologically induced changes in morphology may be useful for biostratigraphic correlation as well.

3.8 Multivariate methods

Multivariate methods of correlation, using sample by sample matrices of similarity, or distance coefficients, seek clustering of samples (Q-mode) as a function of comparative fossil content. In the final dendrogram, the level of clustering of samples may be selected according to a value which is a function of the degree of association of the original taxa observed. Biostratigraphic fidelity is a simple numerical expression of the preference of a species for a particular cluster (zonal) unit. Depending on the similarity coefficient and weighting procedure selected, multivariate cluster analysis and expression of biostratigraphic fidelity for taxa in the final dendrogram will define assemblage type zonations. Excellent reviews were given by Hazel (1977), Brower et al. (1978) and Millendorf et al. (1978). Individual dendrogram clusters may be either of paleoecologic or stratigraphic significance, or both. The same is true for multivariate clustering on species by species matrices (R-mode). The latter may be insensitive to rare and scattered first and last occurrences of taxa, but such may be an advantage for robust correlation. R-mode clustering may be successfully applied to small data sets. Multivariate methods have been reviewed by Brower (1985a). For applications to chemical determinations and borehole logs, see Reyment and Sturesson (1987).

Methods of multivariate analysis including principal components analysis, factor analyses, multidimensional scaling, correspondence analysis and cluster analysis are firmly based on relatively simple statistical theory (Kendall, 1975b). Computer programs are widely available for these techniques which are used extensively mainly outside the earth sciences. Hohn (1978, 1985) used principal components for stratigraphic correlation. Order of stratigraphic events in time is not necessarily preserved when multivariate statistical methods are applied. For example, Brower (1985a) obtained four clusters (A, B, C and D) for a data set of Upper Cretaceous Foraminifera from the Western Interior Seaway of the United States. These clusters clearly identify assemblages of similar fossils but their order in the dendrogram (A, C, B, D) is not according to their order in relative geological time which is A, B, C, D. Nevertheless, the clusters are useful for lateral tracing.

Palynologists have developed a method of stratigraphically constrained cluster analysis which has proved particularly satisfactory for pollen frequency data (Grim, 1987). As opposed to ordinary, unconstrained analysis, only stratigraphically adjacent clusters are considered for merging. Grim's (1987) computer program CONISS for stratigraphically constrained cluster analysis uses the method of incremental sum of squares. As an option, this program will also perform an unconstrained analysis which can be useful for comparison because this option can indicate re-occurrence if a pollen assemblage higher up in the sequence.

Another recent example of application of multivariate analysis in biostratigraphy is provided by Bonham-Carter et al. (1986). Foraminiferal data from 36 offshore wells on the Labrador Shelf, Grand Banks, and Scotian Shelf were analyzed statistically for biostratigraphic correlation and for systematic trends in distribution related to paleobiogeography. Ranking and Scaling (RASC) of the data allowed the recognition of reliable assemblage zones, grouped for this analysis into six well-defined time slices. Subsequent application of correspondence analysis using Hill's (1979) computer program DECORANA (for DEtrended CORrespondence ANAlysis) showed clearly geographic trends in faunal distribution, differing according to latitude. About one-half of the taxa are planktonic; many of these restricted to southern and more offshore wells that were influenced by the presence of a proto-Gulf Stream. The remaining taxa are predominantly benthonic, and may be allocated broadly to two groups, one with widespread species occurring throughout the region, and a smaller group that is restricted to northern wells on the Labrador Shelf, possibly favored by the influence of terrigenous sediment supply. This threefold effect of southern planktonics, ubiquitous benthonics, and minor northern benthonics is recognized throughout the Cenozoic, with minor fluctuations. During Middle-Late Eocene, relatively many taxa are restricted northerly benthonics, reflecting the fossiliferous, thick terrigenous mudstone sequence in northern wells. During Early-Middle Miocene, the southerly restricted planktonics predominate, reflecting Gulf Stream influence during climatic warming. In the late Neogene, a small group of benthonics are relatively ubiquitous due to the onset of the shelfbound Labrador current. In this study the combined use of RASC and correspondence analysis provided a good tool for unscrambling the influence of both time and paleoenvironment on the dataset.

Burroughs and Brower (1982) applied Wilkinson's (1974) method of seriation to order a data matrix consisting of the presence/absence of *m* taxa taken from *n* samples in *p* stratigraphic sections. The objective of seriation is to arrange the data into a range chart with the taxa in the columns and the samples in the rows. This is accomplished by concentrating the presences of the taxa along the main diagonal of the matrix so that the range zones are minimized. Bonham-Carter et al. (1986) showed that Wilkinson's seriation method may give results similar to Hill's method of correspondence analysis. Brower (1985b) has pointed out that seriation was originally developed by archaeologists who only rarely possess information on the sequence of the taxa in individual sections. Burroughs and Brower (1982) found that ordinary seriation generally yields solutions in which the originally observed relative stratigraphic position of the samples within the individual sections has been lost. They proposed a new method of constrained seriation in which the order relationships of the samples in the sections is preserved in the final solution. Bonham-Carter et al. (1986) approached the same problem, by subdividing their events into six separate time slices on the basis of prior stratigraphic analysis with RASC. The relative position of events within any particular time slice remains uncertain so that clusters of events were more appropriate than a complete stratigraphic ordering of each event in their study.

3.9 Research on time-scales

The construction of good regional and global time-scales provides a key theme for further research in quantitative chronostratigraphy. During the last few years of existence of IGCP Project 148, participants began work along these lines, because it was realized that an ultimate goal in stratigraphic correlation is isochron contouring.

Time-scale research falls into two categories:

1. Calibration and linkage of biostratigraphic and other unique geological events to a common chronostratigraphic scale;

2. Stretching of the (relative) chronostratigraphic scale, along the time axis, to create a geological time scale measured in Ma (10^6y) units.

In the absence of direct radiometric estimates for many chronostratigraphic boundaries, geological and statistical techniques have to be developed to allow reliable inferences on the numerical age of stage boundaries. The use of such indirect methods to construct Mesozoic and Cenozoic scales, applicable both in local basin sequences and in general, became an important activity in IGCP Project 148.

The relative ordering of events in Earth history is a primary concern of geologists. On a regional basis, spatial relationships of separate or overlapping rock volumes are used for accomplishing this goal. The simplest type of relative time scale is a sequence of ordered events. From the variable amounts of overlap between rock volumes, or by making assumptions on rates of sedimentation, it may be possible to estimate intervals between events along a relative time axis. For correlation over large distances between regions or when the rate of change of geological processes in time is being considered, it is necessary to use the numerical time scale which is largely based on radiometric ages of variable precision.

In 1982 two time scales were published (Odin 1982; Harland et al. 1982). There is general agreement on the ages along most of these time scales. The largest discrepancies amount to about 10 percent of the ages estimated (also see Section 1.6). Harland et al. (1982) estimated 144 Ma for the Jurassic-Cretaceous boundary and 590 Ma for the Precambrian-Cambrian boundary, and Odin (1982) 130 Ma and 530 Ma, respectively. Such differences are related to the nature of the materials used for dating.

Although they are helpful for pointing out the existence of significant discrepancies (see e.g. Gradstein et al., 1988), statistical methods cannot be used to resolve difficulties related to the nature of the materials used for dating. Neither can they solve the problem of choosing decay constants in order to avoid bias in radiometric dating. However, any radiometric method is subject to a measurement error which increases with age and is usually much greater than the uncertainties associated with the relative ordering of events using methods of stratigraphic correlation (e.g. biostratigraphic or magnetopolarity methods). The problem of having to estimate the age of stage and chronozone boundaries from relatively imprecise isotope determinations remains even if all sources of bias related to these methods could be eliminated.

Cox and Dalrymple (1967) have developed a statistical approach for estimating the age of boundaries between polarity chronozones in the Cenozoic (Brunhes, Matuyama, Gauss and Gilbert Chronozones). A slightly modified version of their method was used in Harland et al. (1982) for estimating the ages of boundaries between the stages of the Phanerozoic geological time scale.

This statistical approach is as follows. Suppose that t_e represents an assumed trial or "estimator" age for the boundary between two stages. Then the n measured ages t in the vicinity of this boundary can be classified as t_y (younger) or t_o (older than the assumed stage boundary). Each age determination t_{yi} or t_{oi} has its own standard deviation s_i.

Because these standard deviations are relatively large, a number (n_a) of the age determinations may be inconsistent with respect to the estimator t_e. Only the n_a inconsistent ages t_{ai} with $t_{oi} < t_e$ and $t_{yi} > t_e$ were used for estimation by Cox and Dalrymple (1967). These inconsistent ages may be indicated by letting i go from 1 to n_a.

In Harland et al. (1982) a quantity E^2 with

$$E^2 = \sum_{i=1}^{n_a} (t_{ai} - t_e)^2 / s_i^2 \tag{3.6}$$

was plotted against t_e in the chronogram for a specific stage boundary.

Such a plot usually has a parabolic form, and the value of t_e for which E^2 is a minimum was used as the estimated age of the stage boundary.

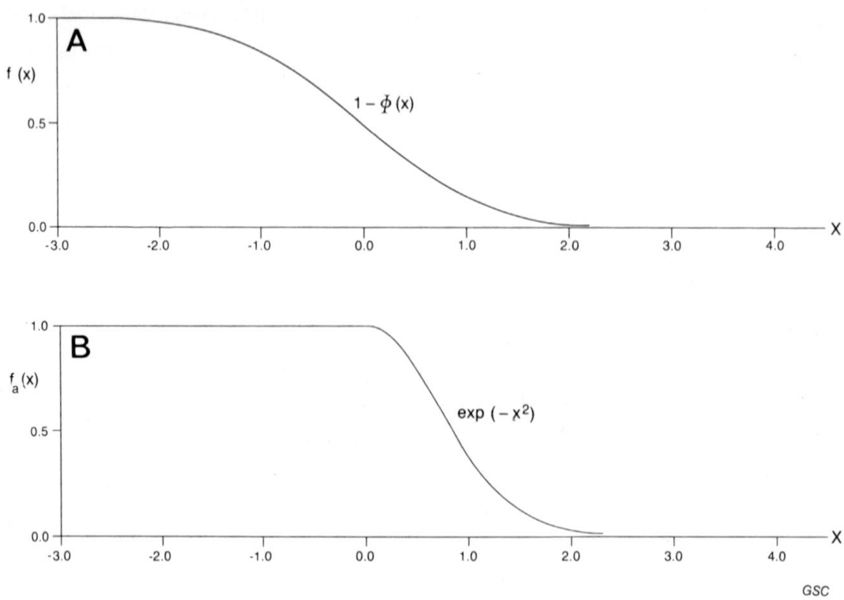

Fig. 3.12 Weighting functions on basis of which likelihood function can be estimated. A. The function $f(x)$ follows from assumption that every age determination is sum of random variables for (1) uniform distribution of (unknown) true ages, and (2) Gaussian distributions for measurements. B. The function $f_a(x)$ is for inconsistent ages only. Its log-likelihood function is $-E^2$.

The statistical model originally proposed by Cox and Dalrymple (1967) may be formulated as follows. Suppose that a stage with upper age boundary t_1 and lower boundary t_2 is sampled at random. This yields a population of ages $t_1 < t < t_2$ with uniform frequency density function $h(t)$. Suppose that every age determination is subject to an error which is normally distributed with unit variance. In general, the frequency density function $f(t)$ of measurements of which the errors satisfy the density function ϕ for the normal distribution in standard form satisfies:

$$f(t) = \int_{-\infty}^{\infty} \phi\,(t-x)\,h\,(t)\,dx \tag{3.7}$$

Because $h(t)$ is uniform, this becomes

$$f(t) = \frac{1}{t_2-t_1}\int_{t_1}^{t_2} \phi\,(t-x)\,dx \tag{3.8}$$

or:

$$f(t) = \frac{1}{t_2-t_1}\left[\Phi\,(t-t_1) - \Phi\,(t-t_2)\right] \tag{3.9}$$

where Φ represents the cumulative distribution function of the normal distribution in standard form. For this derivation, the unit of t was set equal to the standard deviation of the errors. Alternatively, the duration of the stage can be kept constant whereas the standard deviation (σ) of the measurements is changed. Suppose that t_2 - t_1 = 1, then Equation (3.9) becomes

$$f(t) = \Phi\left(\frac{t-t_1}{\sigma}\right) - \Phi\left(\frac{t-t_2}{\sigma}\right) \tag{3.10}$$

Graphical representations of $f(t)$ for different values of σ were given by Cox and Dalrymple (1967; Fig. 7, p. 2611). It could be argued that $h(x)$ is not necessarily uniform and departures from uniformity would affect $f(t)$. However, one would need very large samples of age determinations before the choice of a different model for $h(x)$ would be justified.

Suppose now that the true age τ_e of a single stage boundary is to be estimated from a sequence of estimator ages t_e by using n measurements of variable precision on specimens which are known to be either younger or older than the age of this boundary. This problem can be solved if a weighting function $f(x)$ is defined. The boundary is assumed to occur at the point where $x = 0$. If one is only interested in the lower boundary of a stage, $\Phi\{(t - t_1)/\sigma\}$ can be set equal to one yielding the weighting function $f(x > t_e) = 1 - \Phi(x)$ which is graphically shown in Figure 3.12A. Alternatively, this weighting function can be derived directly: If all possible age above the stage boundary have an equal chance of being represented, then the probability that their measured age assumes a specific value is proportional to the integral of the Gaussian density function for the errors. In terms of the definitions given, any inconsistent age t_y greater than t_e has $x > 0$ whereas consistent ages with $t_y < t_e$ have $x < 0$. It is assumed that standardization of an age t_{yi} or t_{oi} can be achieved by dividing either $(t_{yi} - t_e)$ or $(t_{oi} - t_e)$ by its standard error s_i yielding $x_i = (t_{yi} - t_e)/s_i$ or $x_i = (t_{oi} - t_e)/s_i$.

Suppose that x_i is a realization of a random variable X. The weighting function $f(x)$ then can be used to define the probability $P_i = P(X_i = x_i) = f(x_i)\,\Delta x$ that x will lie in a small interval Δx about x_i. The method of maximum likelihood for a sample of n values x_i consists of finding the value of t_e for which the product of the probabilities P_i is a maximum. Because Δx can be set equal to an arbitrarily small constant, this maximum occurs when the likelihood function

$$L = L(x|t_e) = \prod_{i=1}^{n} \left[f(x_i) \right] \tag{3.11}$$

is a maximum. The so-called log-likelihood function is obtained by taking the logarithm at both sides of this equation. For the model of Figure 3.12A,

$$\log L(x|t_e) = \sum_{i=1}^{n} \log_e \left[1 - \Phi(x_i) \right] \tag{3.12}$$

If the log-likelihood function is written as y and its first and second derivatives with respect to t_e as y' and y'', respectively, then the maximum likelihood estimator $\hat{t}_e$ occurs at the point where $y'=0$ and its variance is $-1/y''$ (cf. Kendall and Stuart, 1966, p. 43). The log-likelihood function becomes parabolic in shape when n is large. Suppose that the equation of this parabola is written as $y = a + bt_e + ct^2_e$. Then the maximum likelihood estimate $\hat{t}_e$ satisfies $\hat{t}_e = -b/2c$ with variance $s^2(\hat{t}_e) = -1/2c$. It will be shown by computer simulation experiments that for most chronograms in Harland et al. (1982) n is sufficiently large and yields good estimates $\hat{t}_e$ of the ages of the stage boundaries with corresponding standard deviations. It can be shown (see Agterberg, 1988) that a chronogram using E^2 represents the maximum likelihood solution for a filter with equation

$$f_a(x|t_e) = exp\left[-(x-t_e)^2\right] \tag{3.13}$$

where $x > t_e$ because n_a inconsistent ages are used only. This weighting function is shown in Figure 3.12B. If the corresponding likelihood function is written as L_a, it follows that $E^2 = -\log_e L_a$.

For example, the quantity E^2 is plotted in the vertical direction of Figure 3.13 for the Caerfai–St. David's boundary example taken from Harland et al. (1982, Fig. 3.7i). The data on which this chronogram is based are shown along the top. Values of E^2 were calculated at intervals of 4 Ma and a parabola was fitted to the resulting values by using the method

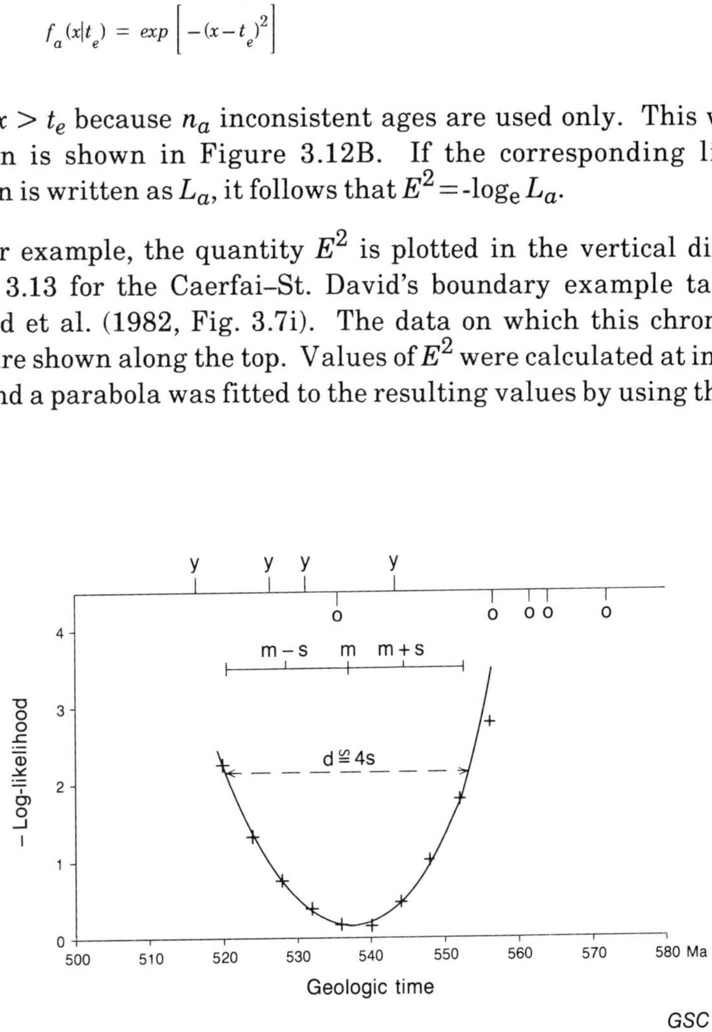

Fig. 3.13 Chronogram for Caerfai-St. David's boundary example and parabola fitted by method of least squares. E^2 = - log-likelihood is plotted in vertical direction. Dates belonging to stages which are older and younger than boundary are indicated by o and y, respectively. Standard deviation follows from d representing width of parabola for E^2 equal to its minimum value augmented by 2.

of least squares. If the log-likelihood function is parabolic, with E^2 satisfying

$$E^2 = -a - bt_e - ct_e^2 \tag{3.14}$$

it follows that the maximum likelihood estimator is normally distributed with mean $\hat{\tau}_e = b/2c$ and variance $s^2(\hat{\tau}_e) = 1/2c$. It will be shown in the next paragraph that graphically $s(\hat{\tau}_e)$ might be determined by taking one fourth of the width of the parabola at the point where E^2 exceeds its minimum value by 2.0 (see Fig. 3.13). The latter result applies to parabolas based on L_a and L. Harland et al. (1982) defined the error of their estimate by taking one-half the age range for which E^2 does not exceed its minimum value by more than 1.0. This yields a standard deviation that is $\surd 2$ times as large as the one resulting from L_a.

A simple proof of the validity of the modified error-range method illustrated in Figure 3.13 is as follows. According to the theory of mathematical statistics (Kendall and Stuart, 1961, pp. 43-44), the likelihood function is asymptotically normal:

$$e^y = \frac{1}{\sigma\sqrt{2\pi}} \exp\left(-t^2/2\sigma^2\right) \tag{3.15}$$

In this expression $e^y = L(x|t_e)$ and $t = t_e - \tau$; σ represents the standard deviation of this normal curve centered about $\tau = 0$. Taking the logarithm at both sides gives the parabola:

$$y = \max - t^2/2\sigma^2 \tag{3.16}$$

where max represents the maximum value of the log-likelihood function. Setting $y = \max - 2$ gives $t = 2\sigma$. This means that the width of the parabola at 2 units of y below its maximum value is equal to 4σ. The parabola shown in Figure 3.13 (and subsequent illustrations) is assumed to provide an approximation of the true log-likelihood function. The standard deviation obtained from the fitted curve is written as s. In Figure 3.13, the y-axis has been inverted so that $-y = E^2$ points upwards in order to facilitate comparison with the chronograms in Harland et al. (1982).

Figure 3.14 shows estimates based on L. The resulting parabola is almost equal to the one in Figure 3.13 which was based on L_a instead of L.

The estimated ages of the Caerfai – St. David's boundary and their standard deviations obtained for L_a and L also are similar. This conclusion will be corroborated by a more detailed comparison of the weighting functions for L and L_a at the end of this section, and by computer simulation experiments to be described in the next section. However, L_a does not provide a good approximation of L when inconsistent ages are missing.

A parabolic chronogram is more readily obtained when the consistent ages are used together with the inconsistent ages as in the method discussed here. A numerical example of the kinds of differences in results obtained is as follows. An age estimate based on the chronogram of Harland et al. (1982, Fig. 3.4h, p. 57) for the Norian-Rhaetian boundary would be approximately 213 Ma. The corresponding standard error as reported by Harland et al. (1982) is 9 Ma. The maximum likelihood method using the same set of 6 data gives an estimated age of 215.5 Ma with corresponding standard error of 4.2 Ma.

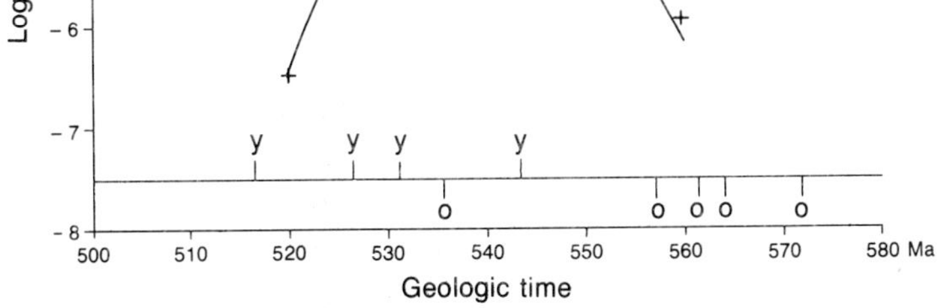

Fig. 3.14 Caerfai-St. David's boundary example. Age (m) estimated by maximum likelihood method using L. Standard deviation (s) and width of 95 percent confidence interval are approximated closely by results shown in Figure 3.13.

The chronogram interpreted as an inverted log-likelihood function

The approach taken is this section differs slightly from the one originally taken by Cox and Dalrymple (1967) as will be discussed in more detail now. The basic assumptions that the dates are uniformly distributed through time and subject to measurement errors are made in both methods of approach. Cox and Dalrymple (1967, see their Fig. 4 on p. 2608) demonstrated that, under these conditions, the inconsistent dates for younger rocks have probability of occurrence P_{Iy} with:

$$P_{Iy}(t) = \tfrac{1}{2}\,\mathrm{erfc}\left[\frac{(t-\tau)}{\sigma_m\sqrt{2}}\right] \tag{3.17}$$

where erfc denotes complementary error function and τ represents true age of the chronostratigraphic boundary (boundary between geomagnetic polarity epochs in Cox and Dalrymple's original paper). The standard deviation for the measurement errors is written as σ_m. Setting $\tau = 0$ and using the relationship $\tfrac{1}{2}\,\mathrm{erfc}\,(z/\sqrt{2}) = 1 - \Phi\,(z)$ it follows that:

$$P_{Iy}(t) = 1-\Phi\left(\frac{t}{\sigma_m}\right) = f\left(\frac{t}{\sigma_m}\right) \tag{3.18}$$

If t/σ_m is replaced by x, the weighting function shown in Figure 3.12A is obtained. Consequently, this weighting function can be interpreted as the probability that an inconsistent age t_a is measured for younger rocks. Likewise, $P_{Io}(t) = f(-t/\sigma_m)$ can be defined for older rocks.

Cox and Dalrymple (1967) next introduced the trial boundary age t_e and defined a measure of dispersion of all inconsistent dates t_a with respect to t_e satisfying:

$$D^2(t_a - t_e) = \int_{-\infty}^{\infty} (t-t_e)^2 P_I(t)\,dt \tag{3.19}$$

where $P_I(t) = P_{Iy}(t)$ if $t \geq 0$; and $P_I(t) = P_{Io}(t)$ if $t \leq 0$. For $t_e = \tau$, this quantity is a minimum (see Cox and Dalrymple, 1967, Fig. 5 on p. 2608). A normalized version of E^2 can be directly compared to the theoretical curve for $D^2(t_a - t_e)$ when the number of inconsistent dates is large. This normalization consisted of dividing E^2 by average number of dates per unit time interval. It is noted that $P_I(t)$ does not represent a probability density function, because it can be shown that

$$\int_{-\infty}^{\infty} P_I(t)\,dt = (2/\pi)^{\frac{1}{2}} = 0.80 \text{ if } t_e = \tau \tag{3.20}$$

In this section, E^2 is not interpreted as approximately proportional to $D^2(t_a - t_e)$. Instead of this, it is regarded as the inverse of a log-likelihood function with Gaussian weighting function. For very large samples, good estimates can be obtained using the inconsistent dates only. For small samples, however, significantly better results are obtained by using the consistent dates also and by replacing the Gaussian weighting function by $f(x)$.

All Gaussian weighting functions provide the same mean age of a chronostratigraphic boundary when the maximum likelihood method is used. However, the standard deviation of this mean depends on the choice of the constant p in $\exp(-px^2)$. For example, $p=1.0$ for $f_a(x)$ in Figure 3.12B. Assuming that $f(x)$ of Figure 3.12A represents the correct weighting function, one can ask for which p the Gaussian function $\exp(-px^2)$ provides the best approximation to $f(x)$ with $x \geq 0$. Let u represent the deviation between the two curves, so that

$$\log_e\{1 - \Phi(x)\} = -px^2 + u \tag{3.21}$$

Minimizing Σu^2 for $x_i = 0.1\,k$ $(k = 1,2..., 20)$ by the method of least squares gives $\hat{p} = 1.13$. Because of the large difference between the two curves near the origin, $\hat{p}$ increases when fewer values x_i are used. It decreases when more values are used. Letting k run to 23 and 24 yields $\hat{p}$ equal to 1.0064 and 0.9740, respectively. These results confirm the conclusion reached before that a Gaussian weighting function with $p = 1.0$ provides an excellent approximation to $f(x)$.

3.10 Computer simulation experiments on estimation of the age of chronostratigraphic boundaries

Computer simulation experiments were performed by Agterberg (1988) in order to attempt to answer the following questions: (a) does the theory of the preceding section remain valid even when the number of available dates is very small; (b) how do estimates obtained by the method of fitting a parabola to the log-likelihood function compare to estimates obtained by the method of scoring which is commonly used by statisticians

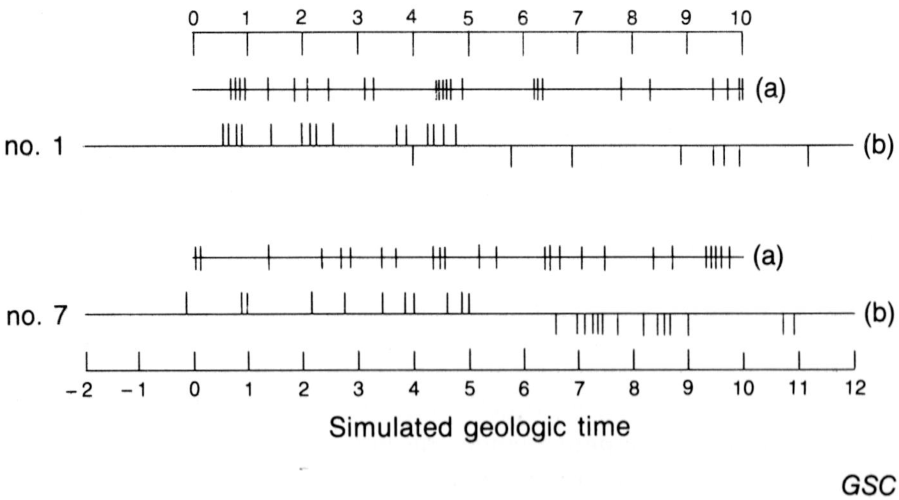

Fig. 3.15 Two examples of runs (Runs No. 1 and No. 7) in computer simulation experiment. True dates (a) were generated first, classified and increased (or decreased) by random amount. Younger and older ages are shown above and below scale (b), respectively.

(see e.g. Rao, 1973); and (c) how do results derived from the chronograms in Harland et al. (1982) compare to those obtained by the maximum likelihood method.

Figure 3.15 and Table 3.2 illustrate the first type of computer simulation experiment performed. Twenty-five random numbers were generated on the interval [0, 10]. These numbers with uniform frequency distribution can be regarded as true dates (τ) without measurement errors. The stage boundary was set equal to 5 (=mid-point of interval). Values of τ less than 5 belong to the younger stage A, and those greater than 5 to the older stage B (see Table 3.2). The measurement error was introduced by adding to τ a normal random number with zero mean and standard deviation equal to one. As a result of this, each value of τ was changed into a date t. Some values of t ended up outside the interval [0, 10], like 11.197 in the first example (Run No. 1 in Table 3.2 and Fig. 3.15), and were not used later. In Run No. 1, a single date for the younger state (A) has $t>5$, and a date for B has $t<5$. Suppose now, for example, that the trial age of the stage boundary t_e is set equal to 4.6. Then there are 3 inconsistent ages for Run No. 1 and these are marked by asterisks in Table 3.2. Each normalized date $x = t - t_e$ was converted into a z-value (=fractile of normal distribution in standard form) by changing its sign if it belongs to the younger stage A. The value of z was transformed into a probability

$P = \Phi(z)$ for values of t on the interval $[t_e - 3, t_e + 3]$ where $\Phi(z)$ denotes cumulative frequency of the normal distribution in standard form. The frequency corresponding to 3 is equal to 0.999 of which the natural logarithm is equal to -0.001. For this reason, values outside the interval $t_e \pm 3$ yield probabilities which are approximately 1 (or 0 for the log-likelihood function) and these were not used for further analysis. Thus a natural window is provided screening out dates that are not in the vicinity of the age of the chronostratigraphic boundary to be estimated. Most probabilities are greater than 0.5. Only inconsistent dates (asterisks in Table 3.2) give probabilities less than 0.5. The value of the log-likelihood

TABLE 3.2

Run 1 for computer simulation experiment. True dates τ were classified as younger (A) or older (B) than true age of stage boundary (=5). Dates t with measurement error are compared to trial age (t_e = 4.6). Inconsistent ages are indicated by asterisks. z = $-x$ for younger rocks (A) and z = x for older rocks (B). Standard normal z-value is fractile of probability P. Total of logs of P gives value of log-likelihood function for t_e = 4.6.

τ		t	x (= t-4.6)	z	P	$\log_e$ P
4.587	A	4.380	-0.220	0.220	0.5871	-0.5325
7.800	B	8.048	3.448	3.448		
2.124	A	2.193	-2.407	2.407	0.9920	-0.0081
0.668	A	2.239	-2.361	2.361	0.9909	-0.0092
6.225	B	5.802	1.202	1.202	0.8853	-0.1218
9.990	B	9.945	5.345	5.345		
4.896	A	4.574	-0.026	0.026	0.5102	-0.6730
4.606	A*	6.487	1.887	-1.887	0.0296	-3.5211
0.796	A	0.553	-4.047	4.047		
1.855	A	2.526	-2.074	2.074	0.9810	-0.0192
6.292	B	6.923	2.323	2.323	0.9899	-0.0101
3.280	A	1.998	-2.602	2.602	0.9954	-0.0046
2.422	A	1.435	-3.165	3.165		
1.397	A	0.912	-3.688	3.688		
4.538	A	4.365	-0.235	0.235	0.5928	-0.5230
0.830	A	0.803	-3.797	3.797		
6.194	B*	4.033	-0.567	-0.567	0.2854	-1.2540
4.545	A	3.930	-0.670	0.670	0.7490	-0.2890
4.774	A*	4.814	0.214	-0.214	0.4154	-0.8786
0.905	A	0.713	-3.887	3.887		
9.763	B	11.197				
8.285	B	8.902	4.302	4.302		
3.131	A	3.676	-0.924	0.924	0.8224	-0.1955
9.987	B	9.435	4.835	4.835		
9.442	B	9.620	5.020	5.020		
					Total =	-8.0397

TABLE 3.3

Values of log-likelihood functions estimated for Run 1 and predicted values for parabola fitted by method of least squares. Initial guesses of extreme values are indicated by asterisks.

TIME	LOG-LIKELIHOOD (Σ log P)	SUM OF SQUARES (E^2)	PREDICTED LLF	PREDICTED E^2
3.0	-15.58	10.86		
3.1	-14.41	9.37		
3.2	-13.30	8.00		
3.3	-12.27	6.75		
3.4	-11.31	5.63		
3.5	-16.98	13.54		
3.6	-15.83	12.07		
3.7	-14.75	10.73		
3.8	-13.75	9.52		
3.9	-12.81	8.43		
4.0	-11.94	7.46		
4.1	-11.13	6.59		
4.2	-10.39	5.84		
4.3	-9.72	5.21		5.11
4.4	-9.10	4.69		4.69
4.5	-8.54	4.27		4.32
4.6	-8.04	3.93	-7.98	3.99
4.7	-7.59	3.65	-7.57	3.71
4.8	-7.20	3.44	-7.21	3.47
4.9	-6.87	3.27	-6.89	3.28
5.0	-6.58	3.15	-6.61	3.14
5.1	-6.35	3.06	-6.38	3.04
5.2	-6.16	3.02	-6.19	2.99
5.3**	-6.02	3.01**	-6.05	2.98**
5.4	-5.93	3.05	-5.95	3.01
5.5	-5.88	3.13	-5.89	3.09
5.6*	-5.88*	3.24	-5.88*	3.22
5.7	-5.92	3.40	-5.91	3.39
5.8	-6.00	3.59	-5.98	3.61
5.9	-6.13	3.84	-6.10	3.88
6.0	-6.29	4.15	-6.26	4.18
6.1	-6.49	4.51	-6.46	4.54
6.2	-6.73	4.94	-6.71	4.94
6.3	-7.01	5.42	-7.00	5.38
6.4	-7.33	5.97	-7.33	
6.5	-7.69	6.57	-7.71	
6.6	-8.08	7.23	-8.13	
6.7	-8.50	7.91		
6.8	-8.97	8.65		
6.9	-9.47	9.43		
7.0	-10.01	10.24		

function for t_e is the sum of the logs of the probabilities as illustrated for $t_e = 4.6$ in Table 3.2.

Log-likelihood values for Run No. 1 are shown in Table 3.3 with t_e ranging from 3 to 7 in steps of 0.1. The largest log-likelihood value is reached for $t_e = 5.6$ and this value was selected as the first approximation $\hat{t}_{e1}$ of the age of the stage boundary. In total, 21 values of $\hat{t}_e$ with $| t_e - \hat{t}_{e1} | \leq 1.0$ were used for fitting a parabola as shown in Figure 3.16. The fitted parabola is more or less independent of number of values used (=21) and width of neighborhood (=2). However, the neighborhood should not be made too wide because of random fluctuations (local minima or maxima) near $t_e = 3$ or 7 (see e.g. Table 3.3). These edge effects should be avoided.

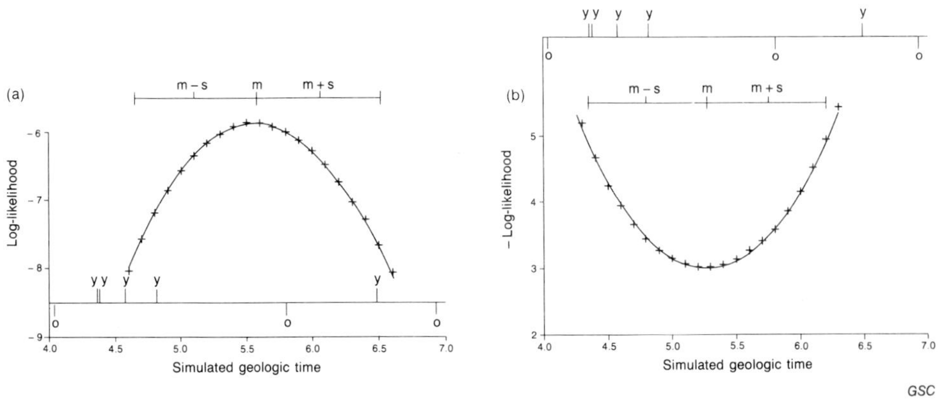

Fig. 3.16 Maximum-likelihood method used for estimating mean of age of stage boundary in Run 1 (data as in Fig. 3.15). Standard deviation (s) and 95 percent confidence interval also are shown. A. Likelihood function L was used. B. Chronogram for Run 1 (using L_a instead of L). Note similarity of s and 95 percent confidence interval in Figs. 3.16A and B.

They are due to the fact that the initial range of simulated time was arbitrarily set equal to 10 in the computer simulation experiment. The peak of this parabola provides the second approximation $m = \hat{t}_{e2}$ of the estimated age. The standard deviation (s) of the corresponding normal distribution can be used to estimate the 95 percent confidence interval $m \pm 1.96s$ also shown in Figure 3.16.

The sum of squares E^2 for L_a, using inconsistent dates only, is also shown in Table 3.3 as a function of t_e. The first approximation of its minimum value is 5.3. The corresponding parabola is shown in Figure 3.16. The mean age resulting from L_a is about 0.3 less than the mean based on L and its standard deviation is nearly the same. It is fortuitous that the mean based on L_a is closer to the population mean (=5) than that based on L. On the average, the original maximum likelihood (L) method gives better results (see results for 50 runs given at the end of this section).

Younger and older ages generated in each of the first 10 (unit variance) computer simulation runs are shown in Figure 3.17 together with their estimated mean and 95 per cent confidence interval using L. Theoretically, each population mean (=5) is contained within the 95 percent confidence interval around the sampling mean with a probability of 95 percent. The means and standard deviations used for

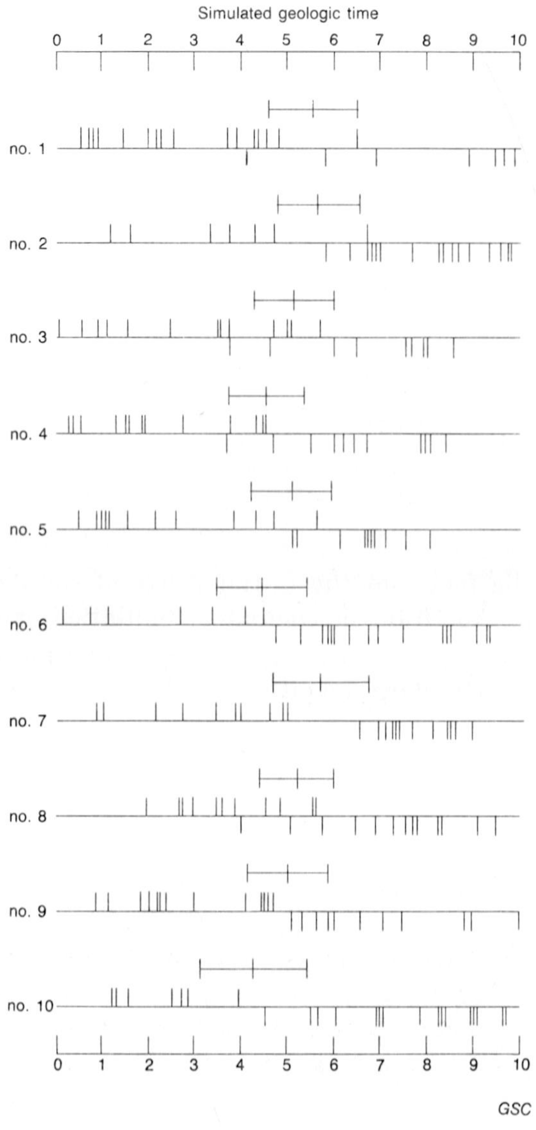

Fig. 3.17 Dates generated in first 10 runs of computer simulation experiment (cf. results for No. 1 and No. 7 shown in Fig. 3.15). Mean and 95 percent confidence interval estimated by maximum-likelihood method are shown for comparison with true mean (=5).

Figure 3.17 are listed in Table 3.4 (Maximum likelihood method with parabola). Also listed in Table 3.4 are the corresponding results for L_a (Gaussian weighting function with parabola). The means based on L_a are close to those for L. The estimated standard deviations tend to be either

TABLE 3.4

First 10 runs of computer simulation experiment. Comparison of results obtained by fitting parabola and scoring method, respectively. Standard deviations marked by asterisks are too large (cf. Fig. 3.18B).

	Maximum Likelihood Method					Gaussian Weighting Function				
Run		Parabola		Scoring			Parabola		Scoring	
No.	Mid-point	Mean	S.D.	Mean	S.D.	Mid-point	Mean	S.D.	Mean	S.D.
1	5.6	5.582	0.479	5.554	0.481	5.3	5.269	0.470	5.260	0.500
2	5.7	5.632	0.481	5.663	0.489	6.3	6.190	0.480	6.264	0.500
3	5.1	5.153	0.420	5.142	0.423	4.8	4.884	0.335	4.828	0.316
4	4.5	4.506	0.447	4.507	0.452	4.2	4.321	0.395	4.216	0.354
5	5.1	5.070	0.461	5.089	0.466	5.3	5.217	0.482	5.293	0.408
6	4.4	4.419	0.502	4.448	0.505	4.6	4.625	0.749*		
7	5.7	5.710	0.531	5.728	0.542	5.8	5.767	3.924*		
8	5.2	5.205	0.406	5.200	0.411	5.0	5.025	0.364	5.017	0.408
9	5.0	5.022	0.417	5.018	0.419	5.0	4.966	0.614*		
10	4.2	4.231	0.609	4.232	0.623	4.3	4.248	1.001*		

slightly smaller or much greater. It can be seen from the results for Run No. 7 shown in Figure 3.18 that the greater standard deviations are due to a break-down of this particular method of estimation.

Results obtained by means of the method of scoring (see e.g. Rao, 1973, p. 366-374) also are shown in Table 3.4. In our application of this method, the following procedure was followed. As before, the log-likelihood was calculated for 0.1 increments in t_e and the largest of these values was used as the initial guess. Suppose that this value is written as y. Two other values x and z were calculated representing log-likelihood values close to y at small distances -10^{-4} and 10^{-4} along the t_e-axis. The quantities $D_1 = 0.5(z - x)\cdot 10^4$ and $D_2 = (x - 2y + z)\cdot 10^8$ were used to obtain a second approximation of the mean by substracting D_1/D_2 from the initial guess. The procedure was repeated until the difference between successive approximations became negligibly small. Then the standard deviation of the estimate is given by $SD = 1/|D_2|$.

For L, the scoring method generally yields estimates of SD which are slightly greater than those resulting from the parabola method. However, the difference is negligibly small (Table 3.4). For L_a, the scoring method provided an answer in only 6 of the 10 experiments of Table 3.4.

Similar results were obtained for runs in a second type of computer simulation experiment using variable measurement error (see Agterberg, 1988, for details). In total, 50 runs were made for each of the two types of

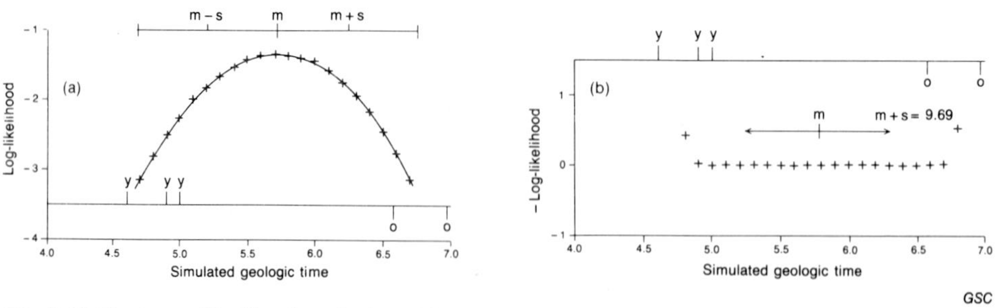

Fig. 3.18 Maximum-likelihood method used for estimating mean age of stage boundary in Run 7 (data as in Fig. 3.15). A. Likelihood function L was used. B. Likelihood function L_a did not give good result.

experiments. For constant variance of measurement errors, the parabola method for L gave an overall mean equal to 4.9287 and standard deviation 0.4979 as calculated from 50 means. The corresponding numbers for the second type of experiment were 4.9442 and 0.5160. The Gaussian weighting scheme gave overall means equal to 4.9213 and 4.9414 for the two types of experiments, and corresponding standard deviations equal to 0.5790 and 0.6541, respectively. If the parabola did not provide a good fit to the function E^2, because of zero values around its minimum, the mean was approximated by the mid-point of the range of zero values in these calculations. The results of the 50 runs for the two types of experiments confirm the earlier results described in this section. Additionally, they show that the Gaussian weighting function (using L_a) provides results which are almost as good as the method of maximum likelihood (using L).

3.11 Smoothing of time-scales with the aid of cubic spline functions

When the ages of a number of successive chronostratigraphic boundaries have been estimated, they can be further improved by smoothing with the aid of cubic smoothing splines (cf. Section 3.6). The ages shown in Table 3.5 and Figure 3.19 will be used for example. They were derived from chronograms in Harland et al. (1982) with the following relatively minor modifications: (a) if the chronograms for the two boundaries of a stage are the same, indicating absence of dates for that stage, the estimate was assigned to a single point mid-way between the stage boundaries; (b) imprecise estimates for 6 successive Jurassic stages were not used; (c) when inconsistent dates are missing, the estimated age was set equal to the mid-point of the range for missing data in the

TABLE 3.5

Ages and estimated standard deviations used for fitting spline-curve No. 1 shown in Figure 3.19.

Lower boundary of stage		Age	S.D.
1	Maastrichtian (Maa)	72	1.41
2	Campanian (Cmp)	84	1.59
3	Santonian (San)	87.5	1.59
4	Coniacian (Con)	88.5	0.88
5	Turonian (Tur)	91	0.88
6	Cenomanian (Cen)	97.5	0.70
7	Albian (Alb)	113	1.41
8	Aptian (Apt)	122	3.18
9	Barremian (Brm)		
10	Hauterivian (Hau)	124	2.83
11	Valanginian (Vlg)	135	1.77
12	Berriasian (Ber)		
13	Tithonian (Tth)	145	4.24
14	Kimmeridgian (Kim)	151	2.12
15	Oxfordian (Oxf)	158	5.30
16	Callovian (Clv)		
17	Bathonian (Bth)		
18	Bajocian (Baj)		
19	Aalenian (Aal)		
20	Toarcian (Toa)		
21	Pliensbachian (Plb)		
22	Sinemurian (Sin)		
23	Hettangian (Het)	212	4.95
24	Rhaetian (Rht)	213	6.36
25	Norian (Nor)	218	2.83
26	Carnian (Crn)	228	7.78
27	Ladinian (Lad)	238	3.54
28	Anisian (Ans)	242	7.43
29	Scythian (Scy)		
30	Tatarian (Tat)	246	7.07
31	Kazanian/Ufimian (Kaz-Ufi)	253	8.13
32	Kungurian (Kun)		
33	Artinskian (Art)	268	4.24
	Sakmarian/Asselian (Sak-Ass)		

chronogram; and (d) the standard deviation was set proportional to the age range listed in the summary time scale (Harland et al., 1982, pp. 52-55) with constant of proportionality equal to $\frac{1}{2}\sqrt{2}$.

The fourth modification (d) is based on the earlier considerations corroborated by the computer simulation experiments proving that the parabola for L_a provides an excellent approximation to the parabola for L.

A cubic spline-curve was fitted to the data in Figure 3.19 for the following reasons. A spline-curve is very smooth because there are no abrupt changes in the rate of change of its slope; the principle of least squares is used; and deviations between observed values (crosses in

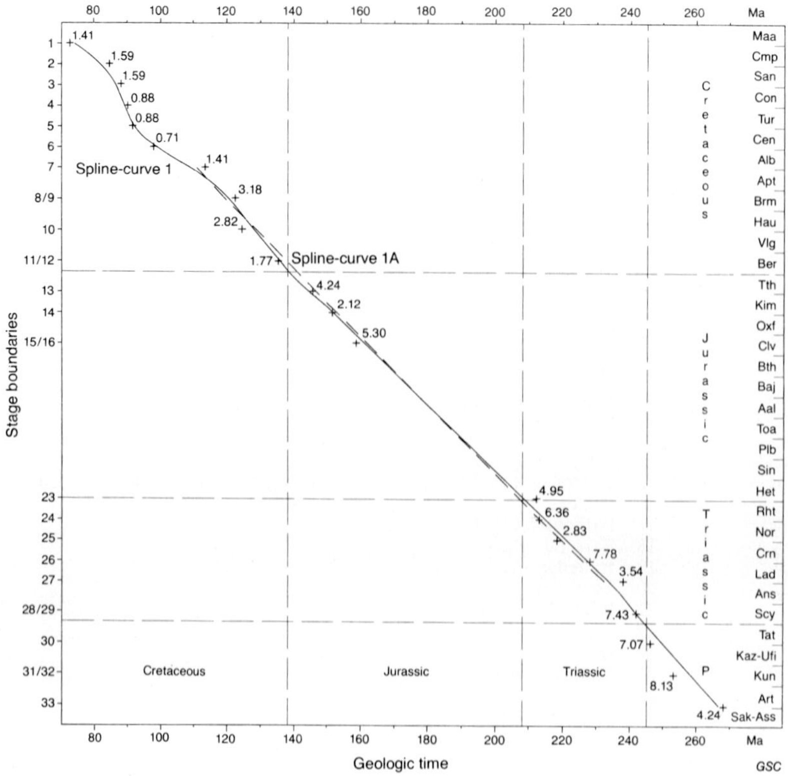

Fig. 3.19 Spline-curves fitted to ages of stage boundaries listed in Table 3.5. Spline-curve 1A was fitted to data for stage boundaries numbered 7 to 27 only.

Fig. 3.19) and spline-curve are permitted to exist but the sum of squares of these deviations can be regulated; a weight can be assigned to each observed value. This weight is inversely proportional to the variance of the observed value.

Let the vertical and horizontal axes in Figure 3.19 represent observations written as x_i, y_i $(i = 1,\ldots, n)$, respectively. Then the smoothing spline-function to be constructed minimizes

$$\int_{x_1}^{x_n} g''(x)^2 \, dx \tag{3.22}$$

among all functions $g(x)$ under the condition that:

$$\sum_{i=1}^{n} \left[\frac{y_i - g(x_i)}{s(y_i)} \right]^2 \leq S \tag{3.23}$$

Here the $s(y_i)$ are the standard deviations of the values y_i. The sum of standardized deviations S is a random variable approximately distributed as chi-squared with n degrees of freedom and variance equal to $2n$. The expected value of S, which is equal to n, was used in the applications of this section.

It can be seen in Figure 3.19 that the fitted spline-curve No. 1 tends to follow the stage boundaries in the Cretaceous more closely because these are relatively precise. In places where the uncertainity is great, the spline-curve tends to become a straight line. Spline-curve No. 1A shown also in Figure 3.19 was fitted to points for stage boundaries between the Anisian and Cenomanian. It is nearly straight and closely approximates Spline-curve 1.

Because the intervals between stage boundaries in the vertical direction of Figure 3.19 are equally spaced, a straight line in this type of plot would agree with the hypothesis of equal duration of stages. Harland et al. (1982) applied linear interpolation between relatively precise stage boundaries (tie-points). The boundaries numbered 1 to 7, 27 and 33 were used as tie-points. Because the crosses for boundaries No. 7 and 27 fall slightly to the right of the fitted spline-curves, the estimates

TABLE 3.6

Ages used for fitting spline-curve No. 2 based on equal duration of Hallam's ammonite zones in the Jurassic; without and with tie-points, respectively.

i	Stage		n_i	x_i	Age	S.D.
13	Tithonian	(Tth)	8	13.4	145	4.24
14	Kimmeridgian	(Kim)	4	14.1	156	0.00
15	Oxfordian	(Oxf)	7	15.4	158	5.30
16	Callovian	(Clv)	6	16.4		
17	Bathonian	(Bth)	7	17.7		
18	Bajocian	(Baj)	7	18.9		
19	Aalenian	(Aal)	3	19.5		
20	Toarcian	(Toa)	6	20.5		
21	Pliensbachian	(Plb)	5	21.4		
22	Sinemurian	(Sin)	6	22.5		
23	Hettangian	(Het)	3	23.0	208	0.00

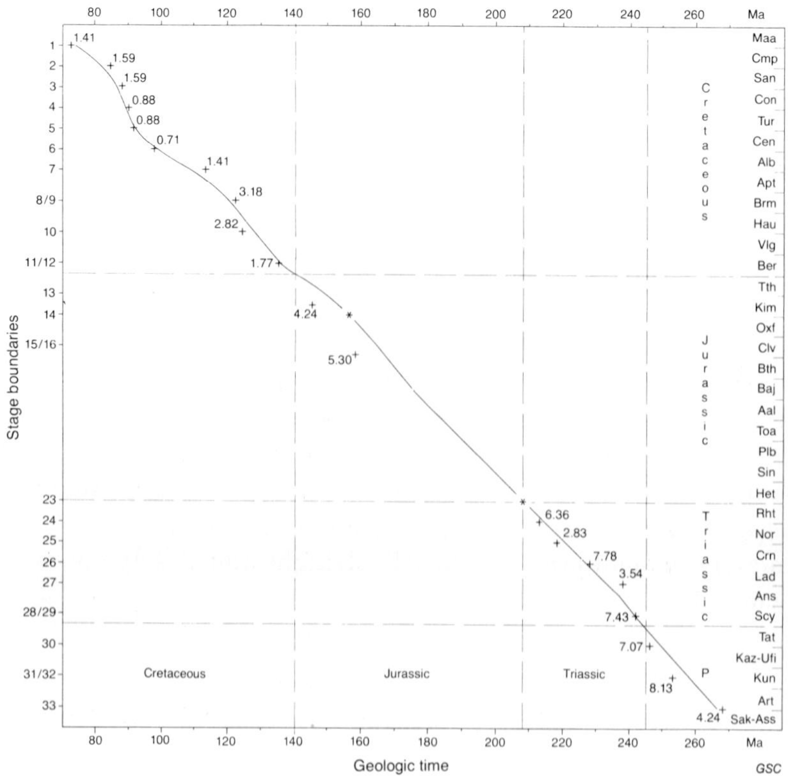

Fig. 3.20 Spline-curve fitted to ages of stage boundaries for Jurassic listed in Table 3.6. This cubic smoothing spline passes exactly through two tie-points with SD = 0.

obtained by spline-interpolation are younger than those of Harland et al. (1982) as will also be shown later (see Fig. 3.21).

With respect to the Jurassic time scale, Kent and Gradstein (1985, 1986) have argued that it is more reasonable to assume equal duration of zones than equal duration of stages. They used Hallam's (1975) ammonite zones for spacing the stage boundaries in the Jurassic between tie-points at the base of the Kimmeridgian and Hettangian, respectively. On the basis of other evidence including data on rates of seafloor spreading in the Late Jurassic and Early Cretaceous between marine magnetic anomalies M25 and M0, Kent and Gradstein assumed ages of 156 Ma and 208 Ma for these two stage boundaries (No. 14 and No. 23), respectively.

The values of x_i used for constructing the spline-curve of Figure 3.19 can be modified by using n_i for number of ammonite zones per stage (see Table 3.6). The new values x_i shown in Table 3.6 satisfy

$$x_i = x_{i-1} + cn_i$$

$$x_{12} = 12; \quad i = 13, ..., 23$$

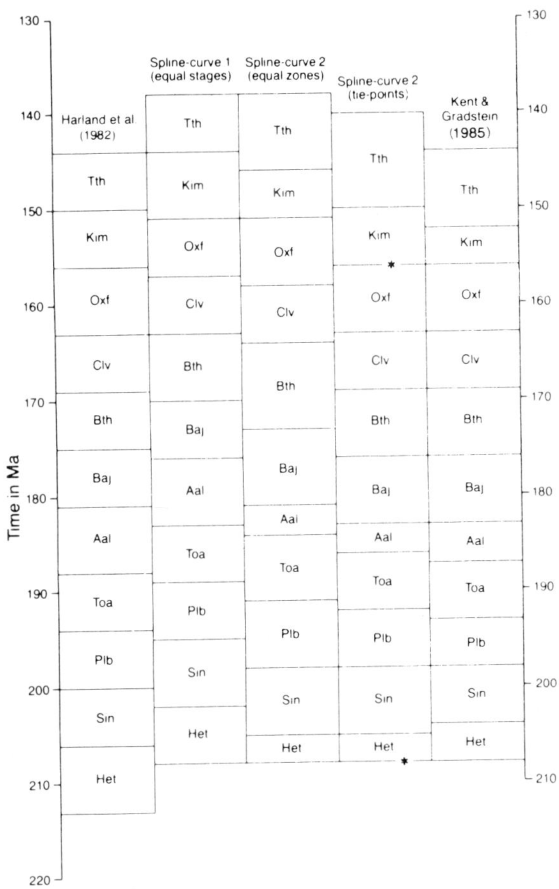

Fig. 3.21 Comparison of spline-curve ages (rounded off to nearest integer Ma values) for Jurassic to ages estimated by Harland et al. (1982) and by Kent and Gradstein (1985). The asterisks in column 4 denote key ages of tie-points through which the spline-curve solution was forced to pass. For further information see Agterberg (1988).

where $c = 11/62 = 0.1774$ represents the ratio of total number of stages (= 11) and zones (= 62) in the Jurassic.

The input for spline-curve fitting was further modified by using as tie-points 156 Ma instead of 151 Ma for the Oxfordian-Kimmeridgian and 208 Ma instead of 212 Ma for the Triassic-Jurassic boundary, respectively, setting the standard deviations of these ages equal to zero. As demonstrated in Agterberg (1988, Appendix 2), the spline-curve has the property of passing exactly through points of which the standard deviation is zero. Spline-curve No. 2 with tie-points is shown in Figure 3.20. The ages of stage boundaries (rounded off to 1 Ma) obtained by three methods of cubic spline-fitting are shown in Figure 3.21 for comparison with the other age estimates. Ages for the modified spline-curve (No. 2) for equal duration of zones but without use of tie-points are shown between those based on Figures 3.20 and 3.21. The spline-curves all gave 208 Ma for the age of the Triassic-Jurassic boundary which is younger than estimate of 213 Ma in Harland et al. (1982) although the same original age determinations were used.

The spline-curves yield ages of 138 Ma and 140 Ma for the Jurassic-Cretaceous boundary which are younger than the 144 Ma age in Harland et al. (1982) and Kent and Gradstein (1985). This relatively young age is mainly due to the effect of (a) a relatively young Oxfordian glauconite age listed as 148.22 Ma in Harland et al. (1982) and as 145 ± 3 Ma in Armstrong (1978) who, in turn, extracted it from Gyji and McDowell (1970), and (b) 4 other relatively young glauconite ages listed in Harland et al. (1982) for the Tithonian. If these 5 dates would not be used, the spline-curves would also give an age of approximately 144 Ma for the top of the Jurassic. In the beginning of Section 3.9 it was pointed out that Odin (Editor, 1982) using more glauconite dates estimated a much younger age (130 Ma) for this boundary. The problem of estimating the age of the Jurassic-Cretaceous boundary also will be considered in the next section.

3.12 Statistical significance of ages

The book on a geological time scale by Harland et al. (1982) differs from earlier publications on the same subject in that it contains tables with all dates that were used and detailed description of results (e.g. chronograms) obtained by systematic treatment of the data. In the last three sections it has been shown that statistical estimation of the ages of chronostratigraphic boundaries in the geological time scale can be improved in two ways: (a) the maximum likelihood method can be used for estimation of the age of individual chronostratigraphic boundaries, and (b) after estimating the ages of a set of successive boundaries by the method of maximum likelihood, these can be further improved by using a cubic spline-curve for smoothing. The resulting methodological improvements, however, are small in comparison with changes that result from changing the input data. Harland et al. (1982) used high-temperature dates mainly. If low-temperature dates are used (cf. Odin, Editor, 1982) significantly younger ages are obtained, for some stages, especially those near the Jurassic-Cretaceous and Proterozoic-Phanerozoic boundaries.

Haq et al. (1987) provided a new sea level and sedimentary cycles chart, calibrated to a new geological time scale for which they used mixtures of low- and high-temperature dates. This procedure was criticized by Gradstein et al. (1988) partly because it can be shown that the low-temperature (glaucony) ages are systematically younger. Odin (Editor, 1982) had pointed out for one sample (NDS2) that its glauconite age of 39.6$\pm$1.8 Ma is a minimum age and that 1.5 to 2 Ma should be added to it "bearing in mind the long time necessary for the evolution of the dated glaucony". Similar corrections may have to be applied to other glauconite dates as well.

The following statistical experiments performed by the author was briefly described in Gradstein et al. (1988). In total, 19 low-temperature and high-temperature dates listed by Harland et al. (1982; Table 3.1, p. 61) were used to estimate three different ages of the Jurassic-Cretaceous boundary. The 7 high-temperature dates in this group of 19 dates are plotted along the top of Figure 3.22, and the 12 low-temperature dates along the bottom. The maximum likelihood method was applied taking the high- and low-temperature dates separately, and to the combined group of 19 values. Best-fitting parabolas are shown in Figure 3.22. Trial ages t_e at intervals of 4 Ma were used. Detailed calculations are shown in Table 3.7 for t_e=132 Ma for high-temperature dates only. The parabola fitted to the log-likelihood values of the high-temperature dates shows a relatively poor fit mainly because these values are determined, to a large extent, by a single Jurassic date (153.32$\pm$5.00 Ma). The other older date

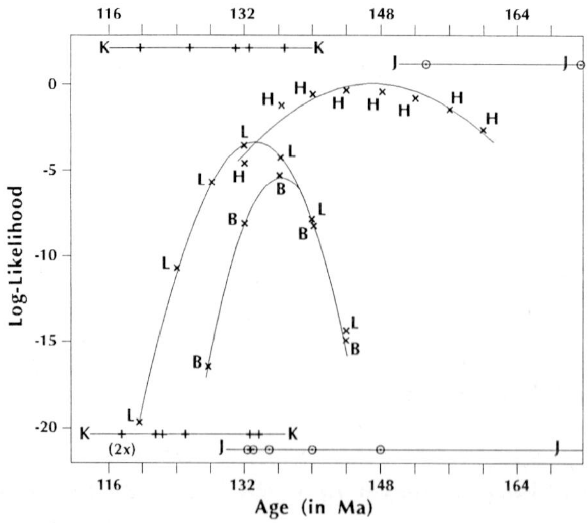

Fig. 3.22 Maximum likelihood method used for estimating age of Jurassic-Cretaceous boundary. See text for further explanation.

(171.66±9.80 Ma) is too far removed from the Jurassic-Cretaceous to make a significant difference.

The glaucony dates separately give a mean age of 133.2±2.3 Ma (error is one standard deviation) which is close to Haq et al.'s (1987) estimate of 131 Ma for the Jurassic-Cretaceous boundary. The high-temperature dates give 147.3±5.4 Ma which is close to the estimates of 144 Ma by Harland et al (1982) and Kent and Gradstein (1985). The estimate based on all 19 dates is 136±1.8 Ma. It is close to Harland et al.'s (1982) chronogram age of 135 Ma. Harland et al. rejected this chronogram age in favor of their 144 Ma age for the Jurassic-Cretaceous boundary because of the former's relative lack of precision. The 144 Ma estimated was obtained by linear interpolation between tie-points for the Aptian-Albian (=113 Ma) and the Anisian-Ladinian (=238 Ma) boundaries.

The difference between the 133.2±2.3 Ma low-temperature and the 147.3±5.4 Ma high-temperature estimates of Figure 3.22 has its own normal distribution with mean of 14.1 Ma and standard deviation of 5.8 Ma. In the absence of bias, this mean difference would be approximately zero. Its standardized value (14.1/5.8=2.93) exceeds the 99% confidence limit (=2.33) of the *z*-test for testing a difference between two means for statistical significance. Statistically, it is therefore 99% certain that the glauconite-based maximum likelihood age is different and younger than the one based on the high-temperature isotope ages in agreement with other comparisons reported in Gradstein et al. (1988).

As pointed out in Section 3.9, Harland et al. (1982) gave a quantitative estimate of the error in the age obtained from a chronogram by taking this error as half the age range for which the error did not exceed its minimum value by more than 1.0. They pointed out that the significance of this error is readily seen where only two identical ages determine a boundary, one of these being from the youngest stage, the other from the older stage. From Equation (3.6) for computing E^2, this quantity is zero at the boundary and rises to 1.0 on both sides of the boundary when the trial age differs from the experimental age by the quoted error. By using the concept of maximum likelihood it was shown that the error of Harland et al. is approximately $\sqrt{2}$ times larger than the standard error, provided that the number of dates is sufficiently large so that the chronogram has become parabolic in shape.

The following slight modification of the preceding argument by Harland et al. also results in a modified estimate of the standard deviation. Two identical ages at a boundary, one from the younger and the other from the older stage, can be averaged to provide a single estimate of the age of this boundary. If the standard deviations of the two age determinations are equal, their average will have a standard deviation

TABLE 3.7

Calculation of logs of probabilities (P) for trial age of 132 Ma using 7 high-temperature dates only. The sum of these values is one of the values plotted in Figure 3.22 and used to fit the parabola for high-temperature dates. Procedure is similar to the one followed in the example of Table 3.2. However, every z-value for an age was obtained after dividing the deviation from the trial age by the measurement error (s) which previously was equal to unity for all deviations in Table 3.2. A and B represent Cretaceous and Jurassic material, respectively.

	Age	s	z	P	$\log_e P$
A	119.66	4.00	-3.09	0.001	-0.001
A	125.26	6.00	-1.12	0.131	-0.140
A	132.51	12.00	0.04	0.516	-0.726
A	136.50	2.50	1.80	0.964	-3.324
A	130.87	4.35	-0.26	0.397	-0.506
B	153.32	5.00	-4.26	0.000	-0.000
B	171.66	4.80	-8.26	0.000	-0.000

which is $\sqrt{2}$ times smaller than the errors of the individual ages. This result is in agreement with the maximum likelihood approximation of L by L_a.

Various authors have assigned different meanings to the error on the Mesozoic and Paleozoic time scales of Harland et al. (1982). For example, Carr et al. (1984) assumed that Harland et al. (1982), by stating that this error is 2.5 Ma, estimated the age of the Jurassic-Cretaceous boundary and 95% confidence interval as 144±2.5 Ma. On the other hand, Menning (1989) quotes "confidence limits" for this boundary as 144±5 Ma. The standard error corresponding to the error of 2.5 Ma estimated by Harland et al. is ($2.5/\sqrt{2}=$) 1.77 Ma. Multiplication of this standard error by 2 gives a statistically-based estimate of 144±3.5 Ma for the 95% confidence interval. This width is between those of Carr et al. (1984) and Menning (1989), respectively.

In order to estimate the precision of the ages of chronostratigraphic boundaries, it is important to have good estimates of the errors of the isotopic dates on which these age estimates are based. Harland et al. (1982) found that although most determinations quote an error, a significant number do not. Errors for these determinations were estimated by fitting a linear regression line to the available error/time data.

For those isotopic ages that have published errors, it may not be immediately obvious whether these are standard deviations or 95% confidence limits. For example, Harland et al. (1982) used a number of Ordivician and Silurian fission track ages from McKerrow et al. (1980) with quoted errors of about 10 Ma. In Gale et al. (1980), these same ages are tabulated with errors "at the 2σ level" that are twice as large (about 20 Ma). From this, it can be inferred that the age determination errors in Harland et al. (1982) are indeed standard deviations, although they were not identified as such in McKerrow et al. (1980).

If errors are standard deviations, it generally can be assumed that there is 68 percent probability that the unknown true value occurs within the error interval reported. By taking error limits that are twice as large this probability is increased to 95 percent. It should be kept in mind that statements of this type imply that the error distributions are Gaussian or "normal".

CHAPTER 4

CODING AND FILE MANAGEMENT OF STRATIGRAPHIC INFORMATION

4.1 Introduction

During the past five years it has become common practice to use microcomputers for the creation, updating and quantitative analysis of stratigraphic information. Lists of fossils and stratigraphic events observed in wells or outcrop sections can be coded and stored together with measurements on their position. The resulting files can be readily submitted to various types of data processing. In the Microsoft Disk Operating System (DOS), for example, files are identified by filenames which are from one to eight characters long. These filenames may be followed by extensions consisting of a period followed by one, two or three characters.

In order to illustrate data management in biostratigraphy, a number of datasets ranging from small and simple, to large and complex will be introduced in this chapter. Later, these same datasets will be used to illustrate automated stratigraphic correlation techniques. The primary purpose of the data management required is to create various types of sequence files for different stratigraphic sections which can later be systematically compared with one another in preparation of automated stratigraphic correlation. Before presentation of the datasets, five types of files are defined which will be used in the examples. For convenience, the different types of files are indicated by three-letter extensions as in Microsoft DOS.

4.2 Five basic types of files

The five basic types of files to be distinguished are: DIC, DAT, SEQ, PAR, and DEP files.

A dictionary file (DIC) is an ordered list of names of taxa or events. The sequence position numbers of the items in the list provide unique identifiers for coding purposes. Data (DAT) files contain coded stratigraphic information for taxa using formats which closely reflect original data collection procedures. Sequence (SEQ) files are lists of successive or coeval stratigraphic events which can either be coded directly or derived automatically from DAT files. Parameter (PAR) files contain the settings of switches and values of parameters required for running the RASC computer program for RAnking and SCaling or other data analysis procedures. Depth (DEP) files contain stratigraphic data for individual wells or sections, augmented by regional time-scale information for automated stratigraphic correlation.

As input, the RASC computer program requires a DIC file for stratigraphic events and a SEQ file for their superpositional relations within individual sections. Although SEQ files can be coded from original data records, it is usually more convenient to create DAT files instead of SEQ files, especially if the information is to be extracted from large databases. Depth data can be extracted from a DAT file if automatic stratigraphic correlation between sections is to be performed on the basis of probable dephts derived by analysis of DEP files.

DIC files

Dictionary (DIC) files contain lists of fossil names (or event names). They include all names to be used for a regional study. The order of the names in the DIC files is arbitrary when the file is created. The names may be initially ordered according to a system selected by the user. For example, the alphabetic order of taxa can be used, taxa can be grouped according to families, with alphabetic order within families, or use can be made of the order in which different taxa are identified in one or more relatively complete stratigraphic sections for a region.

Microsoft DOS permits rapid alphabetic sorting of names. (It also is possible to obtain alphabetic lists by means of RASC.) However, most stratigraphers prefer other types of order for their lists. When a list of fossil names, alphabetic or otherwise, is available for a region, the names can be automatically numbered for the DIC files. The assigned sequence numbers will later be used as codes for the taxa. It is convenient to enter only one name per taxon in the original DIC file for a region. In exploratory drilling, when well cuttings are used to determine highest occurrences of taxa (and lowest occurrences are not used because of downhole contamination), the DIC file initially created for taxa, can be used for the highest occurrences as well. If both highest and lowest occurrences of taxa are used, it may be necessary to create a new DIC file for events from the DIC file for taxa. A simple procedure for this is to automatically replace each taxon dictionary number i (i= 1,2,...,n) by two numbers (2i-1) and (2i). The odd numbers (2i-1) may be used for lowest occurrences and even numbers (2i) for highest occurrences. In the RASC computer program for this procedure the same taxon name is used for highest and lowest occurrences. They are distinguished in the event dictionary by preceding them with the indicators HI and LO, respectively.

DAT files

Data (DAT) files contain information on all events in all sections to be used for the study of a region. Different formats can be used. These formats may emulate data entry procedures of the paleontologist. DAT files consist of separate lists of samples corresponding to the separate stratigraphic sections or wells for a region. Examples of formats are as follows: For exploratory wells, the paleontologist often works with cuttings which successively become available while proceeding in the stratigraphically downward direction. For each well, the depth of a sample, e.g. as measured from sealevel, can be entered , followed by the highest occurrences of all taxa identified for this sample. For outcrop sections, the paleontologist usually works in the stratigraphically upward direction. The distances measured in the stratigraphic direction (perpendicular to bedding) may be measured for each region from the base of each section upwards. Consequently, every section has its own scale. The origins of these scales which are set at the stratigraphically lowest points in the sections usually do not occur in the same bed. A common procedure of coding the information consists of entering the name of a taxon followed by its lowest and highest occurrence measured along the scale for the section. This scale may be in meters or feet, or may be a sequence of numbers representing beds counted in the stratigraphically upward direction. If beds without highest or lowest occurrences are skipped in the counting, the numbers represent so-called "event levels". DAT files can automatically be changed into SEQ and preliminary DEP files. The depth files that can be created from a DEP file are preliminary because information on probable depths of events in wells (or probable locations of events in outcrop sections) which is needed for automated stratigraphic correlation only can be added after application of ranking and scaling to the SEQ file.

SEQ files

Sequence (SEQ) files consist of sequences of all stratigraphic events in all sections to be used for the study of a region. The events are positioned according to their relative stratigraphic position, usually proceeding in the stratigraphically downward direction. Normally, SEQ files are automatically created from DAT files, replacing them by superpositional or equipositional (coeval) relations. The relative event levels are used for indicating order in the SEQ files. The information in a SEQ file is sufficient to ascertain for any pair of events (A, B) in a section whether A was observed to occur stratigraphically above or below B, or whether A and B were observed to be coeval in this section. SEQ files will be used for ranking and scaling of the events in the region. In the optimum sequence for a region, each event will obtain a rank above or below other events. In the scaled optinum sequence there will be different intervals between successive events. Zero interval between successive events along the RASC scale would indicate that the events are coeval on the average for the study region.

PAR files

Parameter (PAR) files contain the settings of switches and values of parameters needed to run the RASC computer program. For example, the user may decide to only use events that occur in k_c or more sections. The value of the parameter k_c then has to be set in the PAR file. In some versions of RASC (e.g. micro-RASC, see Chapter 10), the parameters have default values which can be changed interactively by the user.

DEP files

Depth (DEP) files contain information on the depths (in meters or in terms of event levels) of stratigraphic events measured in the stratigraphically downward direction for single sections. This information is compared to the average positions of the events expressed either as ranks or as RASC distances. Ranks and RASC distances are obtained by ranking and scaling applied to a SEQ file. If the age (in Ma) is known for a sufficiently large subgroup of the events used for a region, the RASC scale can be transformed into a numerical time scale. This may facilitate interpretation and allows isochron contouring (e.g. automated construction of lines of correlation for multiples of 10 Ma). Then the estimated age (in Ma) must be entered into the DEP file. For many types of applications it may seem to be hazardous to convert scaling results to the numerical time-scale. It is not necessary to change RASC scale into a numerical time scale for automated stratigraphic correlation. Also, even if this transformation is applied, the automated stratigraphic correlation between sections actually remains based on the RASC scale because the same regional time scale transformation is applied to all sections. The RASC scale is subjected to local stretching or shrinking to change it into a numerical time scale. In general, the same pattern is obtained for the lines of correlation based on transformed RASC distances (in Ma) or original RASC distances. For specific stratigraphic events, it does not matter whether their probable locations in the sections are based on the RASC scale or on a numerical time scale derived from it.

Fig. 4.1 Locations of sections of the Sullivan database.

4.3: Hay example as derived from the Sullivan database: Lower Tertiary nannoplankton in California

In his original article on probabilistic stratigraphy, Hay (1972) used stratigraphic information on calcareous nannofossils from sections in the California Coast Ranges for example (see Fig. 4.1 for locations). These sections had originally been studied by Sullivan (1964; 1965) and Bramlette and Sullivan (1961). The distribution of Lower Tertiary nannoplankton described in the latter three papers also was used by Davaud and Guex (1978) and Guex (1987) for testing other types of quantitative stratigraphic correlation techniques. The original paper by Hay (1972) resulted in extensive discussions (e.g. Edwards, 1978; Harper, 1981) and applications of other techniques to the Hay example (e.g. Hudson and Agterberg, 1982). For these reasons, the Hay example will be used again here.

Hay (1972) restricted his example to Lower Tertiary nannofossils for samples shown on Sullivan's (1965) correlation chart augmented by stratigraphic information on the Lodo Gulch section from Bramlette and Sullivan (1961). Several of the nannofossil taxa selected for the example are known to occur in older Paleocene strata in the Media Agua Creek and Upper Canada de Santa Anita sections (see Sullivan, 1964). Addition of this other information to the example changes the relative order of the lowest occurrences in these two sections. In general, care should be taken to minimize bias due to lack of sampling older or younger rocks containing fossils of which the highest and lowest occurrences are recorded for a section. This source of bias will be discussed on the basis of the Hay example. It arises only when the time-span for the example has a length which is comparable to those of the ranges of the taxa studied. The problem is almost entirely avoided in datasets which deal with periods, rather than ages (see later).

Tables 4.1 and 4.2 are DIC files for the Hay dataset and larger Sullivan dataset originally coded by Davaud and Guex (1978). Hay (1972) selected for his examples the lowest occurrences of 9 taxa and the highest occurrence of one taxon (*Discoaster tribrachiatus*). The DIC file of Table 4.1 can directly be used as a RASC input file. On the other hand, the DIC file of Table 4.2 is for taxa only and a DIC file should be created from it before RASC can be used. Agterberg et al.(1985) automatically replaced the number (i) of each taxon by a pair of numbers ($2i$-1) and $2i$ for its lowest and highest occurrence, respectively. For example, taxon 89 (*Discoaster*

TABLE 4.1

Dictionary (DIC file) for Hay example. LO and HI represent lowest and highest occurrences of nannofossils, respectively.

1	LO DISCOASTER DISTINCTUS
2	LO COCCOLITHUS CRIBELLUM
3	LO DISCOASTER GERMANICUS
4	LO COCCOLITHUS SOLITUS
5	LO COCCOLITHUS GAMMATION
6	LO RHABDOSPHAERA SCABROSA
7	LO DISCOASTER MINIMUS
8	LO DISCOASTER CRUCIFORMIS
9	HI DISCOASTER TRIBRACHIATUS
10	LO DISCOLITHUS DISTINCTUS

tribrachiatus) was replaced by event 177 (LO *Discoaster tribrachiatus*) and event 178 (HI *Discoaster tribrachiatus*). Thus, event 9 in Table 4.1 represents the same stratigraphic event as event 178 in the RASC input DIC file based on Table 4.2.

TABLE 4.2

Fossil name file (preliminary DIC file) for Sullivan database coded by Davaud and Guex (1978) and Agterberg et al. (1985). A RASC input DIC file was obtained automatically from this file (see text).

1	CHIPHRAGMALITHUS CRISTATUS	27	DISCOLITHUS FIMBRIATUS	53	RHABDOSPHAERA TRUNCATA	79	DISCOASTER BINODOSUS
2	CHIPHRAGMALITHUS ACANTHODES	28	DISCOLITHUS OCELLATUS	54	RHABDOSPHAERA INFLATA	80	DISCOASTER DEFLANDREI
3	CHIPHRAGMALITHUS CALATUS	29	DISCOLITHUS PANARIUM	55	ZYGODISCUS SIGMOIDES	81	DISCOASTER DELICATUS
4	CHIPHRAGMALITHUS DUBIUS	30	DISCOLITHUS PUNCTOSUS	56	ZYGODISCUS ADAMAS	82	DISCOASTER DIASTYPUS
5	CHIPHRAGMALITHUS PROTENUS	31	DISCOLITHUS SOLIDUS	57	ZYGODISCUS HERLYNI	83	DISCOASTER DISTINCTUS
6	CHIPHRAGMALITHUS QUADRATUS	32	DISCOLITHUS VESCUS	58	ZYGODISCUS PLECTOPONS	84	DISCOASTER FALCATUS
7	COCCOLITHUS BIDENS	33	DISCOLITHUS VERSUS	59	ZYGOLITHUS CONCINNUS	85	DISCOASTER LODOENSIS
8	COCCOLITHUS CALIFORNICUS	34	DISCOLITHUS PERTUSUS	60	ZYGOLITHUS CRUX	86	DISCOASTER MULTIRADIATUS
9	COCCOLITHUS EXPANSUS	35	DISCOLITHUS EXILIS	61	ZYGOLITHUS DISTENTUS	87	DISCOASTER NONARADIATUS
10	COCCOLITHUS GRANDIS	36	DISCOLITHUS DUOCAVUS	62	ZYGOLITHUS JUNCTUS	88	DISCOASTER STRADNERI
11	COCCOLITHUS SOLITUS	37	DISCOLITHUS INCONSPICUUS	63	ZYGRHABLITHUS SIMPLEX	89	DISCOASTER TRIBRACHIATUS
12	COCCOLITHUS STAURION	38	CYCLOLITHUS ROBUSTUS	64	ZYGRHABLITHUS BIJUGATUS	90	DISCOASTER CRUCIFORMIS
13	COCCOLITHUS GIGAS	39	ELLIPSOLITHUS MACELLUS	65	BAARUDOSPHAERA BIGELOWI	91	DISCOASTER GERMANICUS
14	COCCOLITHUS DELUS	40	ELLIPSOLITHUS DISTICHUS	66	BAARUDOSPHAERA DISCULA	92	DISCOASTER LENTICULARIS
15	COCCOLITHUS CONSUETUS	41	HELICOSPHAERA SEMILUNUM	67	MICRANTHOLITHUS FLOS	93	DISCOASTER MARTINII
16	COCCOLITHUS CRASSUS	42	HELICOSPHAERA LOPHOTA	68	MICRANTHOLITHUS INAEQUALIS	94	DISCOASTER MINIMUS
17	COCCOLITHUS CRIBELLUM	43	LOPHODOLITHUS NASCENS	69	MICRANTHOLITHUS VESPER	95	DISCOASTER SEPTEMRADIATUS
18	COCCOLITHUS EMINENS	44	LOPHODOLITHUS RENIFORMIS	70	MICRANTHOLITHUS BASQUENSIS	96	DISCOASTER SUBLODOENSIS
19	CYCLOCOCCOLITHUS GAMMATION	45	LOPHODOLITHUS MOCHOLOPHORUS	71	MICRANTHOLITHUS CRENULATUS	97	DISCOASTER HELIANTHUS
20	CYCLOCOCCOLITHUS LUMINIS	46	RHABDOSPHAERA CREBRA	72	MICRANTHOLITHUS AEQUALIS	98	DISCOASTER LIMBATUS
21	DISCOLITHUS PECTINATUS	47	RHABDOSPHAERA MORIONUM	73	CLATHROLITHUS ELLIPTICUS	99	DISCOASTER MEDIOSUS
22	DISCOLITHUS PLANUS	48	RHABDOSPHAERA PERLONGA	74	RHOMBOASTER CUSPIS	100	DISCOASTER PERPOLITUS
23	DISCOLITHUS PULCHER	49	RHABDOSPHAERA RUDIS	75	POLYCLADOLITHUS OPEROSUS	101	DISCOASTEROIDES KUEPPERI
24	DISCOLITHUS PULCHEROIDES	50	RHABDOSPHAERA SCABROSA	76	SPHENOLITHUS RADIANS	102	DISCOASTEROIDES MEGASTYPUS
25	DISCOLITHUS RIMOSUS	51	RHABDOSPHAERA SEMIFORMIS	77	FASCICULOLITHUS INVOLUTUS	103	HELIOLITHUS KLEINPELLI
26	DISCOLITHUS DISTINCTUS	52	RHABDOSPHAERA TENUIS	78	DISCOASTER BARBADIENSIS	104	HELIOLITHUS RIEDELI

Figure 4.2 (after Hay, 1972, Fig. 2, p.261) shows stratigraphic information for the 10 events of Table 4.1 which occur in the nine sections

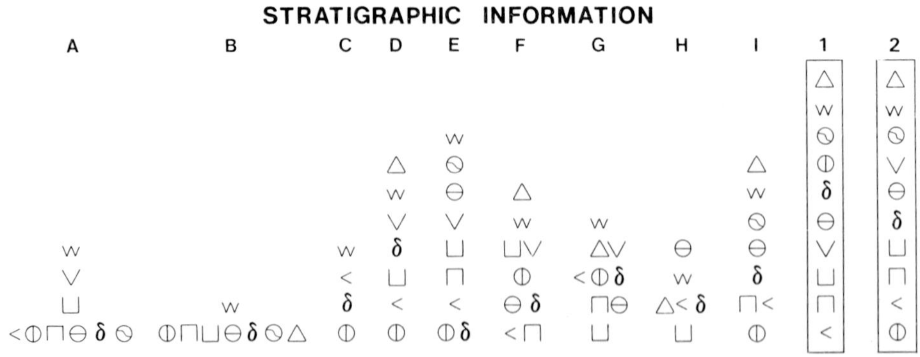

Fig. 4.2 Hay example. Highest and lowest occurrences of Lower Tertiary nannofossils selected by Hay (1972) from the Sullivan database. The 10 events are represented by symbols (cf. Fig. 5.1) which correspond to numbers in Tables 4.1 and 4.3. δ = lowest occurrence of *Coccolithus gammation*; ⏀ = lowest occurrence of *Coccolithus cribellum*; ⊖ = lowest occurrence of *Coccolithus solitus*; V = lowest occurrence of *Discoaster cruciformis*; < = lowest occurrence of *Discoaster distinctus*; ⊓ = lowest occurrence of *Discoaster germanicus*; ⊔ lowest occurrence of *Discoaster minimus*; w = highest occurrence of *Discoaster tribrachiatus*; Δ = lowest occurrence of *Discolithus distinctus*; ⊘ = lowest occurrence of *Rhabdosphaera scabrosa*. See Fig. 4.1 for locations of the 9 sections (A-I). The columns on the right represent a subjective ordering of the events and Hay's original optimum sequence, respectively.

TABLE 4.3

Two SEQ files for Hay example. Minus signs (or hyphens) denote coeval events (cf. Fig. 4.1). The last entry for a section is followed by -999. **Left side**: SEQ file for stratigraphically downward direction. **Right side**: SEQ file for stratigraphically upward direction.

```
A
   9   8   7   6  -5  -4  -3  -2  -1-999
B
   9  10  -6  -5  -4  -7  -3  -2-999
C
   9   1   5   2-999
D
  10   9   8   5   7   1   2-999
E
   9   6   4   8   7   3   1   5  -2-999
F
  10   9   8  -7   2   5  -4   3  -1-999
G
   9   8 -10   5  -2  -1   4  -3   7-999
H
   4   9   5  -1 -10   7-999
I
  10   9   6   4   5   1  -3   2-999
```

```
A
   1  -2  -3  -4  -5  -6   7   8   9-999
B
   2  -3  -7  -4  -5  -6 -10   9-999
C
   2   5   1   9-999
D
   2   1   7   5   8   9  10-999
E
   2  -5   1   3   7   8   4   6   9-999
F
   1  -3   4  -5   2   7  -8   9  10-999
G
   7   3  -4   1  -2  -5  10  -8   9-999
H
   7  10  -1  -5   9   4-999
I
   2   3  -1   5   4   6   9  10-999
```

of Figure 4.1. One or more symbols on the same level in a section in Figure 4.2 indicate that the events they represent cannot be separated. Column 1 on the right side is a subjective ranking based on visual inspection of some of the more complete sections. Column 2 represents Hay's original optimum sequence. The order of the events in column 2 is based on pairwise comparison of the events in the nine sections. An event is placed above other events if it occurs more frequently above than below these other events in the sections. This is one of several possible methods for ranking events (see Chapter 5).

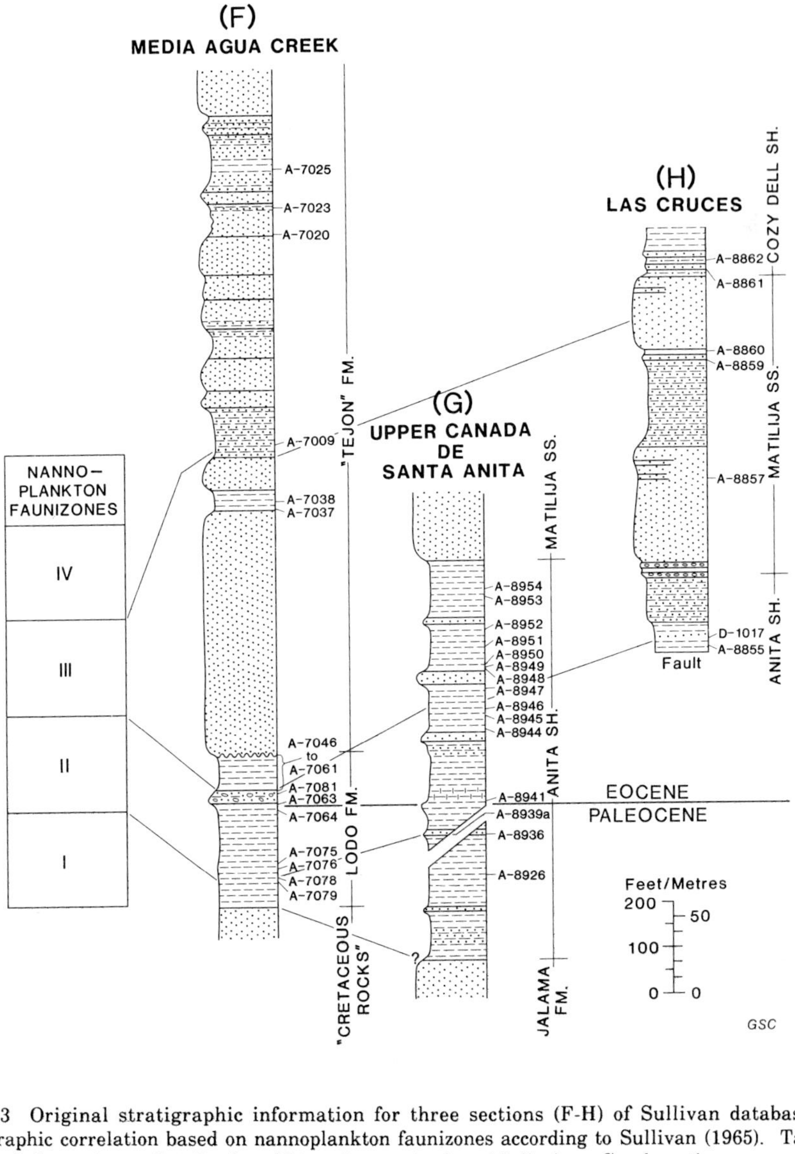

Fig. 4.3 Original stratigraphic information for three sections (F-H) of Sullivan database with stratigraphic correlation based on nannoplankton faunizones according to Sullivan (1965). Table 4.4 contains information on distribution of 9 taxa in samples from Media Agua Creek section.

Table 4.3 shows two possible SEQ files for the stratigraphic information of Figure 4.2. They are for the stratigraphically downward and upward directions, respectively. For reasons to be discussed in Chapter 5, the RASC computer program may give slightly different results for the upward and downward directions. It will be instructive to run the program on both SEQ files of Table 4.3 in order to illustrate the minor changes brought about by inverting the order. Such minor changes are usually much smaller than those resulting from altering the dataset by resetting switches or parameters in the PAR file (see later). Unless stated otherwise, we will use SEQ files for the stratigraphically downward direction which is also the direction in which results are printed out in tables and graphical displays.

The SEQ files of Table 4.3 contain all information represented in Figure 4.2. Coeval events are shown by hyphens in the SEQ files. The RASC computer program reads these hyphens as minus signs. There is one-to-one correspondence between the SEQ files of Table 4.3 and the graphical representation of Figure 4.2 in that the latter can be reconstructed from the former and vice versa. No use was made of a DAT file in order to obtain the SEQ files from Figure 4.2. This stage can be skipped for the Hay example because the stratigraphic information is of a simple nature. Normally, the stratigrapher will wish to construct a DAT file from which the SEQ file is extracted automatically. This procedure will be illustrated in the next section.

4.4 Partial DAT file for the Hay example

Figure 4.3 shows three of the sections with positions of samples studied by Sullivan (1964, 1965). For example, a partial DAT file will be created for section F (Media Agua Creek section) only. Table 4.4 contains the original stratigraphic information for nine of the ten taxa selected by Hay (see Table 4.1). Only *Rhabosphaera scabrosa* was not observed in the Media Agua Creek section. Hay (1971) used Sullivan's (1965) Eocene information only, for samples extending up to 88 feet below the base of "Tejon" Formation. According to Sullivan (1964), the Paleocene-Eocene boundary occurs about 111 feet below the base of the "Tejon" Formation. Table 4.5 shows two partial DAT files (for Section F only) which were obtained from the information contained in Table 4.4. The first partial DAT file (Table 4.5A) shows taxon identification numbers followed by

TABLE 4.4

Stratigraphic distribution of nine taxa of fossil nannoplanton for individual samples in the Media Agua Creek area, Kern County, California (according to Sullivan, 1964, Table 3, and Sullivan, 1965, Table 6). Stratigraphic distance (D) in feet measured upward and downward from base of "Tejon" Formation; Paleocene-Eocene boundary occurs between 103 and 118 feet. Fossil (F) numbers in first column as in Table 4.2; A-abundant; C-common; 0-few; x-rare. Single bar indicates stratigraphic events E1 to E10 used in Table 4.1 and Figure 4.3 (as defined for samples extending up to 88 feet below base of "Tejon" Formation); relative superpositional relations are changed by using lowest occurrences of four taxa in Paleocene shown in lower part of the table (also see Table 4.5). Level (L) as in Guex (1987, p. 228).

D	218	138	108	693	522	514	1	2	6	10	14	23	34	37	43	48	57	62	66	69	72	81	83	86	88	
F83					0		0	0	0	0	0	0	0	0	0	C	0	0	0	0	x	x		x	x̂	E1
F17								x				0	x				x		x				x̂			E2
F91																	x							x	x̂	E3
F11			x	x	0	x		x				x					x		x	x	x			x̂		E4
F19					0	0	0	0	0	0	0	0	0	x	x	x	x	0	0	0	0	x	x	x̂		E5
F94																	x				x̂					E7
F90						x		x									x	x			x̂					E8
F89					?	?	?				?					x̂	x	x			x	C	x	x	0	E9
F26					0		C	C	0	0	0	x	x̂													E10
L	18	18	17	16	15	15	15	14	14	13	12	11	10	9	9	9	9	9	9	9	9	8	7	7	7	

D	101	103	118	125	129	137	146	154	156	167	171	194	220	228	236	241	257	261	278	
F83			x				x													E1
F17	x																x			E2
F91																				E3
F11																				E4
F19	x		x	x	x	0	x	x	x	x	x	x	x	x	0	0	0			E5
F94																				E7
F90																0				E8
F89	C	C	C	C	C	C	C	C	C	C	C	C	C	A	A	A	A			E9
F26																				E10
L	6	6	5	5	5	5	5	4	4	3	3	3	3	3	2	2	2	1	1	

depths in feet of highest and lowest occurrences. The second file (Table 4.5B) has different depths for the lowest occurrences of five taxa because the data from the Paleocene also were used. Partial SEQ files automatically constructed from the data in Table 4.5 are shown in the first two rows of Table 4.6. The first row (Eocene only) duplicates the row for Section F in Table 4.3 (stratigraphically downward direction). The SEQ file in the second row is different from the initial result. It is more realistic because events 1, 2, 5, and 8 already existed before the Eocene. As mentioned before, continued use will be made of the original Hay example of Figure 4.2 and Table 4.3 for historical reasons. The extended SEQ file incorporating the Paleocene data shown in Table 4.6 will be employed as well. Differences between the SEQ files of Tables 4.3 and 4.6 are restricted

TABLE 4.5

Examples of partial DAT files for Media Agua Creek section of Table 4.4. Distances (in feet) measured downward from base of "Tejon" Formation. Guex Levels are shown as L in bottom row of Table 4.4.

A. Fossil Number	Distances		Guex Levels	
	LO	HI	LO	HI
83	88	-522	7	15
17	83	2	7	14
91	88	57	7	9
11	86	-1080	7	17
19	86	-522	7	15
94	72	57	9	9
90	72	-514	9	15
89	88	48	7	9
26	34	-522	10	15
B. Part A modified to consider Eocene and Paleocene				
83	146	-522	5	15
17	257	2	2	14
91	88	57	7	9
11	86	-1080	7	17
19	257	-522	2	15
94	72	57	9	9
90	241	-514	2	15
89	257	48	2	9
86	34	-522	10	15

to sections F and G because these are the only sections with additional data not used by Hay (1972).

Artificial truncation of the observed ranges of some of the nannoplankton taxa may occur when the coding and analysis are restricted to relatively narrow time intervals, e.g. for one or two ages. Such artificial truncation effects should be avoided as much as possible in practice. It is likely that the relatively large number of coeval events at the base of sections A and B in Figure 4.2 is in part also due to artificial truncation. It is noted that Hay (1972) ignored coeval events in his original method of obtaining an optimum sequence thus counteracting the possible truncation effect. In the RASC method, coeval events will always be considered. Although some ranking methods give the same results whether or not observed coeval events are considered, the scaling methods make extensive use of coeval events and these should not be ignored. The truncation drawback of the Hay example will be avoided in most other datasets to be discussed later.

The lowest and highest occurrences in the DAT and SEQ files for the Hay example are based on rare occurrences within samples. Sullivan (1965) adopted the widely used semi-quantitative method of categorizing abundance (rare, few, common, abundant) in order to improve upon coding presences and absences only without following the laborious and possibly counter-productive, route of actually counting large numbers of individual fossils. His charts normally show uninterrupted sequences for the "abundant" and "common" categories (A's and C's in Table 4.5), whereas the sequences for the "rare" and "few" categories (x's and o's in Table 4.5) are interrupted. As pointed out by Hay (1972), the only reasonable explanation for the gaps in the sequences of x's and o's is that the presence or absence of a rare taxon is the realization of a random variable (also see Section 3.3). All taxa were rare when they first and last appeared in a

TABLE 4.6

Partial SEQ files in stratigraphically downward direction for Media Agua Creek section as derived from partial DAT files of Table 4.5. Event code numbers as in Table 4.1.

Eocene 1 (Distances)	10	9	8	-7	2	5	-4	3	-1
Eocene 2 (Guex levels)	10	9	-8	-7	2	-5	-4	-3	-1
Eocene and Paleocene 1	10	9	7	4	3	1	8	-2	-5
Eocene and Paleocene 2	10	9	-7	4	-3	1	8	-2	-5

basin. Some taxa (e.g. F 17 in Table 4.4) never became abundant contrary to others (e.g. F 89 in Table 4.4) which were abundant as well as rare.

Stratigraphic events can be defined on the basis of rare occurrences as well as abundant occurrences of a taxon. For example, Doeven et al. (1982) applied ranking to a mixture of events in order to construct a nannofossil range chart for Cretaceous nannofossils along the Canadian Atlantic margin. This mixture included subtops (last consistent occurrences) and superbottoms (fist consistent occurrences) as well as the tops (last observed occurrences) and bottoms (first observed occurrences) for selected nannofossils. Definition of more than two events for these taxa helped to improve the range chart. In general, subtops and superbottoms are less subject to random variability in time than first and last occurrences (also see Doeven, 1983).

4.5 DAT files constructed by Guex and Davaud

As mentioned in Section 4.3, Guex and Davaud have used Sullivan's database for the testing of other types of quantitative stratigraphic correlation techniques. Their "Unitary Associations" method aims to emulate the Oppel zones of biostratigraphy. Oppel (1856) had proposed construction of a regional standard consisting of a succession of different zones later called "Oppel zones". Each zone of this type is characterized by one or more taxa, or by a unique assemblage of taxa (also see Fig. 2.1 and previous discussion in Section 2.2). Identification of individual Oppel zones in individual sections provides a vehicle for biostratigraphic correlation. As explained in Section 3.5, Guex (1987) used graph theory to construct Unitary Associations which have essentially the same properties as Oppel zones. Systematic insertion of supposedly missing data in order to establish coexistence of taxa is a guiding principle of this approach. This aim is already reflected in the type of coding stratigraphic information performed before the Unitary Associations are constructed.

It is reasonable to assume that, apart from disturbances such as reworking, each taxon existed continually between the time equivalent of its observed first and last occurrences in a section. This is the well-known "range-through" method (cf. Section 2.1) which usually leads to assumed coexistences of taxa which may not have been observed together within a single bed. The range-through assumption is made in explicit or implicit form in most quantitative stratigraphic correlation techniques including

RASC and the Unitary Associations method. However, in the latter method, the following, additional assumption is made before the data are coded. Adjoining samples are combined into levels representing "maximal horizons" (cf. Guex, 1987, p. 20; also see Guex, 1988) as illustrated for the Media Agua Creek example in the bottom row of Table 4.4.

Davaud and Guex (1987, p. 587) estimated that the number of "maximal horizons" is less than 30 percent of the total number of samples for the Sullivan-Bramlette database. Figure 4.4 illustrates how this type of level was constructed. Each maximal horizon corresponds to a separate clique in the interval graph (cf. Section 3.5) for the section that is being studied. The observed range chart for the section is interpreted as the interval assignment for this interval graph. The seven taxa in the example of Figure 4.4 have only three maximal horizons corresponding to the cliques (1, 2, 3), (2, 3, 4) and (3, 4, 5, 6, 7) respectively. These maximal horizons are separated by horizons with fewer taxa on the range chart for the section.

Individual samples can be represented by lines drawn perpendicular to the ranges. In Figure 4.4 the taxa whose ranges are intersected by such a line would coexist in the corresponding sample. All samples containing taxa of a particular clique are combined with one another as a first step towards constructing the Unitary Associations. If sampling proceeds in the stratigraphically upward direction, a new combination of taxa leading to a new maximal horizon is started as soon as one or more taxa of the next clique are encountered in a sample.

An interval assignment of an interval graph is schematic in that there is no one-to-one correspondence between these two models. In general, it is not possible to reconstruct the range chart for a section from its interval graph. For example, when moving from the right to the left in the range chart of Figure 4.4, one successively encounters 6, 3, 7, 5, and 4 for the end points of the five taxa in the largest clique. Such detailed information obviously does not exist in the interval graph.

The eighteen levels "L" in Table 4.5 were based on maximal horizons for all (= 82) taxa occurring in the Media Agua Creek area. The 44 samples of this section were combined into 18 levels by Guex (1987) with loss of information on the relative order of first and last occurrences. Many pairs of events were made coeval during the coding, although they had a distinct order in the section before the cliques were determined. For

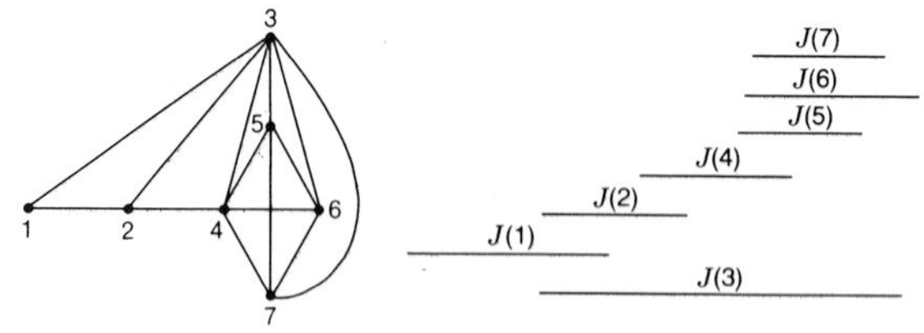

Fig. 4.4 Example of interval assignment $J(i)$, $i=1, 2, \ldots$ for undirected graph (after Roberts, 1976). If applied to a single stratigraphic section, each clique represents a maximal horizon or Guex level.

ranking and scaling generally, it is recommended that all observed superpositional relations for pairs of events in sections are preserved by entering this type of information in the DAT files from which SEQ files will be derived automatically. Table 4.6 shows a partial SEQ file for the Media Agua Creek section of the Hay example based on Guex levels (line 2) in comparison with that based on all samples (line 1). The number of hyphens for coeval events is increased when event levels are combined with one another using the maximal horizons method. For Eocene nannoplankton only, the number of event levels would be reduced from 6 to 3 in Table 4.6, and for the Paleogene (combined Eocene and Paleocene) from 7 to 5. Later Guex (1987) added the information for the Paleocene to the Sullivan data base for the (Media Agua Creek and Upper Canada de Santa Anita sections. Lines 1 and 2 for Eocene and Paleocene in Table 4.6 show the effect of this change with respect to lines 1 and 2 for the Eocene used in the original Hay example.

It is noted that Agterberg et al. (1985) made use of the Sullivan database as originally coded by Davaud and Guex (1978) which did not use Sullivan's (1964) data for the Paleocene, and in which the number of levels had been reduced by adoption of the maximal horizons method.

4.6 Gradstein – Thomas database: Cenozoic Foraminifera in Canadian Atlantic Margin wells

The RASC model for ranking and scaling of stratigraphic events was originally developed during a project on Cenozoic foraminiferal stratigraphy of the northwestern Atlantic margin (Gradstein and

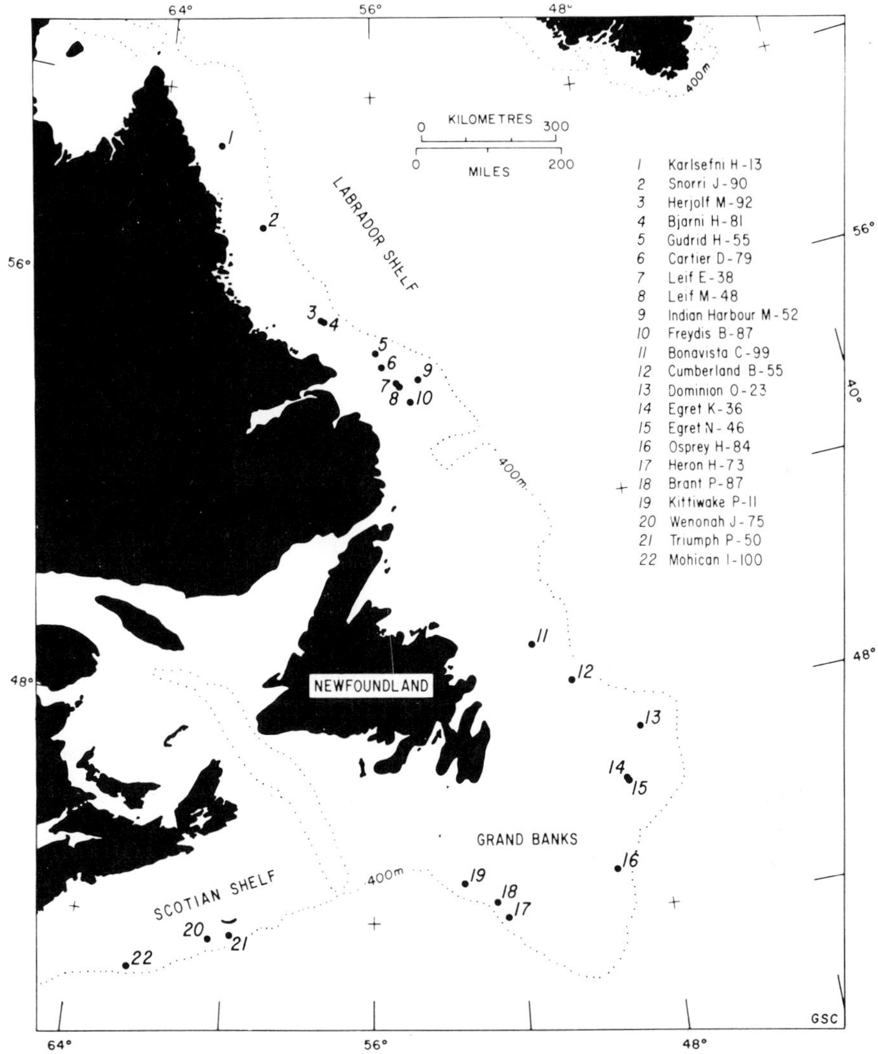

Fig. 4.5 Location of 22 wells along Eastern Canadian margin used for Cenozoic foraminiferal stratigraphy by Gradstein and Agterberg (1982). Original samples were obtained from Eastcan and others: Karlsefni H-13 (1760-12 990'), Snorri J-90 (1260-9950'), Herjolf M-92 (3030-7800'), Bjarni H-81 (2760-6060'), Gudrid H-55 (1660-8580'), Cartier D-79 (1950-6070'); Tenneco and others: Leif E-38 (1210-3557'); Eastcan and others: Leif M-48 (1300-5620'); BP Columbia and others: Indian Harbour M-52 (1740-10 480'); Eastcan and others: Freydis B-87 (1000-5260'); BP Columbia and others: Bonavista C-99 (1860-11 940'); Mobil Gulf: Cumberland B-55 (920-11 830'), Dominion O-23 (1380-10 260'); Amoco Imp Skelly: Egret N-64 (1060-2070'), Egret K-36 (860-2270'), Osprey H-84 (1190-2660'), Brant P-8 (1050-6270'); Amoco Imp: Heron H-73 (970-5800'), Kittiwake P-11 (970-5560'); PetroCanada Shell: Wenonah J-75 (1000-4750'); Shell: Triumph P-50 (990-5490'), Mohican I-100 (1276-5320').

Agterberg, 1982). Figure 4.5 shows the locations of the 22 offshore wells used. They were divided into two groups. Sixteen of these wells are located on the Labrador Shelf and northwestern Grand Banks (northern region). Six occur on the Scotian Shelf and southern Grand Banks (southern region). In total, the highest occurrences (exits) of 206 benthonic and planktonic Foraminifera, were used. Of these 150 and 157 occurred in the northern and southern regions, respectively.

Initial biozonations for the northern and southern regions were based on smaller sets of 41 and 60 data, respectively. The two regions had 14 of these taxa in common. The southern biozonation had 32, mostly Eocene and Miocene index planktonics and the northern zonation 6, essentially Eocene ones. This difference reflects pronounced post-Middle Eocene latitudinal water mass heterogeneity and differential post-Eocene shallowing across the continental margin. The biozonation with relatively many planktonics for the southern region helped to establish the initially largely unknown biozonation for the northern region.

Later, data for 10 wells were added for the northern region, mainly in the vicinity of the Hibernia oil field on the Grand Banks between wells 13 and 14 in Figure 4.5. New taxa were identified and the original dictionary for the 22 wells of Figure 4.5 was updated. The enlarged dictionary is given in Table 4.7 which is part of the Gradstein-Thomas database for 24 wells on the Labrador Shelf and Grand Banks, published in Gradstein et al. (1985, pp. 515-520).

It is noted that not all events in Table 4.7 are highest occurrences of Foraminifera. For example, four seismic events were included in the database. Also, in total there are 238 events in Table 4.7 which is less than the greatest number (=275) assigned to a taxon. Gaps in the numbering are due to revisions made in the identification of taxa. For example, a taxon with one name in Table 4.7 may be the composite of two taxa of which one had a different name which became obsolete after the renaming. In order to preserve the unique identifier of the name that was retained, a dummy code (e.g. xxx) was assigned in the dictionary to the name that was deleted. The advantage of this procedure is that other taxa retain their original dictionary numbers in RASC input and output files regardless of revisions applied to relatively few taxa.

Table 4.8 is a partial DAT file using 4 of the 24 wells. The depths of the samples were measured in feet for earlier wells and in meters for wells

TABLE 4.7

DIC file of Cenozoic Foraminifera in Gradstein-Thomas database for Canadian Atlantic margin.

1 NEOGLOBOQUADRINA PACHYDERMA
2 GLOBIGERINA APERTURA
3 GLOBIGERINA PSEUDOBESA
4 GLOBOROTALIA INFLATA
5 GLOBOROTALIA CRASSAFORMIS
6 NEOGLOBOQUADRINA ACOSTAENSIS
7 GLOBIGERINOIDES RUBER
8 ORBULINA UNIVERSA
9 FURSENKOINA GRACILIS
10 UVIGERINA CANARIENSIS
11 NONIONELLA PIZARRENSE
12 EHRENBERGINA SERRATA
13 HANZAWAIA CONCENTRICA
14 TEXTULARIA AGGLUTINANS
15 GLOBIGERINA PRAEBULLOIDES
16 CERATOBULIMINA CONTRARIA
17 ASTERIGERINA GURICHI
18 SPIROPLECTAMMINA CARINATA
19 GLOBIGERINOIDES SP
20 GYROIDINA GIRARDANA
21 GUTTULINA PROBLEMA
22 COSCINODISCUS SP3
23 COSCINODISCUS SP4
24 TURRILINA ALSATICA
25 COARSE ARENACEOUS SPP.
26 UVIGERINA DUMBLEI
27 EPONIDES UMBONATUS
28 CIBICIDOIDES SP5
29 CYCLAMMINA AMPLECTENS
30 CIBICIDOIDES BLANPIEDI
31 PTEROPOD SP1
32 AMMOSPHAEROIDINA SP1
33 TURBOROTALIA POMEROLI
34 MARGINULINA DECORATA
35 SPIROPLECTAMMINA DENTATA
36 PSEUDOHASTIGERINA WILCOXENSIS
37 ACARININA AFF PENTACAMERATA
38 LENTICULINA SUBPAPILLOSA
39 ALABAMINA WILCOXENSIS
40 BULIMINA ALAZANENSIS
41 PLECTOFRONDICULARIA SP1
42 CIBICIDOIDES ALLENI
43 BULIMINA MIDWAYENSIS
44 CIBICIDOIDES AFF WESTI
45 BULIMINA TRIGONALIS
46 MEGASPORE SP1
47 PLANOROTALITES PLANOCONICUS
48 ANOMALINA SP5
49 OSANGULARIA EXPANSA
50 SUBBOTINA PATAGONICA
51 ACARININA PRIMITIVA
52 ACARININA SOLDADOENSIS
53 UVIGERINA BATJESI
54 SPIROPLECTAMMINA NAVARROANA
55 GAVELINELLA BECCARIIFORMIS
56 GLOMOSPIRA CORONA
57 SPIROPLECTAMMINA SPECTABILIS LCO
58 EPONIDES SP8
59 RZEHAKINA EPIGONA
60 PLANOROTALITES COMPRESSUS
61 SUBBOTINA PSEODOBULLOIDES
62 GAVELINELLA DANICA
63 NODOSARIA SP11
64 CASSIDULINA ISLANDICA
65 COSCINODISCUS SP1
66 COLEITES RETICULOSUS
67 SCAPHOPOD SP1
68 SPIROPLECTAMMINA SPECTABLIS LO
69 NODOSARIA SP8
70 ALABAMINA WOLTERSTORFFI
71 EPISTOMINA ELEGANS
72 CYCLOGYRA SP3
73 EPONIDES SP3
74 EPONIDES SP5
75 LENTICULINA ULATISENSIS
76 CASSIDULINA SP
77 ELPHIDIUM SP
78 UVIGERINA PEREGRINA
79 GLOBIGERINA TRIPARTITA
80 CYCLAMMINA CANCELLATA
81 GLOBIGERINA VENEZUELANA
82 GLOBIGERINA LINAPERTA
83 PLANOROTALITES PSEUDOSCITULUS
84 GLOBIGERINA YEGUAENSIS
85 PSEUDOHASTIGERINA MICRA
86 TURRILINA BREVISPIRA
87 BULIMINA AFF. JACKSONENSIS
88 SIPHOGENEROIDES ELEGANTA
89 MOROZOVELLA SPINULOSA
90 ACARININA DENSA
91 RADIOLARIANS
92 MOROZOVELLA CAUCASICA
93 ACARININA AFF. BROEDERMANNI
94 GLOBIGERINATHEKA KUGLERI
95 ARAGONIA VELASCOENSIS
96 ACARININA INTERMEDIA WILCOXENSIS
100 GLOBIGERINA RIVEROA
109 CASSIDULINA CURVATA
110 GLOBIGERINA BULLOIDES
111 PARAROTALIA SP1
112 MARGINULINA BACHEI
113 GLOBOROTALIA MENARDII GROUP
114 GLOBIGERINOIDES SACCULIFER
115 GLOBOROTALIA OBESA
116 ORBULINA SUTURALIS
117 SPHAEROIDINA BULLOIDES
118 EPISTOMINA SP5
119 SPHAEROIDINELLA SUBDEHISCENS
120 GLOBOROTALIA SIAKENSIS
121 GLOBIGERINA NEPENTHES
122 SPHAEROIDINELLOPSIS SEMINULINA
123 GLOBIGERINOIDES TRILOBUS
124 GLOBOQUADRINA DEHISCENS
125 "GLOBOROTALIA" CONTINUOSA

TABLE 4.7 (continued)

126	GLOBIGERINOIDES OBLIQUUS	211	HANTKENINA SP
127	GLOBIGERINITA NAPARIMAENSIS	213	ARENOBULIMINA SP2
128	GLOBOROTALIA PRAEMENARDII	216	GLOBIGERINOIDES SICANUS
130	SIPHONINA ADVENA	217	GLOBOROTALIA SCITULA
131	CIBICIDOIDES TENELLUS	218	MARGINULINA AMERICANA
132	"GLOBOROTALIA" OPIMA NANA	219	MARTINOTIELLA COMMUNIS
133	LENTICULINA SP3	220	CIBICIDOIDES WUELLERSTORFFI
134	LENTICULINA SP4	221	GLOBIGERINOIDES SUBQUADRATUS
135	GLOBIGERINA SP40	222	GLOBOQUADRINA ALTISPIRA
136	MELONIS BARLEANUM	223	GLOBIGERINA CIPEROENSIS
137	GLOBIGERINOIDES PRIMORDIUS	224	UVIGERINA MEXICANA
138	GLOBIGERINA ANGUSTIUMBILICATA	225	GLOBIGERINA AFF. AMPLIAPERTURA
139	"GLOBOROTALIA" OPIMA OPIMA	226	GLOBIGERINA SENNI
140	ROTALIATINA BULIMINOIDES	227	CIBICIDOIDES AFF. TUXPAMENSIS
141	PLANULINA RENZI	228	CASSIDULINA TERETIS
142	GYROIDINA SOLDANII MAMILLIGERA	230	BULIMINA OVATA
143	UVIGERINA GALLOWAY	231	UVIGERINA RUSTICA
144	GLOBOROTALIA CERROAZULENSIS	232	GLOBIGERINOIDES IMMATURUS
145	ANOMALINOIDES ALLENI	233	CATAPSYDRAX UNICAVUS
146	SUBBOTINA EOCAENA	234	TRUNCAROTALOIDES AFF. ROHRI
147	CATAPSYDRAX AFF. DISSIMILIS	235	SUBBOTINA BOLIVARIANA
148	GLOBIGERINATHEKA INDEX	236	EPONIDES SP4
149	GLOBIGERINATHEKA TROPICALIS	237	LENTICULINA SP8
150	GLOBIGERINA GORTANII	238	CIBICIDOIDES SP7
151	BULIMINA BRADBURYI	239	NONIONELLA LABRADORICA
153	BULIMINA COOPERENSIS	240	ELPHIDIUM CLAVATUM
154	ANOMALINOIDES MIDWAYENSIS	241	GLOBOROTALIA TRUNCATULINOIDES
155	ANOMALINOIDES GROSSERUGOSA	242	GLOBOROTALIA FOHSI GROUP
156	SUBBOTINA FRONTOSA	243	GLOBIGERINA DECAPERTA
157	TRITAXIA SP3	244	GAUDRYINA SP10
158	SUBBOTINA INAEQUISPIRA	245	PRAEORBULINA GLOMEROSA
159	MOROZOVELLA ARAGONENSIS	246	GLOBIGERINATELLA INSUETA
160	ACARININA PSEUDOTOPILENSIS	247	GLOBIGERINOIDES ALTIAPERTURA
161	PLANOROTALITES AUSTRALIFORMIS	248	"GLOBOROTALIA" AFF. INCREBESCENS
162	MOROZOVELLA AEQUA	249	GLOBIGERINATHEKA SEMIINVOLUTA
164	NUTTALIDES TRUMPYI	250	VULVULINA JARVISI
166	MOROZOVELLA SUBBOTINAE	251	ANOMALINA SP4
167	MOROZOVELLA FORMOSA GRACILIS	252	MOROZOVELLA AFF. QUETRA
169	EPISTOMINELLA TAKAYANAGII	253	SUBBOTINA TRILOCULINOIDES
172	PSEUDOHASTIGERINA SP	254	PLANOROTALITES PSEUDOMENARDII
173	ANOMALINA SP1	255	MOROZOVELLA CONICOTRUNCATA
175	ALLOGROMIA SP	256	"MOROZOVELLA" AFF. PUSILLA
176	ALLOMORPHINA SP1	257	CHILOGUEMBELINA SP
177	BOLIVINA DILATATA	258	TAPPANINA SELMENSIS
179	GLOBOROTALIA SCITULA PRAESCITULA	259	AMMODISCUS LATUS
180	GYROIDINA SP4	260	HAPLOPHRAGMOIDES KIRKI
181	CYCLOGYRA INVOLVENS	261	HAPLOPHRAGMOIDES WALTERI
182	PLECTOFRONDICULARIA SP3	262	KARRERIELLA APICULARIS
184	GYROIDINA OCTOCAMERATA	263	AMMOBACULITES AFF POLYTHALAMUS
187	CIBICIDOIDES GRANULOSA	264	KARRERIELLA CONVERSA
188	PLEUROSTOMELLA SP1	265	ASTERIGERINA GURICHI (PEAK)
190	ANOMALINOIDES ACUTA	266	GLOBOROTALIA PUNCTICULATA
191	"GLOBIGERINA" AFF. HIGGINSI	267	GLOBOROTALIA HIRSUTA
194	PLANOROTALITES CHAPMANI	268	GLOBOROTALIA AFF KUGLERI
196	OSANGULARIA SP4	269	NEOGLOBQUADRINA ATLANTICA
201	SEISMIC EVENT #1	270	CIBICIDOIDES GROSSA
202	SEISMIC EVENT #2	271	GLOBOROTALIA INCREBESCENS
203	SEISMIC EVENT #3	272	GLOBOQUADRINA BAROEMOENSIS
204	SEISMIC EVENT #4	273	BULIMINA GRATA
206	EPONIDES POLYGONUS	274	GAUDRYINA AFF HILTERMANNI
210	LOXOSTOMOIDES APPLINAE	275	PARAROTALIA SP2

TABLE 4.8

Partial DAT file for Gradstein-Thomas database. Numbers in brackets below well names are for rotary table height and water depth, respectively (M = meters; F = feet). Depths (first column) are followed by highest occurrences.

Hibernia P-15 (M 11.3; 80.2)			Adolphus D-15 (F 98.0; 377.0)			Bjarni H-81 (F 40.0; 456.0)			Indian Harbour M-52 (F 98.0; 649.0)		
255	17		1140	10		2860	16		1740	1	3
275	18	265	1410	71		3360	67		1740	4	5
310	16		1500	218		3460	20	21	1740	8	
410	20	100	1590	16	136	3560	18	69	1890	9	10
550	26		1680	18		3560	70	71	1950	269	
620	201		1980	20		4060	15		2090	2	7
695	15		2700	179		4260	24		2130	6	18
720	71		2900	201		4860	25		2460	15	20
915	72		3060	26		5060	34		2460	16	
945	69		3660	15	81	5360	29	265	2550	17	
960	202		3660	69		5560	42	74	3600	24	25
975	81		4200	24	33	5560	41	32	4140	26	27
1005	27		4200	202		5560	30	264	4140	28	
1035	147		4440	259	25	5560	75		5400	259	
1075	24		4562	263		5960	57		5590	261	
1125	25	32	4920	82		6060	46		5780	30	
1125	57	259	4950	85	261	6590	56		6370	260	32
1125	260		5400	203					6970	33	
1185	261		5420	147	260				7660	34	35
1195	29		5550	68					7760	263	36
1200	203		5778	32					7760	39	
1315	53	263	5896	90					7860	29	40
1345	40		6018	30					7860	41	42
1375	45		6200	49	29				7960	86	
1400	204		6646	144	90				8140	37	38
			6646	156	37				8230	44	
			6646	89					8860	45	46
			6975	234					8860	47	
			7596	160	93				9130	49	
			7917	36					9560	57	54
			8020	161	164				9560	50	52
			8258	50	230				9940	55	56
			8384	54					10090	59	
			8520	57	56				10230	60	61
			8700	55					10230	62	
			8726	194	95						

TABLE 4.9

SEQ file for 24 wells of Gradstein-Thomas database for Labrador Shelf and Grand Banks.

```
BJARNI H-81
  16  67  20 -21  18 -69 -70 -71  15  24  25  34  29-261  42 -74 -41 -32  30-264
 -75  57  46  56-999
CARTIER D-70
  16  18  15  21 -70  67  69  24-172  25 259  34 260-261 118 -85 -29-263  46 -42
 -32  35  41 -57  54  56 175 -59-999
FREYDIS B-87
  16 181 -67 -21 -18  20  69 -27  15 -70  25 190 -34-206 -42 -74 260  29-261 -45
  33 -81 -41 -75-210 -32 211 -85 -94  57 -88 -86 -30 -46 -35  56  54 213 -55  59
-999
GUDRID H-55
  10 -17 265  20 -21 -18 -16  24  15 -25  33 259  40 -34  84 -90 -36  37-260-261
  29  35  45 -74  42  57 -88 -30  32  46 -50  56 -59 -54  55-999
INDIAN HARBOUR M-52
   1  -3  -4  -5  -8   9 -10 269   2  -7   6 -18  15 -20 -16  17  24 -25  26 -27
 -28 259 261  30 260 -32  33  34 -35 263 -36 -39  29 -40 -41 -42  86  37 -38  44
  45 -46 -47  49  57 -54 -50 -52  55 -56  59  60 -61 -62-999
KARLSEFNI H-13
 228  67  25  41-118  69 260-261  68 -39  53-206  29  86 -30 -63 -34  46-264 230
 -44 -42  96 -36 164 -50  52  45 -54  56  55 -62  61-253 258-999
LEIF M-48
 228 -77 -10 181  16 -67  15  20 -21 -18  70  69  85 -24  25-238  42  29 260 -34
  57 -74-118-263  30 -41  46 -56 -54-999
LEIF E-38
 228 -77-270  17  67 -16  18 -21  20-999
SNORRI J-90
  77 228  16  67  15 -21  18  25  57-263 -32 -34  29-260 -53 -41 -30 -36  27 -46
 118 264 230  86 -63  42  45  56  59 -54-999
HERJOLF M-92
  67  18 -15 -20 -16  78  70  25-259  85-145 -71 -40  45 -35-263-261 -34  29  41
 -53 -30 -32-264  86  57  54  46 190  47-154 -56  55  60  59-999
BONAVISTA C-99
  76 -77  10  17 -16  21  25 -20  18  79 -15 259  24 -26  81 -33  82  83  40  84
 -27  29-261  32-263  85 -86 -87-264  41 -34  57  88 -42 -90  89 159 -92 -93 -94
  56 -50 -30  47 -96 -36  46-999
DOMINION O-23
 177-109-169  11  -9  17  10-117 -78 112  18 179 -16 -15 -71 122 180  26-123-137
  14-136  27  20  21-181 201  24  25  34 264-260 -38 259 142 -81 184 -82 -30-146
  69-263 202  32  68 187  49-188-147-190-140  29 -40 191-156 151 250-226  36 -44
 194 -90 -57 203  50 -47-158 161 -52 -46  37-159-162 196  45-230 164-999
EGRET K-36
  17  26  16  20 -21 -18 -71 -15  24  27 -42 202  69  82-999
OSPREY H-84
  17  18 -20  15 -16  26-181  81  82  84-147 -69-148  90 -89 -33-187-234 -34-244
  52 -51-162-159-166 -50 -93-999
CUMBERLAND B-55
  76 228  -1  17  10 -11  -9-109 -71 265 -16 -20  18  15-119 117 219  26  24  25
-259 132  42 261  41  84  29  32 226 144  49  57 -36  90  52 -54 161 -93 -96-151
-164-157  46 -50-159  55 -56-254-194-999
EGRET N-46
  11 -16 -18  14 -27 -71  26 -20 202  15 -24 172-999
ADOLPHUS D-50
  10  71 218  16-136  18  20 179 201  26  15 -81 -69  24 -33-202 259 -25 263  82
  85-261 203 147-260  68  32  40  30  49 -29 144 -90-156 -37 -89 234 160 -93  36
 161-164  50-230  54  57 -56  55 194 -95-999
```

TABLE 4.9 (continued)

```
HIBERNIA O-35
  17 201  26  18 -20  16 275  24 -71  72  27 140 202  34 -81 203 259 -29 -25  15
 -28  57-260-261 204  40 -32  91-999
FLYING FOAM I-13
   9 -10  16  71  17 275-265  18-110  70  26 -15 -81 201  24 -20 -27  25 259 202
 263 -32 -34 260-261 264  29 -57-203  54  46  36  41 230-999
BLUE H-28
  77   1   4 267 269 110 -10 -64 266 124-125  -6-113 122  26 -71 268  -2 147 -27
  29-261 -81-150  82 -15-118-138 146 -84  32 -79-172 -53 -68 164-190  42  86-151
  33 -94 -57  37  90 -52-999
HARE BAY H-31
 228-270  77   1  10 136  16  70 -15  24  18 -20 -25 260-263 259  29-233 -69-118
 -32 -81  68  49  41 227  93 -42 -96  50  57  66 -54  55-161 -56  59 253-255 -46
-999
HIBERNIA K-18
 201  16 -18 -20 -71 -72  24 -27  15 -34  81 202 259 147  25 -29-260  30 -57-203
  32 263  36 -40 -63  45 -91-155-230 204-999
HIBERNIA B-08
  17  26  18 -20  16  15 -27 -71  72  81 -25  24 146-259  32 -57-147-260-261-263
  36 -40  45  63  47-144-194 -54 -91-230  56  55 -61  52 -59 -96-253-999
HIBERNIA P-15
  17  18-265  16  20-100  26 201  15  71  72  69 202  81  27 147  24  25 -32 -57
-259-260 261  29 203  53-263  40  45 204-999
```

drilled more recently. Rotary table height and water depth are given separately for each well. For the DEP files to be constructed later for the purpose of automated stratigraphic correlation, rotary table height will be subtracted so that all depths were measured from sealevel downward. Feet will be converted to metres.

Only the relative depths of the samples with respect to one another are used in ranking and scaling. For example, the Adolphus D-15 well has 32 distinct "event levels" for 50 exits. The majority (=19 of 32) of these levels have a single observed exit; there are 10 levels with 2, 2 with 3, and 1 with 5 exits, respectively. The total number of samples studied exceeded the total number of event levels because highest occurrences of microfossils were coded only. The exits in Table 4.8 have the same numbers as the Foraminifera in Table 4.7. The complete SEQ file for all 24 wells in the Gradstein-Thomas database is shown in Table 4.9.

4.7 Characteristic features of Gradstein-Thomas database

The original reasons for applying probabilistic stratigraphy (see Gradstein and Agterberg, 1982) may be summarized as follows. It is well known that the sequence of first and last occurrences of planktonic foraminiferal species in open marine Cenozoic sediments in the low-latitude regions of the world is closely spaced and shows a regular order. As a result, standard planktonic zonations provide a stratigraphic resolution of 30 to 45 zones over a time span of 65 x 10^6y (Blow, 1969; Postuma, 1971; Berggren, 1972; Stainforth et al., 1975). Although several Cenozoic taxa are indigenous to mid-latitudes, the absence of many lower-latitude forms and the longer stratigraphic ranges of mid-latitude taxa cause stratigraphic resolution to decrease away from the lower-latitude belt. In high latitudes (65°N and S), the virtual absence of planktonic foraminiferal taxa makes standard zonations inapplicable.

The northwest Atlantic margin, offshore eastern Canada, spans the mid- to high-latitudinal realms (north of 42°) and although there were temporal northward incursions of lower-latitudinal taxa in Early or Middle Eocene times, there is a drastic overall diminution of the number of biostratigraphically-useful Cenozoic planktonic species (from about 75 to 30) from the Scotian Shelf to the Grand Banks to the Labrador Shelf. A change from a deeper, open marine facies in the Paleogene to nearshore, shallower conditions in the Oligocene to Neogene (Gradstein et al., 1975; Gradstein and Srivastava, 1980) also curtails the number of taxa present in the younger Cenozoic section.

As a consequence, the construction of a planktonic zonation is mainly applicable to the southern Grand Banks and Scotian Shelf where 12 zones have been recognized using species of standard zonations which are not too rare locally to be of practical value in correlation. Similarly, on the northern Grand Banks and Labrador Shelf a 7-fold planktonic subdivision of the Cenozoic sedimentary strata is possible; the regional application is limited but the zonal markers and associated planktonic species improve chronostratigraphic calibration for the benthonic zones.

Independently, the Cenozoic benthonic foraminiferal record also shows temporal and spatial trends in taxonomic diversity and number of specimens. Calcareous benthonic species diversity and number of specimens decreases northward from the Scotian Shelf to the Grand Banks to the Labrador Shelf whereas the early Cenozoic agglutinated species diversity and numbers of specimens drastically increases on the Labrador Shelf. This benthonic provincialism is complicated by incoherent geographic distribution of some taxa, which in part is due to sampling.

Few of the agglutinated taxa, only a dozen out of more than 50 determined, are of biostratigraphic value (Gradstein and Berggren, 1981), but among the hundreds of calcareous benthonic forms determined, more potentially locally-useful or widely-known index species occur. As a consequence of the ecological sensitivity of these bottom dwellers, and because of the long stratigraphic ranges, facies changes can be expected to modify stratigraphic ranges. This is known as the problem of total versus local stratigraphic range. As a result, the benthonic stratigraphic correlation framework based on exits forms the appearance of a weaving pattern of numerous small and a few large-scale cross-correlations. Considerable mismatch in correlation is the result of misidentifiation, reworking, or large differences between local stratigraphic ranges of a taxon. In addition, some correlation lines only transverse part of the combined shelves area.

The previous summary provides insight into some of the constraints on a regional foraminiferal zonation. The most important additional one is sampling method. Only samples of cuttings obtained dominantly over 30-ft. (10-m.) intervals, are available generally from the wells, inferring that instead of entry, relative range, peak occurrence, and exit, only the exit of a taxon is known. Furthermore, downhole contamination in cuttings hinders recognition of stratigraphically-separate benthonic or planktonic homeomorphs. Other limiting factors are that species occur frequently in small numbers and that tests usually are reworked in the younger Neogene section of the Labrador Shelf.

In summary, the Gradstein-Thomas database of Tables 4.7 - 4.9, shows the following properties, ranked according to their importance with respect to stratigraphic resolution:

(1) Samples are predominantly cuttings, which forces use of the highest parts of stratigraphic ranges or of the highest occurrences (tops, exits), and restricts the number of stratigraphically useful taxa.

(2) There is limited application of standard planktonic zonations, due to the mid- to high-latitude setting of the study area and the presence of locally unfavorable facies.

(3) There are minor and major inconsistencies in relative extinction levels of benthonic taxa.

(4) Many of the samples are small which limits the detection of species represented by few specimens; this contributes to factor (3) and to the erratic, incoherent geographic distribution pattern of some taxa.

(5) There is geographic and stratigraphic provincialism in the benthonic record from the Labrador Shelf to the Scotian Shelf which makes representation of details in a general zonation difficult.

Despite the limiting factors, it was possible to erect a zonation based on a partial database. Gradstein and Williams (1976) used four Labrador Shelf/northern Grand Banks wells to produce an 8-fold (benthonics) subdivision of the Cenozoic section. Similar stratigraphic resolution and improved zone delineation was obtained by Gradstein (unpublished) using 9 wells on the Labrador Shelf and northern Grand Banks. Some of the zones were tentative and their ages not well defined. These initial subjective zonations were compared to RASC output (Gradstein and Agterberg, 1982) suggesting that a slightly improved zonation resulted from the latter method.

Increase of the Cenozoic database through incorporation of more wells has clarified the broader correlation pattern and increased the number of chronostratigraphic calibration points based on planktonic foraminiferal occurrences. It also increased noise in the stratigraphic signal (factors 3 and 4) due to more stratigraphic inconsistencies and geographic incoherence of exits.

The RASC method initially was developed in an attempt to optimize stratigraphic resolution based on all observations that could be employed for a zonation. Other benefits of using the computer for ranking and scaling included the following. Obviously reworked highest occurrences of taxa never were included in the database. Such reworking is apparent from anomalous, poor preservation of tests relative to the remainder of the assemblage and from highly erratic stratigraphic position. However, when the database is large, it is difficult to evaluate the possibility of anomalous stratigraphic position for all samples in a systematic manner. The normality test in RASC (cf. Gradstein, 1984; also see Section 6.6 and Chapter 8) allows comparison of the positions of the events in each section with those in the optimum sequence of the biozonation. Events that are either too high or too low in a given section in comparison with their neighbors are flagged in the normality test. Such anomalies then can be scrutinized and excluded from the database if they are due to reworking, contamination or misidentification.

4.8 Frequency of occurrence of taxa of Cenozoic Foraminifera along the northwestern Atlantic margin

In the previous section, it was mentioned that samples obtained during exploratory drilling are small, limiting the chances that microfossils will be detected if present within a zone. It is reasonable to assume that many taxa will not be detected at all in a well. It they are detected, their highest occurrence is likely to be recorded at a stratigraphically lower level. The first kind of statistical analysis performed in the RASC program simply consists of counting for how many different sections (or wells) each taxon has been recorded. Table 4.10 shows such counts for the 150 Foraminifera from the 16 wells in the northern region introduced at the beginning of Section 4.6 (cf. Fig. 4.5). As many as 110 events listed in Table 4.10 have zero counts. Most of these occurred in the southern region only. Some numbers with zero counts represent "dummy" events (see Section 4.6). In total, 56 events occur in a single well only. The following tabulation shows how many events occur in 1, 2, ..., 16 wells of the northern region:

Number of wells:	1	2	3	4	5	6	7	8	9	10	11	12	13	14	15	16
Number of events:	56	26	13	14	11	4	5	2	2	3	4	5	2	1	2	0

This is clearly a skew frequency distribution with relatively few Foraminifera occurring in relatively many wells. The corresponding frequency distribution for the southern region is:

Number of wells:	1	2	3	4	5	6
Number of events:	56	51	29	21	10	6

TABLE 4.10

RASC computer program preprocessing output for number of times that successive events occur in a well; e.g. event 1 occurs in 2 wells and event 2 in 1 well.

TABULATION OF EVENT OCCURRENCES:
DICTIONARY CODE NUMBER VERSUS FREQUENCY OF OCCURRENCE

1- 2	53- 5	105-0	157-2	209-0
2- 1	54- 9	106-0	158-1	210-1
3- 1	55- 6	107-0	159-4	211-1
4- 1	56-12	108-0	160-0	212-0
5- 1	57-12	109-2	161-2	213-1
6- 1	58- 1	110-0	162-2	214-0
7- 1	59- 6	111-0	163-0	215-0
8- 1	60- 2	112-1	164-3	216-0
9- 3	61- 2	113-0	164-0	217-0
10- 5	62- 3	114-0	166-1	218-0
11- 4	63- 3	115-0	167-0	219-1
12- 1	64- 2	116-0	168-0	220-0
13- 1	65- 5	117-2	169-0	221-0
14- 3	66- 0	118-4	170-0	222-0
15-14	67- 8	119-1	171-0	223-0
16-15	68- 0	120-0	172-0	224-0
17- 7	69-10	121-0	173-4	225-0
18-15	70- 7	122-1	174-0	226-1
19- 1	71- 6	123-1	175-1	227-0
20-13	72- 0	124-0	176-4	228-5
21-11	73- 2	125-0	177-1	229-0
22- 6	74- 4	126-0	178-0	230-3
23- 1	75- 3	127-0	179-1	231-0
24- 9	76- 2	128-0	180-1	232-0
25-12	77- 4	129-0	181-4	233-0
26- 7	78- 2	130-0	182-2	234-1
27- 8	79- 1	131-1	183-0	235-0
28- 1	80- 1	132-1	184-1	236-1
29-12	81- 4	133-0	185-0	237-1
30-10	82- 4	134-0	186-0	238-1
31-13	83- 2	135-0	187-1	239-0
32- 4	84- 4	136-1	188-1	240-0
33- 4	85- 5	137-1	189-0	241-0
34-11	86- 5	138-0	190-3	242-0
35- 5	87- 1	139-0	191-1	243-0
36- 7	88- 3	140-2	192-0	244-1
37- 2	89- 2	141-0	193-0	245-0
38- 2	90- 5	142-1	194-2	246-0
39- 2	91- 0	143-0	195-0	247-0
40- 5	92- 1	144-1	196-1	248-0
41-11	93- 3	145-1	197-0	249-0
42-11	94- 2	146-1	198-0	250-1
43- 5	95- 0	147-2	199-0	251-0
44- 3	96- 3	148-1	200-0	252-0
45- 7	97- 0	149-0	201-0	253-1
46-12	98- 0	150-0	202-0	254-1
47- 4	99- 0	151-2	203-0	255-0
48- 1	100- 0	152-0	204-0	256-0
49- 3	101- 0	153-0	205-0	257-0
50-10	102- 0	154-0	206-2	258-0
51- 2	103- 0	155-0	207-0	259-0
52- 5	104- 0	156-1	208-0	260-0

It should be kept in mind that a taxon, if it occurred in a well, may have been observed in several samples. Of these, only the depth of the sample with the highest occurrence was recorded.

Suppose that the number of wells is represented by the index k. It is useful to work with cumulative frequencies expressing how many events occur in k or more wells. The preceding two tabulations then become:

Northern region:

Number of wells:	1	2	3	4	5	6	7	8	9	10	11	12	13	14	15	16
Cumulative frequency:	150	94	68	55	41	30	26	21	19	17	14	10	5	3	2	0

Southern region:

Number of wells:	1	2	3	4	5	6
Cumulative frequency:	157	101	60	31	16	6

The largest cumulative frequency is equal to total number of events in the region considered.

The cumulative distribution provides a simple guide for selecting a threshold parameter k_c in order to retain only those events that occur in k_c or more wells. It will be seen later that results of ranking and scaling may become imprecise if they are based on all events including those that occur in only one or a few wells. The precision of the results increases when only those events are used that occur in at least k_c wells. The events occurring in fewer than k_c wells are filtered out. For example, by setting $k_c = 5$ for the northern region, further analysis was restricted to 41 events. For the southern region, 60 events with $k_c = 3$ were used. Although statistical results become more precise when the minimum sample size k_c is increased, an increasingly large number of events then is deleted. The stratigrapher must make a judicious choice of k_c taking care that not too much information is lost. It is possible that certain key fossils , important for establishing a regional biozonation, occur in one or a few sections only. In the RASC method, such special fossils can be coded as "unique" events.

These occur in fewer than k_c sections. Although unique events are not used for ranking and scaling, they are inserted later on the basis of their superpositional relations with other events in the one or more sections containing them.

The study of the frequency distribution of the events in a region, selection of the threshold parameter k_c and definition of unique events belong to the preprocessing module of the RASC computer program. During this stage, the user should also identify possible "marker horizons". These are stratigraphic events with positions that can be coded with certainty in the k_c or more sections containing them. Marker horizons (e.g. bentonite layers or seismic events) will receive more weight than other events in the scaling part of RASC.

4.9 Artificial datasets based on random numbers

The Gradstein-Thomas database introduced in the previous sections is characterized by the fact that it has information on many microfossils and most of these occur in relatively few sections. Ranking and scaling are based on superpositional relations between stratigraphic events. If there are n events in total, the number of pairs of events is $n(n\text{-}1)/2$. For example, $n=101$ results in 5050 pairs. It means that there are fifty times as many pairs of events as there are individual events. It will be seen in Chapters 5 and 6 that the frequency distributions for pairs of events in the Gradstein-Thomas database have smaller frequencies and are even more skewed than the frequency distributions for counts of events shown in the previous section.

In order to test the statistical models for ranking and scaling to be developed in later chapters it is desirable to have "complete" artificial datasets in addition to the real datasets. Such artificial datasets can be obtained from random numbers. In this section, random normal numbers will be used. In general, it is most convenient to obtain these by means of a pseudo-random number generator on a computer. Table 4.11 shows how artificial sequences of three events (A, B and C) can be created from random normal numbers. The first three columns of Table 4.11 are random normal numbers from Dixon and Massey (1957). Each number is a realization of the same random variable X with "normal", Gaussian distribution and mean (or expected value) $E(X)=2$ and variance $Var(X)=1$. By subtracting 1 from the numbers in column 1 and adding 0.5

TABLE 4.11

Artificial sequences of events A, B and C created from random normal numbers with $E(X)=2$ and Var $(X)=1$ taken from Table A-23 of Dixon and Massey (1957). Event "Distances" were obtained by subracting 1 from random normal numbers in column 1, maintaining column 2, and adding 0.5 to random normal numbers in column 3.

Random Normal Numbers			Event "Distances"			
1	2	3	A	B	C	Sequence
2.422	0.130	2.232	1.422	0.130	2.732	BAC
0.694	2.556	1.868	-0.306	2.556	2.368	ACB
1.875	2.273	0.655	0.875	2.273	1.155	ACB
1.017	0.757	1.288	0.017	0.757	1.788	ABC
2.453	4.199	1.403	1.453	4.199	1.903	ACB
2.274	1.767	1.564	1.274	1.767	2.064	ABC
3.000	1.618	1.530	2.000	1.618	2.030	BAC
2.510	2.256	1.146	1.510	2.256	1.646	ACB
1.233	2.085	2.251	0.233	2.085	2.751	ABC
3.075	1.730	2.427	2.075	1.730	2.927	BAC
1.344	-0.095	2.166	0.344	-0.095	2.666	BAC
1.246	3.860	1.253	0.246	3.860	1.753	ACB
0.889	2.299	2.458	-0.111	2.299	2.958	ABC
1.154	1.401	1.935	0.154	1.401	2.435	ABC
3.031	1.048	0.719	2.031	1.048	1.219	BCA
0.534	1.155	1.705	-0.466	1.155	2.205	ABC
2.230	3.096	0.045	1.230	3.096	0.545	CAB
2.355	1.761	1.816	1.355	1.761	2.316	ABC
1.461	0.947	0.717	0.461	0.947	1.217	ABC
3.034	1.778	2.122	2.034	1.778	1.622	BAC
2.761	0.473	3.726	1.761	0.473	4.226	BAC
1.961	0.965	1.481	0.961	0.965	1.981	ABC
2.639	4.010	1.915	1.639	4.010	2.415	ACB
1.349	2.225	0.644	0.349	2.225	1.144	ACB
2.959	2.797	4.635	1.959	2.797	5.135	ABC
2.440	2.919	1.455	1.440	2.919	1.955	ACB
3.078	3.279	0.352	2.078	3.279	0.852	CAB
1.736	1.968	0.011	0.736	1.968	0.511	CAB
3.275	3.147	2.800	2.275	3.147	3.300	ABC
2.579	2.297	2.030	1.579	2.297	2.530	ABC

TABLE 4.12

Sequences of artificial stratigraphic events A, B and C generated from random normal numbers for subsamples 1 to 5. Sequences for subsample 1 are same as those shown in last column of Table 4.11.

1	2	3	4	5
BAC	ACB	CBA	BAC	ABC
ACB	ACB	ACB	ACB	ABC
ACB	BAC	ABC	ACB	ABC
ABC	ABC	ACB	ACB	ACB
ACB	CAB	BAC	ACB	CAB
ABC	CAB	CBA	ABC	ACB
BAC	ABC	BAC	ACB	ABC
ACB	BCA	ACB	ABC	ABC
ABC	ACB	ACB	ACB	ACB
BAC	BAC	ACB	ABC	ABC
BAC	CBA	ACB	ABC	ACB
ACB	ACB	ABC	ACB	BAC
ABC	ABC	ACB	ABC	ABC
ABC	CBA	ACB	ABC	ABC
BCA	ACB	ACB	BAC	ABC
ABC	BAC	ABC	BAC	CBA
CAB	BCA	ABC	ABC	ABC
ABC	ABC	CAB	ABC	ACB
ABC	ABC	ABC	BAC	ABC
BAC	ACB	ACB	ABC	ACB
BAC	ACB	ABC	BAC	BAC
ABC	ABC	ABC	ACB	CAB
ACB	ABC	ACB	ACB	BAC
ACB	ABC	ACB	CBA	ABC
ABC	CAB	ACB	ACB	ABC
ACB	ABC	ACB	ABC	CAB
CAB	CAB	ABC	BAC	ABC
CAB	BAC	ABC	BAC	ACB
ABC	BAC	BAC	ABC	ABC
ABC	ACB	BCA	ACB	ABC

to the numbers in column 3, artificial "distances" along the real line were created for the events A, B and C which are regarded as realizations of the normal random variables X_A, X_B and X_C, respectively.

On the average, the random numbers for events A, B and C occupy the positions $E(X_A)=1.0$, $E(X_B)=2.0$, and $E(X_C)=2.5$ which follow one another along the real line. Consequently, their expected or average "optimum" sequence is ABC. Each event, however, has variance equal to one. This implies, that in the realizations, simulating separate stratigraphic sections, A may be following B or C instead of preceding them. Thirty "observed" sequences for sections are shown in the last column of Table 4.11. The artificial sequences are of nine different types with the following frequencies:

Sequence:	ABC	ACB	BAC	BCA	CAB	CBA
Frequency:	12	8	6	1	3	0

The optimum sequence is observed in 12 of the 30 sections. Because $E(X_b)=2$ AND $E(X_c)=2.5$ are closer together on the real line than $E(X_A)=1$ and $E(X_B)=2$, it is expected that A in the sections precedes B more frequently than that, for example, B is followed by C. For frequencies of pairs of events,

Sequence:	AB	BA	AC	CA	BC	CB
Frequency:	23	7	26	4	19	11

It can be attempted, by statistical modelling, to estimate the optimum sequence (ABC) and also the relative positions of $E(X_A)$, $E(X_B)$ and $E(X_C)$ along the real line from the frequencies of observed sequences in the sections. Normally such experiments are carried out on a large scale using a pseudo-random number generator on a computer. An advantage of computer simulation experiments similar to the experiment of Table 4.11 is, that predictions can be compared to true values, e.g. to $E(X_B\text{-}X_A)=1.0$, $E(X_C\text{-}X_A)=1.5$, $E(X_C\text{-}X_B)=0.5$. The statistical techniques for making these predictions will be further developed in later chapters.

The experiment of Table 4.11 was repeated on other random normal numbers listed in Dixon and Massey (1957, p.452-453) with the resulting sequences shown in Table 4.12. The final column of Table 4.11 is the first column of Table 4.12. In this new table, the previous experiment is regarded as the first subsample for a set of five experiments, all with $E(X_A)=1$, $E(X_B)=2$, $E(X_C)=2.5$ and $Var(X_A)=Var(X_B)=Var(X_C)=1$. In the first subsample, the frequencies of the ordered pairs BC and CB were 19 and 11, respectively. The relative frequency of BC, therefore, is (19/30=)0.633. The set of relative frequencies for all subsamples is

TABLE 4.13

Sequence file with artificially created superpositional relations for 20 events (numbered 1 to 20) in 25 sections. The interval between expected positions of the events along the linear scale was set equal to 0.5.

1	2	5	4	3	6	10	8	9	11	13	14	12	15	7	17	16	18	19	20
1	4	3	2	7	8	9	6	11	5	12	13	10	15	18	19	16	14	17	20
3	1	2	4	5	6	10	8	7	9	12	11	13	15	16	14	17	18	19	20
5	3	1	2	4	7	6	8	9	10	12	11	13	14	18	19	16	15	17	20
2	1	3	5	6	4	7	8	9	12	10	13	11	14	15	16	19	17	20	18
3	4	5	2	1	6	11	9	7	10	12	8	16	15	14	13	17	18	20	19
2	3	4	1	7	6	9	10	5	12	8	13	14	15	11	16	18	17	19	20
1	3	5	4	9	6	2	7	11	12	8	10	13	16	15	14	17	19	18	20
1	8	3	2	4	6	9	5	12	7	10	11	14	13	15	16	18	17	20	19
2	3	4	1	8	7	6	5	10	12	14	16	11	13	9	15	17	18	19	20
1	5	6	2	3	4	8	7	9	13	10	14	16	11	12	15	17	18	19	20
1	4	6	2	3	5	8	7	9	13	11	14	10	12	15	17	18	16	19	20
2	4	1	5	3	11	6	7	9	8	10	13	14	12	16	15	17	18	19	20
6	3	1	4	2	5	7	8	14	9	11	12	15	16	10	13	17	18	19	20
3	4	2	1	5	7	6	8	9	12	10	11	14	13	16	17	15	19	18	20
3	1	7	6	2	5	4	8	10	15	12	9	13	14	11	17	16	20	19	18
1	2	4	5	7	3	8	6	14	10	9	11	16	12	13	19	18	17	15	20
2	1	4	3	8	6	5	7	9	11	15	14	12	13	10	16	17	20	18	19
1	2	4	7	3	5	6	9	10	11	8	18	13	12	14	15	16	17	19	20
2	1	4	3	6	5	7	11	10	9	8	14	15	16	12	13	18	17	19	20
3	1	5	4	10	6	2	7	8	11	9	12	14	16	13	17	15	18	19	20
1	2	5	3	4	6	8	7	9	11	10	15	14	13	12	16	19	17	18	20
1	5	4	3	6	2	8	7	11	9	12	10	16	14	17	15	18	13	19	20
2	1	7	3	6	5	4	8	13	12	9	10	11	16	18	20	14	15	19	17
4	1	3	2	8	6	5	7	11	9	13	10	12	16	14	15	17	18	20	19

TABLE 4.14

Sequence file with artificially created superpositional relations for 20 events (numbered 1 to 20) in 25 sections. The interval between expected positions of the events along the linear scale was set equal to 0.3.

5	1	4	2	10	3	6	8	11	9	15	13	14	17	12	16	7	19	18	20
1	4	3	7	2	8	9	11	6	12	13	18	15	5	10	19	16	20	17	14
3	1	2	4	5	6	10	12	8	9	11	7	16	15	13	17	14	18	19	20
5	3	1	7	6	4	2	9	8	10	12	13	11	14	18	19	17	20	16	15
2	1	3	5	6	8	7	12	9	4	10	14	13	19	15	11	16	17	20	18
3	4	5	11	9	2	6	1	7	10	12	16	15	14	8	17	13	18	20	19
2	3	4	7	1	10	9	6	12	13	15	14	5	8	16	18	11	17	19	20
1	9	3	5	4	6	2	11	7	12	10	16	8	13	15	14	19	17	18	20
8	3	1	2	4	6	9	12	5	10	7	14	11	15	13	16	18	17	20	19
2	3	4	8	7	1	6	10	5	14	12	16	15	13	11	17	9	18	19	20
1	5	6	2	3	8	7	13	4	9	16	14	10	11	12	17	15	18	19	20
1	4	6	5	3	2	8	7	14	13	9	17	11	15	10	12	18	20	19	16
2	4	5	11	3	1	9	6	7	8	13	10	14	16	12	15	17	18	19	20
6	3	4	1	2	5	14	7	8	11	9	16	12	15	17	13	10	18	19	20
3	4	2	1	5	7	12	9	8	6	11	10	14	16	13	17	19	15	18	20
3	1	7	6	5	15	2	10	8	4	14	12	13	9	11	17	16	20	19	18
1	4	7	2	5	14	8	6	3	10	16	11	9	19	12	18	13	17	15	20
2	8	4	1	3	6	7	5	9	11	15	14	12	13	20	18	16	17	19	10
7	1	4	2	5	3	6	9	18	10	11	13	8	12	14	15	16	17	19	20
2	4	1	6	7	3	5	11	14	10	9	8	16	15	18	17	12	13	19	20
3	10	1	5	6	4	7	2	8	11	9	14	12	16	17	13	15	18	19	20
1	2	5	3	4	8	6	7	9	11	15	10	14	13	19	12	16	17	18	20
5	1	4	6	3	11	2	8	7	9	12	16	17	18	14	10	15	13	19	20
2	7	6	1	3	5	13	8	12	4	16	9	10	20	18	11	14	19	15	17
4	3	8	1	2	6	5	11	7	9	13	12	10	16	14	17	15	18	20	19

TABLE 4.15

Sequence file with artificially created superpositional relations for 20 events (numbered 1 to 20) in 25 sections. The interval between expected positions of the events along the linear scale was set equal to 0.1.

5	10	4	2	1	11	17	15	14	8	9	13	6	3	16	12	19	20	18	7
1	4	7	18	11	19	9	13	8	12	3	15	20	2	6	16	17	10	14	5
3	4	1	2	6	10	12	5	16	11	15	8	9	13	7	17	18	19	20	14
5	7	3	6	9	1	4	8	18	10	19	2	12	13	14	20	11	17	16	15
2	5	12	1	3	19	8	6	7	9	10	15	14	20	16	13	17	4	11	18
11	16	9	5	4	3	10	12	6	15	7	17	2	14	18	1	20	13	19	8
10	15	9	3	7	12	4	2	13	14	6	16	18	1	17	8	5	11	19	20
9	1	5	6	3	4	11	16	12	19	7	15	2	10	13	17	14	18	8	20
8	12	3	9	6	1	15	14	4	2	16	10	13	18	11	7	17	5	20	19
2	8	3	4	7	16	14	10	6	12	15	1	17	13	5	19	18	20	11	9
5	6	1	13	16	8	7	14	9	2	3	4	10	17	12	18	19	11	15	20
4	1	6	5	3	8	2	17	14	13	20	15	19	18	11	7	9	16	12	10
11	4	5	2	9	13	8	7	3	6	10	1	14	16	17	18	19	15	12	20
6	3	14	4	16	11	17	5	15	2	8	1	7	12	9	19	18	20	13	10
3	4	12	5	7	2	9	14	8	1	16	19	17	11	6	10	13	15	18	20
3	1	15	7	6	10	14	8	13	5	12	20	17	2	4	16	11	19	9	18
14	7	16	4	1	19	8	5	2	10	6	18	11	12	17	3	9	13	20	15
2	8	15	20	7	4	6	11	14	9	19	5	18	3	17	1	13	16	12	10
18	7	4	5	1	9	10	11	2	6	13	3	12	14	16	17	15	8	20	19
2	4	14	1	11	6	7	16	10	15	9	5	3	8	18	17	19	20	13	12
10	3	5	6	7	1	4	8	11	16	14	17	12	9	2	19	18	15	13	20
5	1	2	3	4	8	15	14	11	6	7	19	13	9	10	16	18	17	12	20
5	11	6	4	1	3	8	18	16	17	9	7	12	2	14	15	19	20	10	13
7	13	2	20	6	16	12	18	5	8	3	1	19	10	9	4	11	14	15	17
8	4	11	13	3	6	16	5	17	9	1	2	18	12	7	15	14	10	20	19

Subsample:	1	2	3	4	5
Relative frequency:	0.633	0.533	0.433	0.600	0.633

The average relative frequency is 0.5667. One might suspect that the average is a better estimate of the "true" population value because it is based on a sample that is five times larger. For this example, this assumption is not correct, because the true relative frequency is $\Phi(0.5/\sqrt{2})=0.638$. In the latter expression, Φ represents the fractile of the normal distribution in standard form (see later). In general, if the interval between the mean positions of two events along the real line is written as D ($D=0.5$ for the interval between B and C in the example), then the population is equal to $\Phi(D/\sqrt{2})$.

Tables 4.13 to 4.15 form an artificial database consisting of three SEQ files for 20 events in 25 sections. The same set of 20x25=500 normal random numbers was used for each SEQ file. The events are numbered 1 to 20. Because their mean positions follow one another along the real line, the optimum sequence is also 1 to 20 for each SEQ file. The 20 events were given expected values that are equally spaced. The spacing along the real line was 0.5, 0.3 and 0.1 for Tables 4.13, 4.14 and 4.15, respectively.

Relative frequencies for the order of pairs of consecutive events in Table 4.13 are similar to those for B and C in Table 4,12, because the interval D between mean positions is equal to 0.5 in both situations. For example, the relative frequencies for the first five ordered pairs in Table 4.13 are

Sequence:	12	23	34	45	56
Relative frequency:	0.640	0.520	0.600	0.600	0.560

The average of these five relative frequencies is 0.584. The population average of 0.638 (see before) would be increasingly closely approximated by the sample average, if the number of ordered pairs in the sample is enlarged. One of the advantages of computer simulation experiments is that the deviations between estimates of parameters based on relatively small samples and the parameters themselves can be systematically studied. As pointed out before, the true values of parameters generally are not available for comparison in real world applications.

0.0 [illegible] the latter expression [illegible]

normal distribution in standard form (Section [illegible]

between the two positions of two operations along [illegible]

$P(X) = $ [illegible]

good coincidence in the GWD [illegible]

Tables 4.12 to 4.15 give an artificial data base consisting of three [illegible] The first 20 events in 4 sections [illegible] The same set of 20 × 20 = 400 [illegible] random numbers was used in each table. The events are numbered [illegible] to 20. Because their input positions follow one another along the real line the optimum sequence is also 1 to 20 for each table. The 20 events were given expected values that were equally spaced. The spacing along the real line was 0.2, 0.3 and 0.1 for Tables 4.13, 4.14 and 4.15, respectively.

Relative frequencies for the [illegible] of pairs [illegible]

example, the relative frequencies for the [illegible]

4.15 are [illegible]

Sequence 12 23 34 45 [illegible]

Relative frequency 0.510 0.525 0.408 0.500 0.550

The average of these five relative frequencies is [illegible] The population average of 0.528 [illegible] by the sample average [illegible]

analyses [illegible]

that the deviations [illegible]

[illegible]

CHAPTER 5

RANKING OF BIOSTRATIGRAPHIC EVENTS

5.1 Introduction

The purpose of the ranking techniques to be discussed in this chapter is to order, for a region, a number of biostratigraphic events for which the observed superpositional relations in individual stratigraphic sections are mutually inconsistent. During the 1960s and 1970s, several methods already were developed to eliminate such inconsistencies in a systematic manner (Shaw, 1964; Hay, 1972; Rubel, 1978; Davaud and Guex, 1978; Edwards and Beaver, 1978; for reviews, see Hay and Southam, 1978; and Brower, 1981). The order obtained for a region after application of a ranking technique will be called an optimum sequence.

The techniques to be introduced in this chapter and the next (scaling) show similarity to the techniques known as "ranking" of objects in mathematical statistics (cf. David, 1988). According to Kendall (1975), a number of individuals are ranked when arranged in order according to some quality which they all possess to a varying degree. The arrangement as a whole is termed a ranking in which each member has a rank. An important difference between the ranking of objects on the basis of their characteristics and the ranking of stratigraphic events on the basis of superpositional relations is that, generally, only subsets of all stratigraphic events are observed within individual sections. These subsets of stratigraphic events may have sizes which are much smaller than the total number of events considered for the study region.

In this chapter and the next, ranking and scaling techniques will be illustrated using the Hay example introduced at the beginning of the previous chapter. In this example, there are 10 stratigraphic events and 9 sections (see Fig. 4.2; Tables 4.1 and 4.3). The preprocessing of the RASC computer program begins with a tabulation of the number of stratigraphic sections in which each event occurs. For the Hay example, this gives:

Number of sections:	8	8	6	7	9	4	7	5	9	6

The following frequency distribution of the stratigraphic events is obtained from this initial tabulation:

Number of sections:	1	2	3	4	5	6	7	8	9
Frequency of events:	0	0	0	1	1	2	2	2	2
Cumulative frequency:	10	10	10	10	9	8	6	4	2

As explained previously (Section 4.8), this frequency distribution is helpful in selecting the threshold parameter k_c which is set to retain only those events that occur in k_c or more wells. For the Hay example, all events occur in at least 4 sections. Initially, we will set $k_c=1$ (Default value for k_c in micro-RASC, see Chapter 10) so that all events will be retained for further analysis.

5.2 Hay's original method

Hay (1972) began constructing an optimum sequence from the stratigraphic information of Figure 4.2 by modifying the subjective sequence in column 1 on the right side of this diagram. While ignoring coeval events, Hay counted how often each of the 10 events was observed to occur above each of the other events. The resulting counts and corresponding sample sizes are shown in Figure 5.1A. Dividing a count by its sample size produces a relative frequency. Because the initial subjective sequence is not very different from the optimum sequence (column 2 of Fig. 4.2), most relative frequencies in Figure 5.1A are greater than 0.5 if they occur above the diagonal consisting of black boxes. Every relative frequency in the upper triangle of Figure 5.1A has a counterpart in the lower triangle. Together the relative frequency and its counterpart add to one and, consequently, most relative frequencies below the diagonal are less than 0.5.

The optimum sequence is determined by re-evaluating the relations of all pairs which show a fraction greater than 0.5 in the lower right half of the matrix. Inspection of the matrix reveals (see Hay, 1972, p. 262) that V and θ should be reversed, the number in the appropriate square in the upper left hand part of the matrix being 1/4, which is less than 0.5. After making this correction it can be seen that δ should come below both θ and V, these relationships being expressed by the fractions 0/5 and 1/4, respectively. Finally, it is evident that the position of ⏀ in the sequence needs to be changed because its relation to δ is 1/4, to V is 0/5, to ц is 1/5, and to < is 1/5. It must come below any of these symbols, and, in fact, became the lowest event in Hay's original optimum sequence shown in column 2 of Figure 4.2. The revised matrix using Hay's optimum sequence is shown in Figure 5.1B. All values greater than 0.5 now are in the upper left part of the matrix. Note that both the upper part and the lower part contain fractions equal to 0.5. These occur in pairs and signify events that are coeval "on the average".

Before or after creation of the optimum sequence, every fraction in the matrix can be tested for statistical significance by comparing it to 0.5 using the binomial frequency distribution model as explained in Section 3.2. Figure 5.2 shows the difference between 1 and the cumulative probability $P_c\ (k, R)$ that an event occurs k times above another one in a sample of pairs of events with size R. If $1\text{-}P_c\ (k, R)$ exceeds 0.95, the

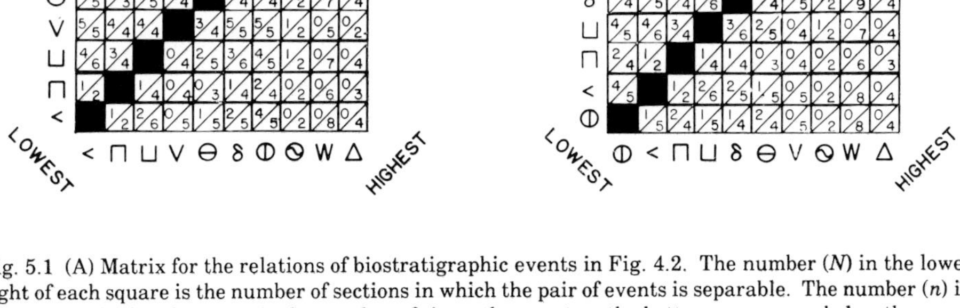

Fig. 5.1 (A) Matrix for the relations of biostratigraphic events in Fig. 4.2. The number (N) in the lower right of each square is the number of sections in which the pair of events is separable. The number (n) in the upper left of each square is the number of times the event on the bottom row occurs below the event on the left side. The sequence from lowest to highest on the bottom and left side of the matrix is that shown in column (1) on right side of Fig. 4.2. (B) Revised matrix in which the ratio n/N has been rearranged so that all values greater than $\frac{1}{2}$ are in the upper left part of the matrix. The lowest-highest sequence along the bottom and left side of the matrix now represents Hay's original optimum sequence also shown as column (2) on right side of Fig. 4.2 (after Hay, 1972).

fraction k/R is greater than 0.5 with a probability of 95 percent. The hypothesis of nonrandom average superpositional relationship can only be accepted for 6 of 45 pairs of events. These are 6 of nine pairs involving the event W which occurs at or near the top of all (9) sections (A to I in Fig. 4.2). In total, two of the values in Figure 5.2 exceed 0.99 They correspond to the facts that (1) W occurs above Φ in eight sections, and (2) W occurs above $<$ in eight sections. These two superpositional relations are statistically significant with a probability of 99 percent.

The binomial model has a drawback for testing whether or not the observed superpositional relation of two events is random, because it ignores the relations of these two events with all other events. For example, the binomial test of Figure 5.2 suggest that W occurs above Φ. On the other hand, the fact that Δ occurs above Φ in 4 out of 4 sections would not be statistically significant, because the sample size is too small. However, W and Δ occur near the top in all sections. In those sections where they coexist, each occurs above the other one 3 out of 6 times. This would suggest that, although the relation between W and Δ remains undecided, both events probably occur above Φ. The relations between these three events are shown graphically in Figure 5.3A. If in addition to

HIGHEST

.8750	.8750	.7500	.8750	.8750	.8750	.5000	.0000	.0000	Δ
.9922	.9922	.9688	.9844	.9961	.8750	.9375	.8750	W	
.5000	.5000	.5000	.0000	.5000	.5000	.0000	⊘		
.9375	.9375	.8750	.8750	.9375	.3750	V			
.0000	.6250	.7500	.0000	.3750	⊖				
.3750	.0000	.3750	.0000	δ					
.6250	.3125	.3750	⊔						
.0000	.0000	⊓							
.6250	<								
Φ									

LOWEST

Fig. 5.2 Values of 1-P where P represents the probability that the sequential relation between two events in nonrandom (cf. Eq. 3.2 for cumulative probability of binomial probability with $p=0.5$; after Hay, 1972).

the relations between these three events (W, Δ and Φ), their relations with a fourth event (V) are also considered, the probability that Δ occurs above Φ is further increased (see Fig. 5.3B). A multivariate statistical test which considers all pairs of events simultaneously and is not subject to the drawback of the binomial test of considering pairs of events in isolation, will be developed in the next chapter on scaling.

5.3 Algorithmic version of Hay's original method

It is obvious that the method of the previous section can be programmed for a digital computer. Slightly different versions have been described in Worsley and Jorgens (1977), Blank (1979), and Agterberg and Nel (1982a). The following changes help to make Hay's method more general.

1. *Choice of initial sequence*

Instead of an initial subjective ranking (e.g. column 1 in Fig. 4.2), one of the sections, if necessary supplemented by information from other sections, can be used as the starting point. Use of Section A in the Hay example gave the event numbers 1 to 9 in Tables 4.1 and 4.3. Only the event Δ (LO *Discolithus distinctus*) does not occur in Section A. It was assigned the number 10. While numbering I moved in the stratigraphically upward direction. However, this decision was arbitrary.

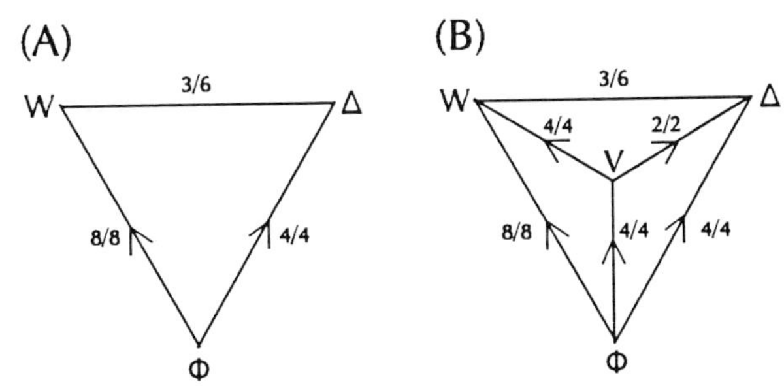

Fig. 5.3 Diagrams to illustrate superpositional relations between (A) three events and (B) four events in the Hay example. Although Δ and Φ both occur in only 4 sections, their superpositional relation is probably nonrandom because of their relations with other events.

TABLE 5.1

A. F-matrix of frequencies of events occurring above or below one another in the sections. The events for the Hay example are labelled 1 to 10 as in Tables 4.1 and 4.3. B. R-matrix of frequencies of coexistence of two events in the same section. Coeval events also were counted.

A	1	2	3	4	5	6	7	8	9	10
1	x	4	1	1	2	0	2	0	0	0
2	1	x	2	2	1	0	1	0	0	0
3	1	2	x	0	1	0	1	0	0	0
4	4	2	3	x	3	0	3	1	1	1
5	3	3	3	1	x	0	3	0	0	0
6	2	2	2	2	2	x	1	1	0	0
7	4	4	3	2	3	1	x	0	0	0
8	5	5	4	3	5	1	4	x	0	0
9	8	8	6	6	9	4	7	5	x	3
10	4	4	3	3	4	1	4	2	3	x

B	1	2	3	4	5	6	7	8	9	10
1	x	7	5	6	8	3	6	5	8	5
2	7	x	6	6	8	4	6	5	8	5
3	5	6	x	6	6	4	5	4	6	4
4	6	6	6	x	7	5	6	4	7	5
5	8	8	6	7	x	4	7	5	9	6
6	3	4	4	4	4	x	3	2	4	2
7	6	6	5	6	7	3	x	5	7	5
8	5	5	4	4	5	2	5	x	5	3
9	8	8	6	7	9	4	7	5	x	6
10	5	5	4	5	6	2	5	3	6	x

One could have started by numbering Δ as 1, followed by W (HI *Discoaster tribrachiatus*) as 2, then moving further downward in Section A.

2. *Matrix notation*

While arranging the information in matrix form, it is customary to number the rows from left to right and the columns from top to bottom. Table 5.1A shows the so-called F-matrix of frequencies which are similar to the counts shown previously in Figure 5.1. The corresponding sample sizes for frequencies of co-existence of two events in the same section are shown in Table 5.1B. Note that the main diagonal goes from the top left to the bottom right side in Table 5.1 .

As already stated in Section 4.3, SEQ files, such as the one shown in Table 4.3A, normally are for the stratigraphically downward direction

(= direction of drilling exploratory wells in sedimentary basins). Table 5.1A corresponds to Table 4.3A in the following sense. Each frequency in Table 5.1A indicates how often the event labelling its column follows the event labelling its row when moving from the left to the right through all the rows of Table 4.3. For example, the first element in the first row of Table 5.1A (after the x on the main diagonal) is equal to 4. This means that event 2 (column label) follows event 1 four times in Table 4.3A. The corresponding sections, in which event 2 is stratigraphically below event 1, are C, D, E and I.

TABLE 5.2

A. *S*-matrix of scores obtained by adding half of the frequencies of ties (shown in Table 5.2B) to the frequencies of the *F*-matrix (see Table 5.1A). B. *T*-matrix of frequencies of ties.

A	1	2	3	4	5	6	7	8	9	10
1	x	5.0	2.5	1.5	3.5	0.5	2.0	0.0	0.0	0.5
2	2.0	x	3.0	3.0	3.0	1.0	1.5	0.0	0.0	0.5
3	2.5	3.0	x	1.5	2.0	1.0	1.5	0.0	0.0	0.5
4	4.5	3.0	4.5	x	4.5	1.0	3.5	1.0	1.0	1.5
5	4.5	5.0	4.0	2.5	x	1.0	3.5	0.0	0.0	1.0
6	2.5	3.0	3.0	3.0	3.0	x	1.5	1.0	0.0	0.5
7	4.0	4.5	3.5	2.5	3.5	1.5	x	0.5	0.0	0.5
8	5.0	5.0	4.0	3.0	5.0	1.0	4.5	x	0.0	0.5
9	8.0	8.0	6.0	6.0	9.0	4.0	7.0	5.0	x	3.0
10	4.5	4.5	3.5	3.5	5.0	1.5	4.5	2.5	3.0	x

B	1	2	3	4	5	6	7	8	9	10
1	x	2.0	3.0	1.0	3.0	1.0	0.0	0.0	0.0	1.0
2	2.0	x	2.0	2.0	4.0	2.0	1.0	0.0	0.0	1.0
3	3.0	2.0	x	3.0	2.0	2.0	1.0	0.0	0.0	1.0
4	1.0	2.0	3.0	x	3.0	2.0	1.0	0.0	0.0	1.0
5	3.0	4.0	2.0	3.0	x	2.0	1.0	0.0	0.0	1.0
6	1.0	2.0	2.0	2.0	2.0	x	1.0	0.0	0.0	1.0
7	0.0	1.0	1.0	1.0	1.0	1.0	x	1.0	0.0	1.0
8	0.0	0.0	0.0	0.0	0.0	0.0	1.0	x	0.0	1.0
9	0.0	0.0	0.0	0.0	0.0	0.0	0.0	0.0	x	0.0
10	1.0	1.0	1.0	1.0	2.0	1.0	1.0	1.0	1.0	x

3. *Incorporation of coeval events*

Coeval events were ignored in Figure 5.1 and Table 5.1A. Although ranking by means of Hay's original method would not be influenced by this modification, two events which are coeval in a section will be scored by adding 0.5 to the two counts for the first event occurring above and below the second event, respectively. Suppose that the elements of the F-matrix of Table 5.1A are written as F_{ij} ($i=1, 2, ..., n; j=1, 2, ..., n$) for n events ($n=10$ in the example). The subscripts i and j indicate rows and columns, respectively. It is noted that these subscripts refer to positions of elements in a matrix. They do not necessarily coincide with the original code

TABLE 5.3

A. P-matrix of relative frequencies obtained by dividing elements of S-matrix by those of R-matrix.
B. P_0-matrix of relative frequencies excluding ties.

A	1	2	3	4	5	6	7	8	9	10
1	x	5.0/7	2.5/5	1.5/6	3.5/8	0.5/3	2.0/6	0.0/5	0.0/8	0.5/5
2	2.0/7	x	3.0/6	3.0/6	3.0/8	1.0/4	1.5/6	0.0/5	0.0/8	0.5/5
3	2.5/5	3.0/6	x	1.5/6	2.0/6	1.0/4	1.5/5	0.0/4	0.0/6	0.5/4
4	4.5/6	3.0/6	4.5/6	x	4.5/7	1.0/4	3.5/6	1.0/4	1.0/7	1.5/5
5	4.5/8	5.0/8	4.0/6	2.5/7	x	1.0/4	3.5/7	0.0/5	0.0/9	1.0/6
6	2.5/3	3.0/4	3.0/4	3.0/4	3.0/4	x	1.5/3	1.0/2	0.0/4	0.5/2
7	4.0/6	4.5/6	3.5/5	2.5/6	3.5/7	1.5/3	x	0.5/5	0.0/7	0.5/5
8	5.0/5	5.0/5	4.0/4	3.0/4	5.0/5	1.0/2	4.5/5	x	0.0/5	0.5/3
9	8.0/8	8.0/8	6.0/6	6.0/7	9.0/9	4.0/4	7.0/7	5.0/5	x	3.0/6
10	4.5/5	4.5/5	3.5/4	3.5/5	5.0/6	1.5/2	4.5/5	2.5/3	3.0/6	x

B	1	2	3	4	5	6	7	8	9	10
1	x	4.0/5	1.0/2	1.0/5	2.0/5	0.0/2	2.0/6	0.0/5	0.0/8	0.0/4
2	1.0/5	x	2.0/4	2.0/4	1.0/4	0.0/2	1.0/5	0.0/5	0.0/8	0.0/4
3	1.0/2	2.0/4	x	0.0/3	1.0/4	0.0/2	1.0/4	0.0/4	0.0/6	0.0/3
4	4.0/5	2.0/4	3.0/3	x	3.0/4	0.0/2	3.0/5	1.0/4	1.0/7	1.0/4
5	3.0/5	3.0/4	3.0/4	1.0/4	x	0.0/2	3.0/6	0.0/5	0.0/9	0.0/4
6	2.0/2	2.0/2	2.0/2	2.0/2	2.0/2	x	1.0/2	1.0/2	0.0/4	0.0/1
7	4.0/6	4.0/5	3.0/4	2.0/5	3.0/6	1.0/2	x	0.0/4	0.0/7	0.0/4
8	5.0/5	5.0/5	4.0/4	3.0/4	5.0/5	1.0/2	4.0/4	x	0.0/5	0.0/2
9	8.0/8	8.0/8	6.0/6	6.0/7	9.0/9	4.0/4	7.0/7	5.0/5	x	3.0/6
10	4.0/4	4.0/4	3.0/3	3.0/4	4.0/4	1.0/1	4.0/4	2.0/2	3.0/6	x

numbers of the events. The resulting modified matrix to be used here is the S-matrix shown in Table 5.2A. Also shown are the symmetrical T-matrix (Table 5.2B) for frequencies $T_{ij} = T_{ji}$ of coeval events (or "ties"). The R-matrix for sample sizes $R_{ij} = R_{ji}$ of pairs of events including ties was already shown in Table 5.1A. Consequently, the scores S_{ij} tabulated in the S-matrix, satisfy the equation: $S_{ij}=F_{ij}+\frac{1}{2}T_{ij}$.

Relative frequencies P_{ij} with $P_{ij}=S_{ij}/R_{ij}$ can be formed by dividing every score by the corresponding sample size in the R-matrix. The resulting P-matrix for relative frequencies is shown in Table 5.3A. Suppose that sample sizes without counting ties are denoted as $R_{0ij}=R_{ij}-T_{ij}$. For comparison, the relative frequencies $P_{0ij}=F_{ij}/R_{0ij}$ are shown in the P_0-matrix of Table 5.3B. These relative frequencies were previously shown in Figure 5.1. Note that any attempt to move all relative frequencies greater than 0.5 to positions above the main diagonal would yield identical results which are independent of whether the P-matrix or the P_0-matrix is used. Later (see Chapter 6), it will be shown that there are advantages to using P instead of P_0 when all superpositional relations between events are considered simultaneously.

4. *Order of checking superpositional relations*

In Hay's original example, the order in which events were selected for comparison with other events was subjective. For an algorithm, it is preferable to proceed in the same way in all applications if possible. The obvious choice is to begin at the beginning of the first row. For example, the first comparison then to be made in the S-matrix of Table 5.2A is for the element $S_{12}=5$ versus $S_{21}=2$. Since S_{12} is greater than S_{21} it is not necessary to reverse the order of events. The next pair of events to be tested is $S_{13}=2.5$ and $S_{31}=2.5$. Again it is not necessary to reverse the order, this time because the two matrix elements are equal to one another. The next pair is $S_{14}=1.5$, $S_{41}=4.5$. Because $S_{41}>S_{14}$ the positions of the first and fourth rows and columns should be interchanged. Table 5.4A shows the revised matrix after the interchange. It now is necessary to return to the first element of the first row for comparison with its counterpart, because the new first row is what originally was the fourth row (with the first element of the original fourth row in the fourth column of the new first row). The original code numbers are shown in parentheses in Table 5.4A.

TABLE 5.4

Illustration of algorithm for systematic checking of superpositional relations in Hay method for constructing optimum sequence. A. Positions of events 1 and 4 were interchanged because in Table 5.2A the element (=1.5) in the fourth column of the first row is less than its counterpart (=4.5) in the lower triangle of the matrix. Original event code numbers are shown in parentheses. B. Positions of events 6 and 4 were interchanged during second iteration. C. Positions of events 9 and 6 were interchanged during third iteration. D. Final order relation matrix after 22 iterations. This matrix has the property that all its elements in the upper triangle are greater than or equal to their counterparts in the lower triangle. Elements in the upper triangle equal to their counterparts are underlined in Table 5.4D. The events corresponding to these elements are coeval on the average. Note that the final (optimum) sequence is nearly the reverse of the original sequence in Table 5.2 because code numbers were assigned to the events while moving in the stratigraphically upward direction (cf. Tables 4.1 and 4.3).

A	1 (4)	2 (2)	3 (3)	4 (1)	5 (5)	6 (6)	7 (7)	8 (8)	9 (9)	10 (10)
1(4)	x	3.0	4.5	4.5	4.5	1.0	3.5	1.0	1.0	1.5
2(2)	3.0	x	3.0	2.0	3.0	1.0	1.5	0.0	0.0	0.5
3(3)	1.5	3.0	x	2.5	2.0	1.0	1.5	0.0	0.0	0.5
4(1)	1.5	5.0	2.5	x	3.5	0.5	2.0	0.0	0.0	0.5
5(5)	2.5	5.0	4.0	2.5	x	1.0	3.5	0.0	0.0	1.0
6(6)	3.0	3.0	3.0	3.0	3.0	x	1.5	1.0	0.0	0.5
7(7)	2.5	4.5	3.5	2.5	3.5	1.5	x	0.5	0.0	0.5
8(8)	3.0	5.0	4.0	3.0	5.0	1.0	4.5	x	0.0	0.5
9(9)	6.0	8.0	6.0	6.0	9.0	4.0	7.0	5.0	x	3.0
10(10)	3.5	4.5	3.5	3.5	5.0	1.5	4.5	2.5	3.0	x

B	1 (6)	2 (2)	3 (3)	4 (1)	5 (5)	6 (4)	7 (7)	8 (8)	9 (9)	10 (10)
1(6)	x	3.0	3.0	2.5	3.0	3.0	1.5	1.0	0.0	0.5
2(2)	1.0	x	3.0	2.0	3.0	3.0	1.5	0.0	0.0	0.5
3(3)	1.0	3.0	x	2.5	2.0	1.5	1.5	0.0	0.0	0.5
4(1)	0.5	5.0	2.5	x	3.5	1.5	2.0	0.0	0.0	0.5
5(5)	1.0	5.0	4.0	4.5	x	2.5	3.5	0.0	0.0	1.0
6(4)	1.0	3.0	4.5	4.5	4.5	x	3.5	1.0	1.0	1.5
7(7)	1.5	4.5	3.5	4.0	3.5	2.5	x	0.5	0.0	0.5
8(8)	1.0	5.0	4.0	5.0	5.0	3.0	4.5	x	0.0	0.5
9(9)	4.0	8.0	6.0	8.0	9.0	6.0	7.0	5.0	x	3.0
10(10)	1.5	4.5	3.5	4.5	5.0	3.5	4.5	2.5	3.0	x

C	1 (9)	2 (2)	3 (3)	4 (1)	5 (5)	6 (4)	7 (7)	8 (8)	9 (6)	10 (10)
1(9)	x	8.0	6.0	8.0	9.0	6.0	7.0	5.0	4.0	3.0
2(2)	0.0	x	3.0	2.0	3.0	3.0	1.5	0.0	1.0	0.5
3(3)	0.0	3.0	x	2.5	2.0	1.5	1.5	0.0	1.0	0.5
4(1)	0.0	5.0	2.5	x	3.5	1.5	2.0	0.0	0.5	0.5
5(5)	0.0	5.0	4.0	4.5	x	2.5	3.5	0.0	1.0	1.0
6(4)	1.0	3.0	4.5	4.5	4.5	x	3.5	1.0	1.0	1.5
7(7)	0.0	4.5	3.5	4.0	3.5	2.5	x	0.5	1.5	0.5
8(8)	0.0	5.0	4.0	5.0	5.0	3.0	4.5	x	1.0	0.5
9(6)	0.0	3.0	3.0	2.5	3.0	3.0	1.5	1.0	x	0.5
10(10)	3.0	4.5	3.5	4.5	5.0	3.5	4.5	2.5	1.5	x

D	1 (9)	2 (10)	3 (6)	4 (8)	5 (4)	6 (7)	7 (5)	8 (1)	9 (3)	10 (2)
1(9)	x	3.0	4.0	5.0	6.0	7.0	9.0	8.0	6.0	8.0
2(10)	3.0	x	1.5	2.5	3.5	4.5	5.0	4.5	3.5	4.5
3(6)	0.0	0.6	x	1.0	3.0	1.5	3.0	2.5	3.0	3.0
4(8)	0.0	0.5	1.0	x	3.0	4.5	5.0	5.0	4.0	5.0
5(4)	1.0	1.5	1.0	1.0	x	3.5	4.5	4.5	4.5	3.0
6(7)	0.0	0.5	1.5	0.5	2.5	x	3.5	4.0	3.5	4.5
7(5)	0.0	1.0	1.0	0.0	2.5	3.5	x	4.5	4.0	5.0
8(1)	0.0	0.5	0.5	0.0	1.5	2.0	3.5	x	2.5	5.0
9(3)	0.0	0.5	1.0	0.0	1.5	1.5	2.0	2.5	x	3.0
10(2)	0.0	0.5	1.0	0.0	3.0	1.5	3.0	2.0	3.0	x

TABLE 5.5

Optimum sequence output of the RASC computer program. Order of events is same as in Table 5.4D.

Sequence Number	Uncertainty Range	Event Code	Event Name
1	0-3	9	HI *Discoaster tribrachiatus*
2	0-3	10	LO *Discolithus distinctus*
3	2-5	6	LO *Rhabdosphaera scabrosa*
4	2-5	8	LO *Discoaster cruciformis*
5	4-6	4	LO *Coccolithus solitus*
6	5-8	7	LO *Discoaster minimus*
7	5-8	5	LO *Coccolithus gammation*
8	7-10	1	LO *Discoaster distinctus*
9	7-11	3	LO *Discoaster germanicus*
10	8-11	2	LO *Coccolithus cribellum*

The step of making one interchange because an element in the upper triangle is less than its counterpart in the lower triangle will be called an iteration. Successive checking of the elements in the first row of Table 5.4A shows that a second iteration is required at the sixth column because $S_{61} > S_{16}$. It means that the first and sixth rows and columns should be interchanged. The result of this second iteration is shown in Table 5.4B. As shown in Table 5.4C one can proceed to the ninth column before the third iteration is required. In Table 5.4C, the situation is finally reached that none of the elements in the first row is less than its counterpart in the first column. It means that one can proceed to the second row. The first element to be tested now is in the third column. The fourth iteration consists of interchanging the positions of the second and fourth rows and columns. In general, once all elements of a given row in the upper triangle have passed the test of comparing them to their counterparts in the corresponding column, then it will not be required to test them again, although they may be moved to other positions within the same row during subsequent iterations. Continuation of the algorithm finally led to the matrix of Table 5.4D, after 22 iterations in total. This is the so-called final order relation matrix. The order of the events in this matrix is considered to be the optimum sequence.

5. Consideration of events which are coeval on the average

A number of elements are underlined in Table 5.4D. They belong to pairs of events which are coeval on the average. In total, there are 6 pairs of this type. The elements of 5 of these 6 pairs are adjoining the main diagonal. If the positions of events which are neighbors in the optimum sequence are interchanged, the sequence remains an optimum sequence because none of its lower triangle elements exceeds 0.5. For example, if events 9 and 10, which are in positions 1 and 2 respectively, are interchanged, all frequencies in the upper triangle remain greater than their counterparts in the lower triangle. This rule does not apply to pairs in the optimum sequence which are coeval on the average but are separated by one or more events with which they are not coeval on the average. For example, events 6 and 7, which are in positions 3 and 6, are separated by events 8 and 4. If events 6 and 7 are interchanged, the resulting sequence is not an optimum sequence because event 7 follows event 4 in most sections, while event 4 follows event 6 in most sections containing both events. Consequently, event 7 must follow event 6 in any optimum sequence.

5.4 Uncertainty ranges for events in the optimum sequence

It is useful to define an uncertainty range for the events in the optimum sequence. Table 5.5 shows the RASC output for the optimum sequence of Table 5.4D. The first column contains the sequence numbers of the events in the optimum sequence. Column 3 gives the original code numbers and the names of the events are shown in the last column. The uncertainty range in the second column of Table 5.5 applies to the sequence number. Its two numbers are less than and greater than the sequence number, respectively. This range was determined by counting, for each event, the number of adjoining events with which it is coeval on the average. For example, because the positions of events 9 and 10 can be interchanged, and there are no other, similar pairs in the vicinity, their uncertainty ranges are 0-3. This indicates that the sequence number of either event could be 1 or 2. It is not possible to decide whether event 9 should come before or after 10 in the optimum sequence. On the other hand, the uncertainty range of event 4 extends from sequence position 4 to 6 indicating that its sequence position (=5) is not, on the average coeval with any other event. Although events 6 and 7 are coeval on the average, it could be established (see before) that event 6 must precede 7 in the optimum sequence. This type of uncertainty does not show up in the uncertainty range.

In general, the uncertainty range provides a quick method for evaluating how firmly an event is positioned between its neighbors in the optimum sequence. Occasionally, the uncertainty ranges of successive events interact with one another and the possible positions of the events are not immediately obvious. For example, in Table 5.5 events 1,3 and 2 have uncertainty ranges 7-10, 7-11 and 8-11, respectively. This means that event 1 or 3 (but not 2) can occupy position number 8. It also means that 2 or 3 (but not 1) can have position 10. Although all three events can occupy position number 9, the preceding conditions imply that 3 must precede 2. This type of conclusion can be drawn more rapidly by inspection of the frequencies in the final order relation matrix shown in Table 5.4D.

Three events A, B and C as a group are mutually inconsistent if, on average, A occurs before B, B before C, and C before A. It will be shown later that if the superpositional relations of 3 or more events are mutually inconsistent, it is not possible to construct an optimum sequence by Hay's original method. Neither can then an optimum sequence be obtained by the algorithm of Section 5.3. A solution can, however, be obtained by ignoring one or more pairs of scores (S_{ij} and S_{ji}) for events participating in inconsistencies involving groups of more than two events. In RASC, ignored pairs of this type will be treated as pairs with equal scores when the uncertainty range is determined.

In general, the scores S_{ij} and S_{ji} are subject to a statistical uncertainty which, in a relative sense, decreases with increasing sample size. R_{ij} $(=S_{ij}+S_{ji})$. If the statistical population from which a sample with size R_{ij} is drawn has fixed probability Π_{ij} that event i is followed by event j, then the difference between the observed proportion P_{ij} $(=S_{ij}/R_{ij})$ and Π_{ij} is relatively large when R_{ij} is small. Binomial theory can be used to quantify the frequency distribution of P_{ij} of which the mean value is Π_{ij}. This dependence on sample size implies that the erroneous observation $S_{ji}>S_{ij}$ (if on the average $S_{ij}>S_{ji}$) will be made more frequently when R_{ij} is small. In RASC, the user has the option of ignoring pairs of scores if sample size is less than a selected threshold value m_{c1}. In the previous example, $m_{c1}=1$ so that all pairs were used. However, if one were to set $m_{c1}=3$, two pairs of events with sample size $R_{ij}=2$, would be ignored in Table 5.4D. These are the pairs (10,6) and (6,8), respectively. For determination of the uncertainty range, pairs of events that are ignored because of the introduction of a threshold value will be treated in the same way as pairs of events that are coeval on the average. By this method, it is possible to consider, to some extent, the statistical uncertainty of event positions in the optimum sequence. Better methods to express the statistical uncertainty of the average position of events can be derived after scaling the events (Chapter 6).

5.5 Other ranking algorithms

In total, 22 iterations were required to produce an optimum sequence (Table 5.5) from the original S-matrix (Table 5.3A). In this section, faster algorithms will be discussed which lead to exactly or approximately the same final product. From a practical point of view, it is not important which one of the algorithms would be selected for this particular example, because there is no significant difference in the computing time required. In other applications, however, hundreds of thousands or more iterations might be required. Then it may become necessary to switch to algorithms by means of which an optimum sequence is produced faster. One method by which the total number of iterations generally can be ranked very quickly, is to set a tolerance value (b_c) greater than zero for the differences S_{ji}-S_{ij}. In the previous algorithm, an iteration is carried out if S_{ji}-$S_{ij}>0$. The user can require that an iteration is only carried and if S_{ji}-$S_{ij}>b_c$ with $b_c>0$. The option of making the tolerance b_c greater than in its default value, which is equal to zero, is available in the RASC computer program. This option reduces the computing time required to obtain an optimum sequence but this accomplished by leaving a variable amount of "noise" in the result.

Use of transposed order relation matrix

It is obvious that a relatively large number (=22) of iterations was required for the example of Table 5.4 because, initially, the majority of the scores in the upper triangle were less than their counterparts in the lower triangle. The transpose of the original S-matrix (Table 5.3A) is obtained by replacing S_{ij} by S_{ji} (and S_{ji} by S_{ij}). The transpose is shown in Table 5.6A. If the algorithm is applied, the first iteration consists of interchanging events 10 and 9 which occupy the first and second position

TABLE 5.6

A. Transposed S-matrix (cf. Table 5.2A). B. Final order relation matrix obtained after 5 iterations.

A	1 (1)	2 (2)	3 (3)	4 (4)	5 (5)	6 (6)	7 (7)	8 (8)	9 (9)	10 (10)
1(1)	x	2.0	2.5	4.5	4.5	2.5	4.0	5.0	8.0	4.5
2(2)	5.0	x	3.0	3.0	5.0	3.0	4.5	5.0	8.0	4.5
3(3)	2.5	3.0	x	4.5	4.0	3.0	3.5	4.0	6.0	3.5
4(4)	1.5	3.0	1.5	x	2.5	3.0	2.5	3.0	6.0	3.5
5(5)	3.5	3.0	2.0	4.5	x	3.0	3.5	5.0	9.0	5.0
6(6)	0.5	1.0	1.0	1.0	1.0	x	1.5	1.0	4.0	1.5
7(7)	2.0	1.5	1.5	3.5	3.5	1.5	x	4.5	7.0	4.5
8(8)	0.0	0.0	0.0	1.0	0.0	1.0	0.5	x	5.0	2.5
9(9)	0.0	0.0	0.0	1.0	0.0	0.0	0.0	0.0	x	3.0
10(10)	0.5	0.5	0.5	1.5	1.0	0.5	0.5	0.5	3.0	x

B	1 (2)	2 (1)	3 (3)	4 (5)	5 (7)	6 (4)	7 (6)	8 (8)	9 (9)	10 (10)
1(2)	x	5.0	3.0	5.0	4.5	3.0	3.0	5.0	8.0	4.5
2(1)	2.0	x	2.5	4.5	4.0	4.5	2.5	5.0	8.0	4.5
3(3)	3.0	2.5	x	4.0	3.5	4.5	3.0	4.0	6.0	3.5
4(5)	3.0	3.5	2.0	x	3.5	4.5	3.0	5.0	9.0	5.0
5(7)	1.5	2.0	1.5	3.5	x	3.5	1.5	4.5	7.0	4.5
6(4)	3.0	1.5	1.5	2.5	2.5	x	3.0	3.0	6.0	3.5
7(6)	1.0	0.5	1.0	1.0	1.5	1.0	x	1.0	4.0	1.5
8(8)	0.0	0.0	0.0	0.0	0.5	1.0	1.0	x	5.0	2.5
9(9)	0.0	0.0	0.0	0.0	0.0	1.0	0.0	0.0	x	3.0
10(10)	0.5	0-.5	0.5	1.0	0.5	1.5	0.5	0.5	3.0	x

in the sequence of columns and rows in Table 5.6A. Table 5.6B shows the final order relation matrix which now was obtained after 5 iterations only. Table 5.7A is RASC output for the optimum sequence of Table 5.6B.

The original SEQ file for this RASC run was shown in Table 4.3B. Because proceeding from left to right in this SEQ file corresponds to moving in the stratigraphically upward direction, the optimum sequence of Table 5.7A is upside down. Table 5.7B is identical to Table 5.7A except for a reversal of the sequence numbers. It is interesting to compare Table 5.7B with the previous result (Table 5.5). The sequence order is different in 4 places. In 3 of these, the order of a pair of two events was reversed. This possibility is expressed by the uncertainty ranges of the events which are identical except for event number 10 which has uncertainty range 8-11 in Table 5.5 and 9-11 in Table 5.7B. This is because the uncertainty ranges of events 8, 9 and 10 interact with one another as explained in Section 5.5. The uncertainty range of 8-11 for event 10 in Table 5.5 is more meaningful than 9-11 in Table 5.7B because event 10 could occur in position 9 provided it would be followed by event 8 in position 10. This illustrates that for a full appreciation of the interaction of uncertainty ranges it may be necessary to inspect the elements of the final order relation matrix.

Use of a transposed order relation matrix is equivalent to reversing the direction for coding the superpositional relations between stratigraphic events. Provided that the uncertainty range is considered, the final optimum sequence is nearly independent of this type of reversal.

Probabilistic ranking

The simple algorithm here termed "probabilistic ranking" was originally added to the RASC computer program as a "presorting option" (Agterberg and Nel, 1982a). It resembles a method earlier proposed by Rubel (1978) which will be discussed in Section 5.6. It will be shown here that, for the Hay example, probabilistic ranking produces the same optimum sequence (Table 5.5) as the algorithm discussed earlier in this chapter. The problem of cycling due to inconsistencies involving more than two events (see Section 5.4) is avoided in probabilistic ranking. Harper (1984) has shown that, in his computer simulation experiments (see Section 7.4), "presorting" consistently gave better results than the modified Hay method which is essentially the same as the algorithm of Section 5.2 with modifications to account for cycling. In Agterberg and Nel (1982a), it was recommended to use presorting followed by the modified Hay method. The new term "probabilistic ranking" reflects that the algorithm previously termed presorting often produces better results than the modified Hay method.

Probabilistic ranking consists of replacing the elements S_{ij} in the S-matrix by $S_{ij}=1$ if $S_{ij}>S_{ji}$, by $S_{ij}=0$ if $S_{ij}>S_{ji}$ and by $S_{ij}=0.5$ if $S_{ij}=S_{ji}$. Table 5.8 shows the A-matrix with elements A_{ij} corresponding to the S-matrix of Table 5.2A. By ordering the row totals A_i according to decreasing magnitude, the optimum sequence of Table 5.9 was obtained.

TABLE 5.7

A. Optimum sequence output of RASC computer program corresponding to Table 5.6B. This result was obtained by using Table 4.3B as SEQ file instead of Table 4.3A. B. Reversed optimum sequence of Table 5.7A. The sequence numbers 1 to 10 for the optimum sequence of Table 5.7A were replaced by new sequence numbers 10 to 1.

A. Sequence Number	Uncertainty Range	Event Code	Event Name
1	0-2	2	LO *Coccolithus cribellum*
2	1-4	1	LO *Discoaster distinctus*
3	0-4	3	LO *Discoaster germanicus*
4	3-6	5	LO *Coccolithus gammation*
5	3-6	7	LO *Discoaster minimus*
6	5-7	4	LO *Coccolithus solitus*
7	6-9	6	LO *Rhabdosphaera scabrosa*
8	6-9	8	LO *Discoaster cruciformis*
9	8-11	9	HI *Discoaster tribrachiatus*
10	8-11	10	LO *Discolithus distinctus*

B. Sequence Number	Uncertainty Range	Event Code	Event Name
1	0-3	10	LO *Discolithus distinctus*
2	0-3	9	HI *Discoaster tribrachiatus*
3	2-5	8	LO *Discoaster cruciformis*
4	2-5	6	LO *Rhabdosphaera scabrosa*
5	4-6	4	LO *Coccolithus solitus*
6	5-8	7	LO *Discoaster minimus*
7	5-8	5	LO *Coccolithus gammation*
8	7-11	3	LO *Discoaster germanicus*
9	7-10	1	LO *Discoaster distinctus*
10	9-11	2	LO *Coccolithus cribellum*

The algorithm for sorting events according to their magnitude is illustrated in Table 5.10. It consists of the following steps. The event with sequence number 1 successively was compared with all following events and its position was interchanged with that of a successor if its magnitude was less. This automatically brings the event (9) with the greatest row total (8.5) to the first position in the optimum sequence. The order of 9 and 10 is not changed because they have the same magnitude. When the event with the largest magnitude is in first position, the algorithm proceeds to

TABLE 5.8

A-matrix to denote average superpositional and coeval relations. Method of probabilistic ranking (or "presorting option") applied to Hay example using *S*-matrix of Table 5.2A as starting point. *F*-matrix of Table 5.1A gives same *A*-matrix. Events will be reordered on the basis of their row totals (A_i).

	1	2	3	4	5	6	7	8	9	10	A_i
1	x	1.0	0.5	0.0	0.0	0.0	0.0	0.0	0.0	0.0	1.5
2	0.0	x	0.5	0.5	0.0	0.0	0.0	0.0	0.0	0.0	1.0
3	0.5	0.5	x	0.0	0.0	0.0	0.0	0.0	0.0	0.0	1.0
4	1.0	0.5	1.0	x	1.0	0.0	1.0	0.0	0.0	0.0	4.5
5	1.0	1.0	1.0	0.0	x	0.0	0.5	0.0	0.0	0.0	3.5
6	1.0	1.0	1.0	1.0	1.0	x	0.5	0.5	0.0	0.0	6.0
7	1.0	1.0	1.0	0.0	0.5	0.5	x	0.0	0.0	0.0	4.0
8	1.0	1.0	1.0	1.0	1.0	0.5	1.0	x	0.0	0.0	6.5
9	1.0	1.0	1.0	1.0	1.0	1.0	1.0	1.0	x	0.5	8.5
10	1.0	1.0	1.0	1.0	1.0	1.0	1.0	1.0	0.5	x	8.5
A_j	7.5	8.0	8.0	4.5	5.5	3.0	5.0	2.5	0.5	0.5	

carry out similar tests for the second position. In Table 5.10 it is shown that it took four iterations to bring event 9 to position 1, followed by five iterations to bring event 10 to position 2. Continuation of the algorithm to find the events for the third and subsequent positions gave the optimum sequence of Table 5.9 after 31 iterations. The new result is identical to that obtained before (Table 5.5). The uncertainty range of an optimum sequence obtained by probabilistic ranking can be determined by using the same method as before (see Section 5.4).

As a further experiment, probabilistic ranking was applied using the SEQ file of Table 4.3B instead the one of Table 4.3A. This is more or less equivalent to ranking the events in ascending order using the column totals A_j of Table 5.8. When the events were first ranked according to descending order of magnitude of their column totals, reversal of the resulting optimum sequence gave an optimum sequence identical to the one shown in Table 5.7 except that event 10 was situated above event 9. The uncertainty ranges resulting from this experiment were identical to those given in Table 5.9.

TABLE 5.9

Optimum sequence output of RASC computer program corresponding to Table 5.8. Events were reordered on the basis of their row totals.

Sequence Number	Code Number	Row Total	Uncertainty Range
1	9	8.5	0-3
2	10	8.5	0-3
3	8	6.5	2-5
4	6	6.0	2-5
5	4	4.5	4-6
6	7	4.0	5-8
7	5	3.5	5-8
8	1	1.5	7-10
9	3	1.0	7-11
10	2	1.0	8-11

Missing data in probabilistic ranking

In practice, the S-matrix may contain pairs of zero elements with $S_{ij}=S_{ji}=0$ because of missing data. The corresponding elements in the A-matrix then can also be set equal to zero ($A_{ij}=A_{ji}=0$). A distinction should be made between a zero whose counterpart is equal to one, and to a zero whose counterpart is zero because it belongs to a pair of zeros for missing information. Suppose that there are B_i zeros of the second type in the i-th row. The row total $\Sigma_j A_{ij}$ may be biased (=too small) because one or more of the missing elements with values equal to 0.0 in reality could be 0.5 or 1.0. The count B_i can be combined with the possibly biased row total to produce the ranking number

$$A_i=(n\text{-}1)(\Sigma_j A_{ij})(n\text{-}1\text{-}B_i)^{-1} \quad (5.1)$$

This is equivalent to rescaling totals for rows with missing information in such a way that the sum of each A_i and its corresponding column total remains equal to (n-1).

Table 5.11A (from Agterberg and Nel, 1982a, p. 74) provides an example of this type of rescaling. Twenty-six highest occurrences of Cenozoic Foraminifera, each occurring in at least $k_c=7$ offshore wells along the northwestern Atlantic margin were subjected to probabilistic

TABLE 5.10

Illustration of computer algorithm used in probabilistic ranking to reorder events on the basis of their row totals in Table 5.8. Final result obtained after 31 iterations is identical to results previously obtained by Hay method (cf. Tables 5.4 and 5.5).

Iteration	1	2	3	4	5	6	7	8	9	10
1	4	2	3	1	5	6	7	8	9	10
2	6	2	3	1	5	4	7	8	9	10
3	8	2	3	1	5	4	7	6	9	10
4	9	2	3	1	5	4	7	6	8	10
5		1	3	2	5	4	7	6	8	10
6		5	3	2	1	4	7	6	8	10
7		4	3	2	1	5	7	6	8	10
8		6	3	2	1	5	7	4	8	10
9		8	3	2	1	5	7	4	6	10
10		10	3	2	1	5	7	4	6	8
11			1	2	3	5	7	4	6	8
12			5	2	3	1	7	4	6	8
13			7	2	3	1	5	4	6	8
14			4	2	3	1	5	7	6	8
15			6	2	3	1	5	7	4	8
16			8	2	3	1	5	7	4	6
17				1	3	2	5	7	4	6
18				5	3	2	1	7	4	6
19				7	3	2	1	5	4	6
20				4	3	2	1	5	7	6
21				6	3	2	1	5	7	4
22					1	2	3	5	7	4
23					5	2	3	1	7	4
24					7	2	3	1	5	4
25					4	2	3	1	5	7
26						1	3	2	5	7
27						5	3	2	1	7
28						7	3	2	1	5
29							1	2	3	5
30							5	2	3	1
31								1	3	2

ranking. The ranking numbers of events 26 and 67 are revised row totals. For this reason, they are not multiples of 0.5 like the other ranking numbers in Table 5.11A. Reordering the 26 events on the basis of the ranking numbers gives the optimum sequence of Table 5.11B.

Probabilistic ranking can be regarded as a primitive kind of scaling method because the events are assigned values along an interval scale.

TABLE 5.11

A. Ranking numbers A_i obtained by method of probabilistic ranking applied to 26 Cenozoic foraminiferal events which occur in $k_c = 7$ or more wells. Original event numbers are shown in column 1. New ranks obtained from ranking numbers A_i are shown in the fourth column. **B**. The ranks are shown in ascending order so that events are in optimum sequence.

A: Event	i	A_i	Rank	**B:** Rank	Event
15	1	19.5	7	1	17
16	2	24.0	2	2	16
17	3	25.0	1	3	67
18	4	21.5	4	4	18
20	5	20.0	6	5	21
21	6	20.5	5	6	20
24	7	15.5	10	7	15
25	8	15.0	11	8	26
26	9	18.2	8	9	70
27	10	14.0	13	10	24
29	11	11.5	15	11	25
30	12	7.0	19	12	69
31	13	12.0	14	13	27
34	14	10.0	16	14	31
36	15	5.5	20	15	29
41	16	9.0	17	16	34
42	17	8.0	18	17	41
45	18	4.5	22	18	42
46	19	3.0	23	19	30
50	20	2.5	24	20	36
54	21	1.0	25	21	57
56	22	0.0	26	22	45
57	23	4.5	21	23	46
67	24	23.9	3	24	50
69	25	14.0	12	25	54
70	26	17.0	9	26	56

Scaling by the averaging of probabilities

Probabilistic ranking gives approximately the same results when the A-matrix is constructed from the F-matrix instead of the S-matrix. The

TABLE 5.12

Ranking numbers obtained by averaging probabilities for the Hay example. See text for further explanation.

	(1)	(2)	(3)	(4)	(5)	(6)
1	15.5	53	10	42	0.292	0.238
2	14.0	55	7	43	0.255	0.163
3	12.0	46	5	32	0.261	0.156
4	24.5	51	18	38	0.480	0.474
5	21.5	60	13	43	0.358	0.302
6	17.5	30	12	19	0.583	0.632
7	20.5	50	17	43	0.410	0.395
8	28.0	38	28	36	0.737	0.778
9	56.0	60	56	60	0.933	0.933
10	32.5	41	28	32	0.793	0.875
Sum	242.0	484	194	388		

only possible difference between outcomes resulting from these two procedures would be due to pairs of locally coeval events which are not considered in the F-matrix. A difference of this type does not arise when probabilistic ranking is applied to the F-matrix of Table 5.1A or the corresponding S-matrix (Table 5.2A).

Suppose that for each row in Table 5.1A or 5.2A, the relative probabilities (shown in Tables 5.3B and 5.3A, respectively) would be added without first replacing these matrices by the A-matrix. Division of its sum by $(n\text{-}1)$ would give an average probability for each event. It can be argued that the probabilities are of variable precision. Their variance is inversely proportional to sample size (= number of pairs). This suggests that it would be advantageous to compute a weighted average of the probabilities in each row using the sample sizes as weights. Multiplication of a probability (e.g. P_{ij}) by its sample size R_{ij} yields the original frequency (e.g. $S_{ij}=P_{ij}\times R_{ij}$). Consequently, the suggested best procedure simply consists of summing the scores in each row of the S-matrix and then dividing the resulting row sums by the corresponding sums for rows of the R-matrix.

Table 5.12 shows ranking numbers obtained by averaging the probabilities P_{ij} (column 5) and P_{0ij} (column 6) for the events of the Hay example, respectively. The average probabilities of column 5 were obtained by dividing the numbers in column 1 by those in column 2 which are row totals for the S-matrix (Table 5.2A) and the R-matrix (Table 5.1B), respectively. The sum of the row totals in column 2 is twice as large as the sum of the row totals in column 1. The numbers in column 3 of Table 5.12 are row totals for the F-matrix (Table 5.1A). These were divided by the numbers of column 4 that represent sample sizes for pairs of events after exclusion of ties (Table 5.2B). The sum for column 4 is twice the sum for column 3.

The optimum sequence obtained after reordering the events on the basis of their ranking numbers in column 5 is identical to the optimum sequences previously given in Tables 5.5 and 5.9. The optimum sequence obtained in column 6 is the same except that event 3 comes below event 2 because it has a lower ranking number. It will be seen in the next chapter that the ranking numbers in columns 5 and 6 of Table 5.12 are very close to the cumulative RASC distances resulting from scaling. There is a natural transition from ranking to scaling as also pointed out by Kemple et al. (1990).

The preceding method of averaging probabilities is a method of probabilistic ranking which is equivalent to a method described by Kendall (1975, p. 151). The method was used for ranking by Blank and Ellis (1982, p. 418) along with a slightly different method to synthesize local range data found among a group of geological sections (Fig. 5.4). The modified average probability values for taxa computed by Blank and Ellis are the same as the ranking numbers of column 6 in Table 5.12, except that a frequency F_{ij} was replaced by F_{ji} if $F_{ji} > F_{ij}$. These modified average probability values cannot be used for ranking or scaling because, on the average, they first decrease from being close to unity near the top to nearly 0.5 in the middle of the composite range chart. Next, continuing to move in the stratigraphically downward direction, they increase to nearly 1.0 toward the bottom of this range chart. Blank and Ellis (1982) found that these modified average probabilities were useful indicators for taxa with mutually inconsistent local range zones. Suppose that the top (highest occurrence) or base (lowest occurrence) of a taxon occupies random position with respect to the tops and bases of other taxa in the sections. The Blank-Ellis average probability of such a random event then would be close to 0.5 (its expected value is slightly grater than 0.5 if tops and bases of the taxa both occur in one or more sections, because the top of a taxon comes above its base). By successively deleting events with the smaller values, Blank

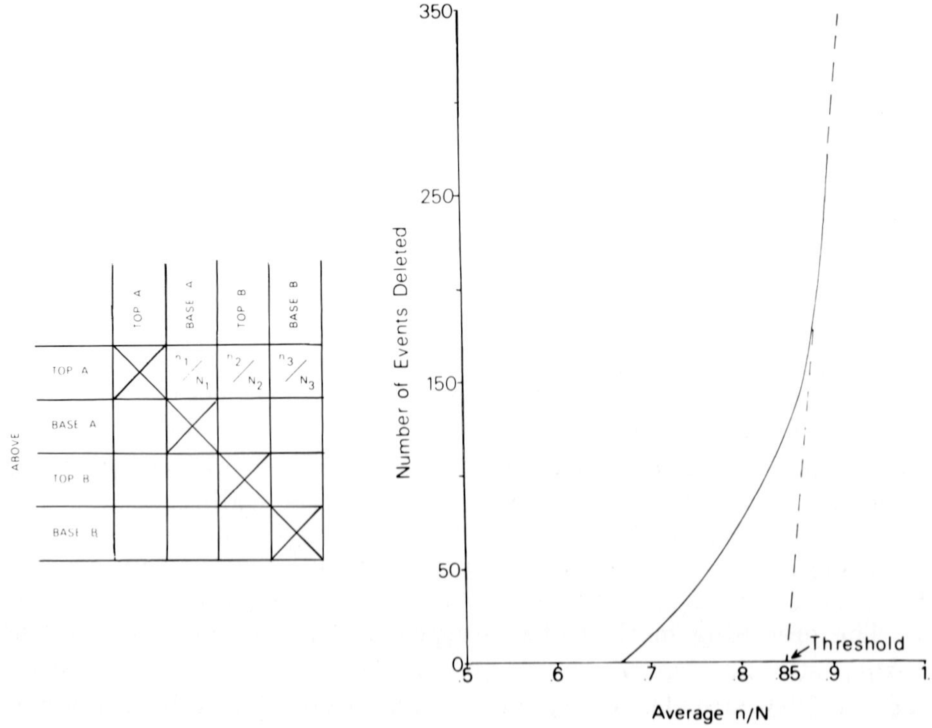

Fig. 5.4 Method of ranking used by Blank and Ellis (1982). **Left side**: The design of the matrix used to synthesize local range data found among a group of geological sections. All taxa range endpoints are identified as being a top or base and are listed at the left and across the top of the matrix. The matrix elements are the ratios *n/N*, and contain the empirical stratigraphic positionings of all endpoints found for a region, taken two at a time. For example, n_2/N_2 is the second matrix element and shows that the Top of taxon A and the Top of Taxon B are found stratigraphically separated in N_2 sections, and the Top of A is found above the Top of B, n_2 times. A row represents an endpoint's total stratigraphic positioning compared to all other endpoints with which it shows a preferred sequence, $n/N > \frac{1}{2}$. Conversely, $n/N < \frac{1}{2}$ also shows a preferred (reversed) stratigraphic sequence and was included in the row total as $1\text{-}n/N$. As the total for a row approaches $\frac{1}{2}$, an endpoint shows a more random stratigraphic positioning, and is not useful in determining biostratigraphic sequence trends. The threshold at which an endpoint is considered randomly distributed with respect to another or with respect to all endpoints with which it is physically associated depends on the level of confidence one is willing to accept. **Right side**: Threshold value determined for the North Atlantic Ocean database of Blank and Ellis (1982). The horizontal axis represents the average *n/N* for a taxon as compared to all other taxa with which it occurs. The vertical axis represents the taxa remaining in the database after successively deleting taxa that fall below a certain value. The relationship defined for the North Atlantic Ocean database in the main body of the figure reveals that at threshold value 0.85, the database maintains a minimum level of confidence and a maximum number of taxa for further analysis. The implication is that taxa falling below the threshold values are less useful in biostratigraphic classification based on sequential similarities (from Blank and Ellis, 1982).

and Ellis determined a threshold value of 0.85 for their very large database of DSDP data (see Fig. 5.4B).

This method must be used with caution because its automated application could result in the rejection of events from the middle of the range about where all events (random and nonrandom) have modified average probability values close to 0.5. Thus other factors should be considered as well when this method is applied.

5.6 Conservative ranking methods

As discussed in Chapter 2, the observed highest occurrences of taxa are probably "too low", and the observed lowest occurrences "too high" in any section.

It may be assumed that, within a study region containing a group of sections, each taxon has unknown true first and last fossilized occurrences. In conservative ranking methods it is attempted to find the relative order of these true stratigraphic events. Different methods have been developed by several authors including Shaw (1964), Edwards (1978) and Guex (1987). A new method for conservative ranking will be introduced later in this book (modified RASC, Chapter 8). Most of these methods use observed positions of events within the sedimentary sequences of the sections as well as their relative order. The conservative ranking method introduced by Rubel (1978) will be used here as an example to illustrate the principles of this approach labelled as "deterministic" by Guex and Davaud (1984) and Rubel and Pak (1984). A comparison with the probabilistic ranking approach also will be made.

Comparison to Rubel's method

Rubel (1978) has proposed the following method: Suppose that, in a stratigraphic section, 12 taxa (numbered 1-12) were observed in 5 consecutive samples. The local ranges of these taxa can be represented as follows:

								9	10	11	12
				5	6	7	8	9	10	11	
		3	4	5	6	7	8	9	10		
	2	3		5	6	7		9	10		
1		3		5				9			

In this tabulation, the taxa are arranged in the order of their disappearance. Table 5.13 is the corresponding matrix of stratigraphic relations between the 12 taxa. Each + in Table 5.13 indicates that the local range of the taxon in the row containing this + is above the corresponding taxon in the column. The counterpart of + is – signifying that the first taxon is below the second taxon. Overlap of local ranges is shown as 0. The three columns in Table 5.13 are for frequencies of +, 0 and – per row. These row tables are written as *a*, *b* and *c*, respectively. They can be used for ordering the taxa. For example, ordering the taxa on the basis of the statistic *a* is equivalent to arranging them in the order of their disappearance. If successive taxa have equal values of *a*, then they are ordered according to their -*c* values.

Table 5.13 resembles the *A*-matrix for probabilistic ranking (cf. Table 5.8) of stratigraphic events. However, the *A*-matrix corresponding to Table 5.13 becomes four times as large if highest and lowest occurrences of all taxa are considered separately as in Table 5.14. Each + in Table 5.13 is equivalent a square block of 4 ones in Table 5.14. Likewise, – becomes a block of 4 zeros. A zero in Table 5.13 is changed into one of 16 possible square blocks with its 4 positions occupied by 1, h($=\frac{1}{2}$) or 0 in Table 5.14. This indicates that Table 5.14 contains more stratigraphic information than Table 5.13. Figure 5.5 shows all these possible configurations together with the relations between the ranges of the taxa they represent. Harper's (1981) eleven possible relative age relations between two taxa (see Fig. 2.5) are all represented. In Table 5.14 and Figure 5.5, there are 6 additional configurations because a separation is made between coexistence of taxa in one or more consecutive samples. Rubel's (1978) example has all possible relations between taxa except the situation (not shown in Fig. 5.5) that two taxa would both occur in one sample only.

TABLE 5.13

Rubel's matrix of stratigraphic relations between 12 taxa in single section (example of local ranges discussed in text). The row totals *a*, *b* and *c* are for +, 0 and -, respectively.

	1	2	3	4	5	6	7	8	9	10	11	12	a	b	c
1	x	+	0	+	0	+	+	+	0	+	+	+	8	3	0
2	-	x	0	+	0	0	0	+	0	0	+	+	4	6	1
3	0	0	x	0	0	0	0	0	0	0	+	+	2	9	0
4	-	-	0	x	0	0	0	0	0	0	+	+	2	7	2
5	0	0	0	0	x	0	0	0	0	0	0	+	1	10	0
6	-	0	0	0	0	x	0	0	0	0	0	+	1	9	1
7	-	0	0	0	0	0	x	0	0	0	0	+	1	9	1
8	-	-	0	0	0	0	0	x	0	0	0	+	1	8	2
9	0	0	0	0	0	0	0	0	x	0	0	0	0	11	0
10	-	0	0	0	0	0	0	0	0	x	0	0	0	10	1
11	-	-	-	-	0	0	0	0	0	0	x	0	0	7	4
12	-	-	-	-	-	-	-	-	0	0	0	x	0	3	8

Suppose that local ranges for the taxa are available for another section. A table similar to Table 5.13 then can be constructed for this other section. The tables for the two sections can be superimposed on one another and combined into a single new table using the following algebra (Rubel, 1978, p. 244): + & + = +, –&–=–, =&0=0 and –&=0. It is implied that 0& + =0 and 0&–=0. If one or both taxa are missing in one of the sections, the matrix element (+, – or 0) for their relation in this section is unknown. Writing x for such an unknown element, the following combinations can be added: +&x= +, –&x=–, 0&x=0 and x&x=x.

It is possible to add more sections to a combination of two sections. The matrix resulting from adding all available sections for a region is independent of the order in which the sections are added to one another. A+ in this final matrix, means that, of the two taxa compared, one occurs above the other in all sections considered. The + is accompanied by a – as its counterpart. A zero means that the two taxa coexisted in at least one sample in at least one section. Great importance is given to coexistences of taxa because the ranges in the composite standard are extended to cover all observed coexistences of taxa. Obviously, this makes conservative ranking methods sensitive to reworking and stratigraphic leaks. Such effects should be eliminated before application of the method.

TABLE 5.14

A-matrix for Rubel's example of 12 local ranges. Each taxon was assigned separate code numbers for its lowest and highest occurrence, respectively. See text for further explanation.

	1 1	2	2 3	4	3 5	6	4 7	8	5 9	10	6 11	12	7 13	14	8 15	16	9 17	18	10 19	20	11 21	22	12 23	24	A_i
1	x	1	1	1	h	1	1	1	h	1	1	1	1	1	1	1	h	1	1	1	1	1	1	1	21.5
2	0	x	1	1	h	1	1	1	h	1	1	1	1	1	1	1	h	1	1	1	1	1	1	1	20.5
3	0	0	x	1	0	1	1	1	0	1	h	1	h	1	1	1	0	1	h	1	1	1	1	1	16.5
4	0	0	0	x	0	1	1	1	0	1	h	1	h	1	1	1	0	1	h	1	1	1	1	1	15.5
5	h	h	1	1	x	1	1	1	h	1	1	1	1	1	1	1	1	1	1	1	1	1	1	1	21.5
6	0	0	0	0	0	x	h	h	0	1	0	1	0	1	h	1	0	1	0	1	1	1	1	1	11.5
7	0	0	0	0	0	1	x	1	0	1	0	1	0	1	h	1	0	1	0	1	1	1	1	1	12.5
8	0	0	0	0	h	h	0	x	0	1	0	1	0	1	h	1	0	1	0	1	1	1	1	1	11.5
9	h	h	1	1	h	1	1	1	x	1	1	1	1	1	1	1	h	1	1	1	1	1	1	1	21.0
10	0	0	0	0	0	0	0	0	0	x	0	h	0	h	0	h	0	1	0	1	h	1	1	1	7.0
11	0	0	h	h	0	1	1	1	0	1	x	1	h	1	1	1	0	1	h	1	1	1	1	1	16.0
12	0	0	0	0	0	0	0	0	0	h	0	x	0	h	0	h	0	1	0	1	0	1	1	1	6.5
13	0	0	h	h	0	1	1	1	0	1	h	1	x	1	1	1	0	1	h	1	1	1	1	1	16.0
14	0	0	0	0	0	0	0	0	0	h	0	h	0	x	0	h	0	1	0	1	h	1	1	1	7.0
15	0	0	0	0	0	h	h	h	0	1	0	1	0	1	x	1	0	1	0	1	1	1	1	1	11.5
16	0	0	0	0	0	0	0	0	0	h	0	h	0	h	0	x	0	1	0	1	h	1	1	1	7.0
17	h	h	1	1	0	1	1	1	h	1	1	1	1	1	1	1	x	1	1	1	1	1	1	1	20.5
18	0	0	0	0	0	0	0	0	0	0	0	0	0	0	0	0	0	x	0	h	0	h	h	h	2.0
19	0	0	h	h	0	1	1	1	0	1	h	1	h	1	1	1	0	1	x	1	1	1	1	1	10.0
20	0	0	0	0	0	0	0	0	0	0	0	0	0	0	0	0	0	h	0	x	0	h	h	h	2.0
21	0	0	0	0	0	0	0	0	0	h	0	1	0	h	0	h	0	1	0	1	x	1	1	1	7.5
22	0	0	0	0	0	0	0	0	0	0	0	0	0	0	0	0	0	h	0	h	0	x	h	h	2.0
23	0	0	0	0	0	0	0	0	0	0	0	0	0	0	0	0	0	1	0	1	0	1	x	1	4.0
24	0	0	0	0	0	0	0	0	0	0	0	0	0	0	0	0	h	h	h	h	h	h	0	x	3.0

In terms of graph theory, Table 5.13 is the adjacency matrix for a local range chart represented as an interval graph. However, after addition of one or more other sections, using the preceding algebra, it may not be

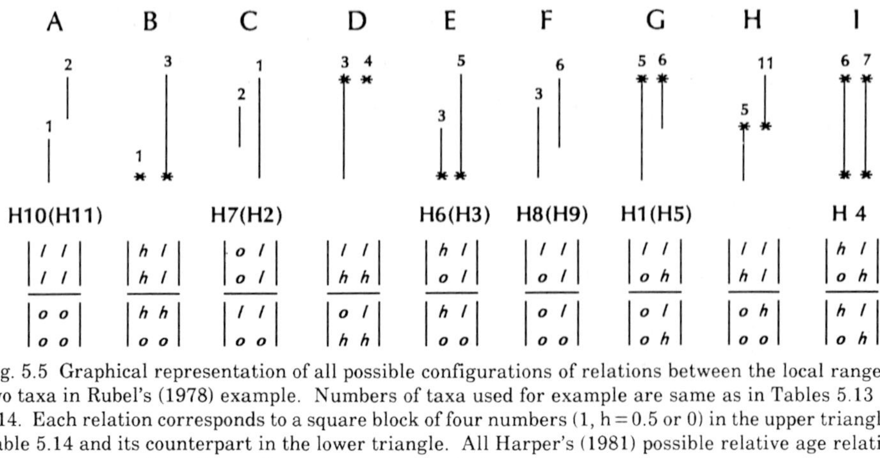

Fig. 5.5 Graphical representation of all possible configurations of relations between the local ranges of two taxa in Rubel's (1978) example. Numbers of taxa used for example are same as in Tables 5.13 and 5.14. Each relation corresponds to a square block of four numbers (1, h = 0.5 or 0) in the upper triangle of Table 5.14 and its counterpart in the lower triangle. All Harper's (1981) possible relative age relations between two taxa (H1 to H11 with numbers as in Fig. 2.5) are represented.

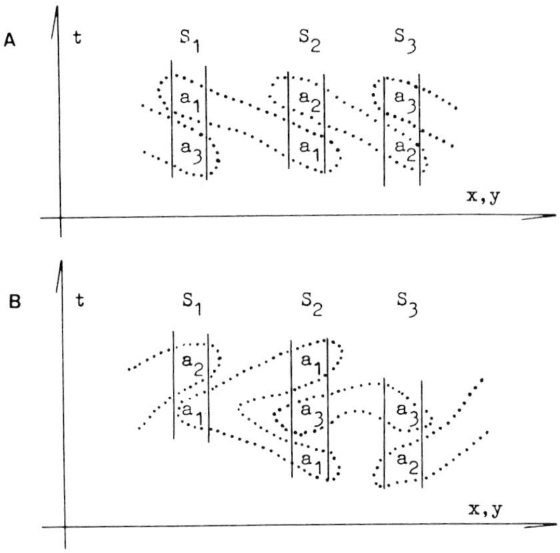

Fig. 5.6 Rubel's (1978) possible explanations of potential inconsistencies for superpositional relations of 3 events in 3 or more sections. In both spatial distribution patterns (A and B), coexistence of the taxa (a_1, a_2 and a_3) cannot be observed in any of the sections (S_1, S_2 and S_3).

possible to directly represent the resulting table as a range chart because it may contain inconsistencies preventing its representation as an interval graph. Figure 5.6 (from Rubel, 1978) shows two inconsistencies of this type. Rubel (1978) would accept such inconsistencies as real phenomena only if their existence is reconfirmed by similar contradictory superpositional relations in other sets of three sections. Unusual superpositional relations in three sections as shown in Figure 5.6 normally will not be preserved in the final table if the latter is based on many sections with other types of superpositional relations for the same three events. It is noted that combining sections by means of the probabilistic ranking method results in an optimum sequence (e.g. Table 5.14) that can be represented as a range chart in which the highest and lowest occurrences of each taxon have average positions with respect to those of all other taxa. As already pointed out in Chapter 2, if the ranges of the taxa in a range chart of this type are plotted along a geological time scale, they are shorter than those in range charts based on conservative ranking methods. This is because superpositional relations with scores less than 0.5 are ignored in probabilistic ranking by setting them equal to zero.

5.7 Three-event cycles

Worsley and Jorgens (1977) have found that the algorithm of Section 5.3 does not necessarily yield an optimum sequence because cyclical inconsistencies may occur in which more than two events are involved. Their original example of cycling events is shown as the first matrix of Table 5.15. When the algorithm is applied, the original S-matrix reoccurs after every set of six consecutive iterations. Hence an optimum sequence could never be determined by means of the preceding algorithm.

In the example of Table 5.15, A occurs more frequently before B ($S_{AB} > S_{BA}$), B before C ($S_{BC} > S_{CB}$), and C before A ($S_{CA} > S_{AC}$). The three events A, B and C are involved in a cyclical inconsistency and are said to form a three-event cycle. It is useful to represent this type of situation by means of a graph. The relationships of Table 5.15 are represented by arrows in the graph shown in Figure 5.7. The three-event cycle involving A, B and C is immediately apparent in Figure 5.7 because the arrows in the triangle ABC point in the same direction at both sides of each of the vertices of this triangle.

If there are no cycles, all inconsistencies can be eliminated by disregarding situations in which $S_{ij} < S_{ji}$. Suppose that each situation $S_{ij} \geq S_{ji}$ is indicated by a + sign for S_{ij} in the upper triangle above the diagonal of the S-matrix where $j > i$ and a − sign for the corresponding element in the lower triangle where $j < i$. Then the S-matrix of Table 5.4D which is a final order relation matrix would be replaced by a matrix with exclusively + signs in the upper triangle and − signs in the lower triangle. If a 3-event cycle occurs, it is not possible to achieve a clear subdivision of this nature as is illustrated in Figure 5.8 for an artificial example. The events of Figure 5.8 are indicated by means of letters. C, F and K form a 3-event cycle. The elements in the first two rows could be tested by means of the previous algorithm. However, iterations would continue indefinitely for the elements in the third row which is for one of the cycling events (C). The event in the margin of the third column of Figure 5.8 can be scanned by putting a "window" on it in the computer algorithm. For the 3-event cycle of C, F and K, this window will begin showing the sequence CKFCKF... which can be readily detected. Once the events involved in a cycle have been identified, the + sign corresponding to the pair of scores with the smallest difference $|S_{ij}-S_{ji}|$ can be allowed to remain in the lower triangle. In the algorithm, this is accomplished by temporary replacement of its scores by zeros. This replacement is

TABLE 5.15

Example of cycling events (initial matrix from Worsley and Jorgens, 1977). Unlike the example of Table 5.4, the algorithm for ordering does not yield an optimum sequence because the initial matrix returns after 6 iterations. Note that event *D* does not participate in the cycling.

(a) *ABCD*	(b) *CBAD*	(c) *BCAD*	(d) *ACBD*	(e) *CABD*	(f) *BACD*	(g) *ABCD*
× 232	× 243	× 511	× 322	× 423	× 151	× 232
1 × 51	5 × 11	2 × 43	4 × 23	3 × 22	2 × 32	1 × 51
42 × 3	32 × 2	23 × 2	15 × 1	51 × 1	24× 3	42× 3
074 ×	470 ×	740 ×	047 ×	407 ×	704 ×	074 ×

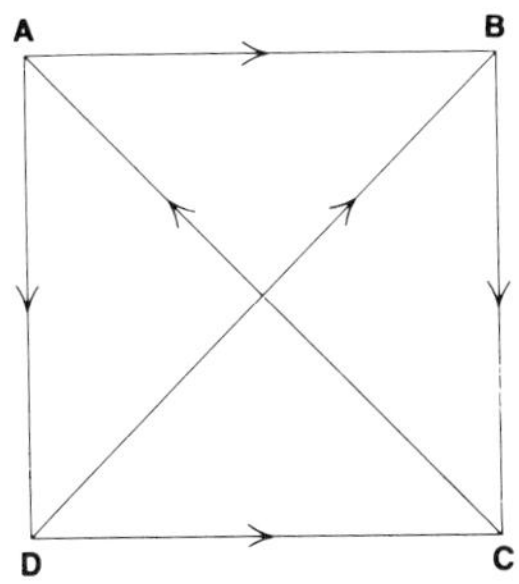

Fig. 5.7 Three-event cycle (ABC) in set of four events is characterized by successive arrows pointing in same direction at both sides of vertices (A, B and C). Arrow between two events indicates that one event precedes other event.

temporary if ranking will be followed by scaling because for scaling, elements in the lower triangle may be larger than their counterparts in the upper triangle.

It is possible that two pairs of scores for events involved in a 3-event cycle have equal smallest difference values, or that all three pairs have equal differences. In those situations only the first pair encountered will be ignored. An example is provided in Table 5.16.

For this example, the data of Table 4.10 were run setting the threshold parameters equal to $k_c=7$ and $m_{c1}=5$, respectively. For n = 26 events , it is possible to make $n(n\text{-}1)/2=325$ comparisons. However, because of the treshold $m_{c1}=5$, forty pairs were not used. The presorting option was used (see Table 5.11) and the 26 events were reordered by means of the modified Hay method using the ranks in the last column of Table 5.11. The final result is shown in Table 5.17. A three-event cycle involving events 25, 27 and 69 was identified with the corresponding output shown in Table 5.16. The event positions printed below the cycling events are temporary and can be used to identify which pair of events (11 and 12) was ignored in order to break the cycle.

In the original input, the three cycling events were encountered together in four wells: Freydis (69, -27, 25), Gudrid (69, 25, 27), Bonavista (25, 27, 69) and Dominion (27, 25, 69). In these expressions, relative order is indicated by means of a comma and coeval events are separated by a comma followed by a hyphen (e.g. in Freydis, 69 and 27 are coeval and both precede 25). For abbreviation, the four expressions can be rewritten as (2-31, 213, 132, 312) where 25, 69 and 27 have been replaced by 1, 2 and 3, respectively. Two of the three events were encountered together in seven wells with relative orders (21, 21, 13, 12, 21, 13, 32). The scores of Table 5.16 can be obtained by counting subsequences for two events (e.g. 21 occurs 5 times while 12 occurs 3 times). All three events

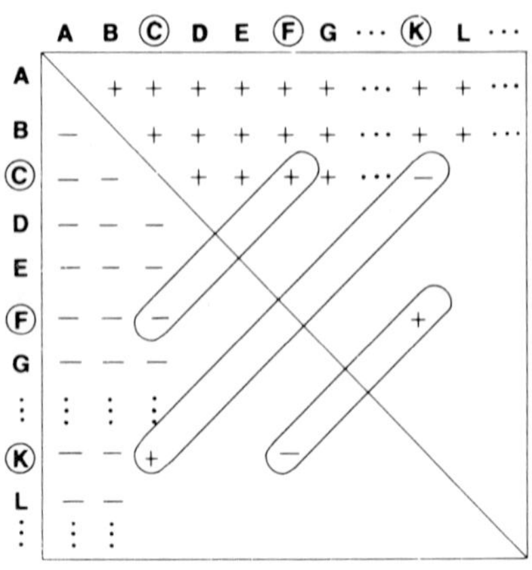

Fig. 5.8 Graphical illustration of algorithm developed to locate three-event cycle. Elements in successive rows of upper triangle are tested proceeding from left to right. Row and column interchanges only take place when element is less than its counterpart in lower triangle. In example, element circled in margin C will be replaced by K which, in turn, will be followed by F. Cycle C K F will repeat indefinitely.

TABLE 5.16

Selected output from RASC program including information on a single 3-event cycle encountered when data of Table 4.10 are run with $k_c = 7$ and $m_{c1} = 5$. See text for explanations.

```
RUN FOR 7 OR MORE OCCURRENCES AND 5 OR MORE PAIRS.

         CYCLING EVENTS:                27          25          69

         EVENT POSITIONS:               11          13          12

         MATRIX ELEMENTS:

         0.0      2.0      3.5

         4.0      0.0      3.0

         1.5      5.0      0.0

C(11, 13) AND C(13, 11) ZEROED

RANKING SOLUTION OBTAINED WITH:

         102 ITERATIONS OUT OF MAXIMUM 9000

         TOLERANCE OF 0.0
```

participate in a cycle because the preferred subsequences 21, 13 and 32 cannot hold true simultaneously.

In this application, the optimum sequence (Table 5.17) is almost equal to the result obtained by means of the presorting option (Table 5.11). In addition to a change in order corresponding to the 3-event cycle, only the events with ranks 21 and 22 have changed places in the sequence. Every cycle is allowed to run 100 times before it is broken. Hence the total number of iterations is 102 instead of 2 in Table 5.16. Extra iterations may be needed to eliminate possible pseudo-cycles which can develop initially before a truly periodic cycle appears. This subject will be explained in the next section which also contains a discussion of the situations in which cycles involving more than three events can develop.

Cycles tend to occur frequently if one or both of the following two conditions are satisfied: (1) many small samples are used (e.g. $R_{ij} < 3$), and (2) the expected values of many of the frequencies $P_{ij} = S_{ij}/R_{ij}$ are close to 0.5. The tolerance parameter (b_c) can be used in the RASC program to reduce the number of cycles. If b_c is set equal to a positive value (e.g. 0.5 or 1.0), scores with $S_{ij} + b_c > S_{ji} > S_{ij}$ will be allowed to occur in the lower triangle ($j < i$) in addition to the values $S_{ji} < S_{ij}$. By leaving a certain

TABLE 5.17

RASC program output of optimum sequence of data of Table 4.10 with $k_c=7$ and $m_{c1}=5$.

Sequence Position	Fossil Number	Range	Fossil Name
1	17	0- 2	Asterigerina gurichi
2	16	1- 3	Ceratobulimina contraria
3	67	2- 4	Scaphopod spl
4	18	3- 6	Spiroplectammina carinata
5	21	3- 6	Guttulina problema
6	20	5- 7	Gyroidina girardana
7	15	6- 8	Globigerina praebulloides
8	26	7-10	Uvigerina dumblei
9	70	7-12	Alabamina wolterstorffi
10	24	8-11	Turrilina alsatica
11	27	10-12	Eponides umbonatus
12	69	11-13	Nodosaria sp8
13	25	12-14	Coarse arenaceous spp.
14	31	13-16	Pteropod spl
15	29	13-16	Cyclammina amplectens
16	34	15-17	Marginulina decorata
17	41	16-18	Plectofrondicularia spl
18	42	17-19	Cibicidoides alleni
19	30	18-20	Cibicidoides blanpiedi
20	36	19-23	Pseudohastigerina wilcoxensis
21	45	19-22	Bulimina trigonalis
22	57	21-23	Spiroplectammina spectabilis
23	46	22-25	Megaspore spl
24	50	22-25	Subbotina patagonica
25	54	24-26	Textularia plummerae
26	56	25-27	Glomospira corona

amount of "noise" in the system, an optimum sequence then is obtained more rapidly requiring less computing time.

5.8 Higher-order cycles and pseudo-cycles

Suppose that four events (A, B, C and D) with $S_{ij}=S_{ji}$ (i=A,B,C,D; j=A,B,C,D; $i \neq j$) are subject to the relationships $S_{AB} > S_{BA}$, $S_{BC} > S_{CB}$, $S_{CD} > S_{DC}$ and $S_{DA} > S_{AD}$.

This situation was in fact shown in Table 5.15. Worsley and Jorgens (1977) assumed that all four events participated in the inconsistency. However, when the algorithm of this paper is applied, only the events A, B and C are involved in what is called a 3-event cycle.

In general, it can be shown that, if $S_{ij}=S_{ji}$ ($i \neq j$) for four events, then there must be two 3-event cycles in the system for the situation defined at the beginning of this section. The scores for A in comparison to C satisfy either $S_{AC} > S_{CA}$ or $S_{CA} > S_{AC}$. If $S_{AC} > S_{CA}$, A, C and D form a 3-event cycle; if $S_{CA} > S_{AC}$, A, B and C form a cycle. Likewise, either A, B and D or B, C and D form a 3-event cycle. If the algorithm is applied, a 3-event cycle (and not a 4-event cycle) will be identified (cf. Table 5.16). When this cycle is broken, the other cycle either remains in the system and would be identified next, or it is broken at the same time as the first cycle. Whether or not two cycles will be identified depends on the relative magnitudes of the differences $|S_{ij} - S_{ji}|$.

A true 4-event cycle with $S_{AB} > S_{BA}$, $S_{BC} > S_{CB}$, $S_{CD} > S_{DC}$, and $S_{DA} > S_{AD}$ arises only if $S_{AC} = S_{CA}$ and $S_{BD} = S_{DB}$ as illustrated in Figure 5.9. Higher-order cycles including the 5-event and 6-event cycles which also are shown in Figure 5.9 only occur if all arrows for arcs on the circumference of the graph point in the same direction while all indirect connections between vertices are undirected with $S_{ij} = S_{ji}$ $(i \neq j;\ j \neq i + 1)$.

Higher-order cycles are identified and eliminated in the same manner as 3-event cycles. It is noted that in Gradstein and Agterberg (1982) all pairs of scores with equal minimum differences were ignored whereas, in the algorithm described here, only the first pair encountered will be ignored. Four-event cycles frequently occur in practice but 5-event cycles are rare. In numerous runs of RASC I have encountered a 6-cycle only twice. The RASC program would identify and break cycles of up to nine events. The problem of dealing with cycles of several stratigraphic events also has been discussed by Salin (1989).

The concept of a pseudo-cycle is illustrated in Figure 5.10. The initial order ABCD is changed into ACDB after four iterations. The sequence ACDB contains a single 3-event cycle (ACD) and reappears with a periodicity of six iterations. When a window is placed on the first event, the observed sequence is ADCBADCADCA... This initially would suggest a 4-event cycle involving all four events. However, this pseudo-cycle is unstable and is automatically replaced by the 3-event cycle for A, C and D.

5.9 The influence of coeval events

In Hay's original method, coeval events are ignored. On the other hand, Davaud and Guex (1978) and Rubel (1978) in their methods assigned more weight to ties (coeval events) than is done in the modified Hay method. In Section 5.3 the practice of several authors including

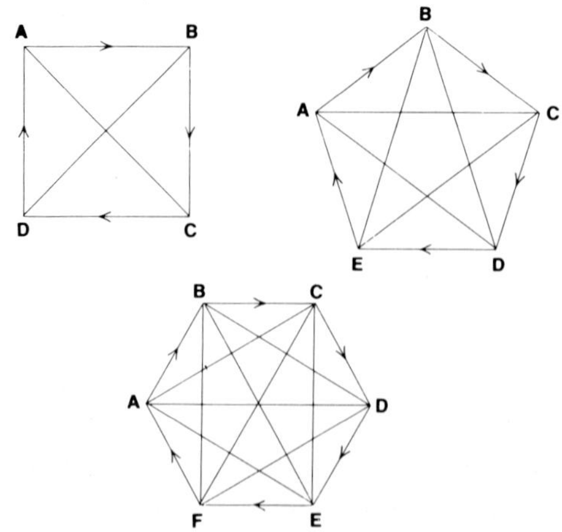

Fig. 5.9 Cycles of more than three events can occur when all events, except those involved in cycle, are pairwise simultaneous (relative frequency P_{ij} is equal to 0.5). Pair of events that are coeval on average have connecting lines without arrows in examples for 4-, 5- and 6-event cycles shown.

	$\hat{A}$	B	C	D
A	x	+	+	-
B	-	x	+	o
C	-	-	x	+
D	+	o	-	x

	$\hat{D}$	B	C	A
D	x	o	-	+
B	o	x	+	-
C	+	-	x	-
A	-	+	+	x

	$\hat{C}$	B	D	A
C	x	-	+	-
B	+	x	o	-
D	-	o	x	+
A	+	+	-	x

	$\hat{B}$	C	D	A
B	x	+	o	-
C	-	x	+	-
D	o	-	x	+
A	+	+	-	x

	$\hat{A}$	C	D	B
A	x	+	-	+
C	-	x	+	-
D	+	-	x	o
B	-	+	o	x

	$\hat{D}$	C	A	B
D	x	-	+	o
C	+	x	-	-
A	-	+	x	+
B	o	+	-	x

	$\hat{C}$	D	A	B
C	x	+	-	-
D	-	x	+	o
A	+	-	x	+
B	+	o	-	x

	$\hat{A}$	D	C	B
A	x	-	+	+
D	+	x	-	o
C	-	+	x	-
B	-	o	+	x

	$\hat{D}$	A	C	B
D	x	+	-	o
A	-	x	+	+
C	+	-	x	-
B	o	-	+	x

	$\hat{C}$	A	D	B
C	x	-	+	-
A	+	x	-	+
D	-	+	x	o
B	+	-	o	x

	$\hat{A}$	C	D	B
A	x	+	-	+
C	-	x	+	-
D	+	-	x	o
B	-	+	o	x

Fig. 5.10 Illustration of pseudo-cycle (ADCB) which initially develops when the algorithm is applied but is automatically replaced by the three-event cycle (ADC). Events with hats are being observed at a "window" and checked for periodicity in the algorithm.

TABLE 5.18

RASC program output of optimum sequence for Hay example after modifications of SEQ file of Table 5.3 (cf. Table 4.6). **A.** Additional information for Paleocene was used. **B.** Guex levels were used for data reduction.

A Sequence Number	Uncertainty Range	Event Code	Event Name
1	0-3	9	HI *Discoaster tribrachiatus*
2	0-3	10	LO *Discolithus distinctus*
3	2-5	6	LO *Rhabdosphaera scabrosa*
4	2-6	8	LO *Discoaster cruciformis*
5	3-6	4	LO *Coccolithus solitus*
6	5-7	7	LO *Discoaster minimus*
7	6-8	3	LO *Coccolithus germanicus*
8	7-9	1	LO *Discoaster distinctus*
9	8-10	5	LO *Discoaster gammation*
10	9-11	2	LO *Coccolithus cribellum*

B Sequence Number	Uncertainty Range	Event Code	Event Name
1	0-2	10	LO *Discolithus distinctus*
2	1-3	9	HI *Discoaster tribrachiatus*
3	2-5	8	LO *Discoaster cruciformis*
4	2-6	6	LO *Rhabdosphaera scabrosa*
5	3-8	7	LO *Discoaster minimus*
6	4-7	4	LO *Coccolithus solitus*
7	6-8	5	LO *Coccolithus gammation*
8	7-10	1	LO *Discoaster distinctus*
9	7-11	3	LO *Discoaster germanicus*
10	8-11	2	LO *Coccolithus cribellum*

Kendall (1975), and Brunk (1960) who scored ties as 0.5 above and below the principal diagonal of the matrix for frequencies. However, arguments that ties should be ignored in some situations have been presented by Hemelrijk (1952) and Tocher (1950). It has already been pointed out that, in the absence of cycling (see Section 5.7), the modified Hay method produces exactly the same optimum sequence as the original Hay method.

In the methods of Davaud and Guex (1978) and Rubel (1978), occurrences of fossil species are considered to be coeval if they are observed to the coeval at least once. For example, even if fossil A is observed to occur above fossil B in several sections, their coexistence in a single section results in the two fossils to co-occur in the standard contructed on the basis of all sections. Clearly, more weight then is assigned to ties than in either the Hay method or modified Hay method. Guex and Davaud (1984) have made extensive use of graph theory in developing their technique. This allowed them to construct an optimum sequence of multiple events which may be subdivided into parts called "Unitary Associations" (see Section 3.5) that can be identified in the original sections and used for correlation.

In Chapter 4 it was pointed out that the results of ranking (and scaling) depend on how the original data are coded. For the Hay example, it was noted that scoring ties for coeval events resulted in bias do to artificial truncation on the stratigraphically lowest levels of some sections. Several of the nannofossils used in the example already existed before the Eocene and their entries with respect to one another in the Paleocene were known for two sections. Use of this information changed the partial SEQ file for the Media Agua Creek section (see Table 4.6). The optimum sequence of Table 5.5 is changed into that of Table 5.18A when a revised SEQ file with data for the Paleocene in the two sections is used. The revisions in the optimum sequence are minor and restricted to the lower part of the optimum sequence.

It also was noted in Chapter 4 that the method of preprocessing by coding events from maximal horizons (cf. Fig. 4.4) gives another type of SEQ file (cf. line 2 in Table 4.6). Table 5.18B shows the optimum sequence obtained for the 10 events of the original Hay example after coding them from Guex levels for all 9 sections. Again the resulting revisions are relatively minor. From the discussions in Chapter 4, it may be concluded that the optimum sequence of Table 5.18A is marginally better than the one of Table 5.5 whereas that of Table 5.18B would be marginally worse. However, for this example, it is not possible to prove whether or not minor revisions of this type are significant. In magnitude they are comparable to the types of changes that arise when one or more of the threshold parameters k_c, m_{c1} and b_c are modified.

CHAPTER 6

SCALING OF BIOSTRATIGRAPHIC EVENTS

6.1 Introduction

The RASC computer program for ranking followed by scaling of stratigraphic events was originally published with documentation in Agterberg and Nel (1982 a, b). Many examples of scaled optimum sequences can be found in Gradstein et al. (1985). The purpose of this chapter is to review the scaling method in detail using relatively small datasets. First the principle of scaling is explained by applying it to simple artificial examples and by approximating the transformation of the relative frequencies P_{ij} into distances Z_{ij}, as performed in RASC, by a linear transformation which is easy to understand.

In the artificial examples of Figure 6.1, observed occurrences of two stratigraphic events (A and B) in 12 sections are compared with one another. An additional event (C) is considered in Artificial Example 4. As a rule, biostratigraphic events are observed only in a subset of the total number of sections (N) in a study region. In Artificial Example 1, $N = 12$ but A occurs only in $N_A = 5$ and B in $N_B = 6$ sections. The number of sections $N_{A,B} = 2$ with both A and B present is even smaller. In these two sections, relative stratigraphic position of A is above that of B. This relation can be quantified by writing $N_{AB} = 2$ and $N_{BA} = 0$, where AB indicates A above B and BA is A below B. In the other examples of Figure 6.1, A-B denotes that A and B were observed to be coeval with frequency $N_{A\text{-}B}$ (e.g. $N_{A\text{-}B} = 4$ in Artificial Example 2).

In total, three threshold parameters have to be set at the beginning of a RASC run: k_c, m_{c1} and m_{c2} with $k_c \geq m_{c2} \geq m_{c1}$. The critical value k_c indicates that an event will only be used for computing if it occurs in at least k_c sections. If one would set $k_c = 6$ in Artifical Example 1, the event A would not be used for ranking and scaling. The parameters m_{c1} and m_{c2} control minimum number of pairs of events to be used for computing optimum sequences in ranking (modified Hay method, see Section 5.4) and scaling, respectively. If $m_{c1} = 1$ and $m_{c2} = 4$ in Artificial Example 1 (with $k_c \geq 5$), A and B would be compared for ranking but not for scaling. If k_c

and m_{c2} are increased, statistical precision of results is improved but fewer events are considered.

The methods of ranking introduced in the previous chapter produce a simple answer for the examples of Figure 6.1. If $N_{AB} > N_{BA}$ as in Artificial Examples 1 and 3, the ranking result is AB. The optimum sequence for the fourth example is ABC, and "undecided" for Artificial Example 2 where a decision cannot be taken. The scaling technique is conceptionally more complex than ranking. Using the frequencies N_{AB}, N_{BA}, $N_{A\text{-}B}$ and $N_{A,B}$, a single relative frequency $P_{AB} = (N_{AB} + 0.5N_{A\text{-}B})/N_{A,B}$ is computed. Obviously, $P_{BA} = 1 - P_{AB}$. The principle of scaling is that the frequency for inconsistencies P_{AB} is transformed into $Z_{AB} = \Phi^{-1}(P_{AB})$ being an estimate of the interval between mean positions of A and B along a distance scale (RASC scale). Φ represents fractile of the normal distribution in standard form. If it is found that $P_{AB} = 1$ for the situation that A and B are relatively close along the RASC scale, $P_{AB} = 1$ is replaced by a probability which is less than 1 and the corresponding interval is set equal to $Z_{AB} = q_c$. In Artificial Example 1, $N_{A,B} = 2$ with $P_{AB} = 1$. If this relation would be used in conjunction with other frequencies (e.g. for "indirect" estimation, see later), we could choose $P_{AB} = 0.90$ with $q_c = 1.282$. The "default" value in RASC is $q_c = 1.645$ for $P = 0.95$.

The transformation Φ^{-1} can be approximated by the linear transformation $Z^*_{AB} = 2.93\ (P_{AB} - 0.5)$ as illustrated in Table 6.1. It is useful to define an interval $Z = Z^* = 0$ for $P = 0.5$ when one is not able to decide whether A should be above or below B in the optimum sequence as in Artificial Example 2. In Artificial Example 3, $P_{AB} = 5/8$ which yields $Z_{AB} = 0.319$ and $Z^*_{AB} = 0.366$. In Artificial Example 4, $P_{AB} = 3.5/5$ which is slightly greater than 5/8 in Example 3. The resulting distance $Z^*_{AB} = 0.59$ ($Z_{AB} = 0.52$) also is slightly greater.

For Example 4, $P_{AC} = 5/6$ with $Z^*_{AC} = 0.98$ ($Z_{AC} = 0.97$), and $P_{BC} = 7/9$ with $Z^*_{BC} = 0.59$ ($Z_{BC} = 0.77$). These three estimates of distance are not mutually consistent. For example, $Z^*_{AB.C} = Z^*_{AC} - Z^*_{BC} = 0.29$ provides an indirect estimate of the distance between A and B which differs considerably from the direct estimate $Z^*_{AB} = 0.59$. This type of inconsistency can be ascribed to small sample sizes and can be eliminated by averaging ; e.g. $\bar{Z}^*_{AB} = 0.5\ (Z^*_{AB} + Z^*_{AB.C}) = 0.38$ which is close to $\bar{Z}_{AB} = 0.36$. Especially when there are many indirect distance estimates, such averages are more precise than direct distance estimates.

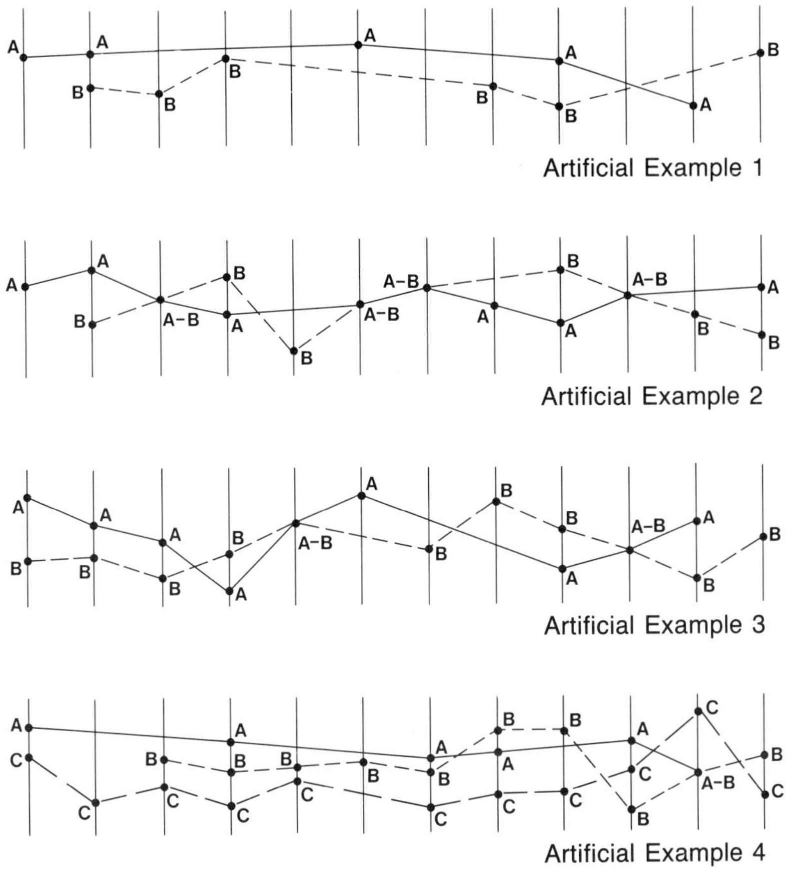

Fig 6.1 Graphical illustration of RASC method for ranking and scaling of stratigraphic events in many stratigraphic sections (shown as vertical lines). Ranking in the stratigraphically downward direction provides optimum sequences AB (A stratigraphically above B) in Examples 1 and 3, A-B (undecided) in Example 2, and ABC in Example 4. Scaling gives distance estimates of intervals between successive events along a linear (RASC) scale. The distance between A and B is estimated as (1) 1.28, (2) 0.00, (3) 0.32 and (4) 0.36 for Artificial Examples 1, 2, 3 and 4, respectively (from Gradstein et al., 1990).

In RASC, the averaging process is refined by considering differences in sample size. For example, $P = 1.5/4$ for $N = 4$ is less precise than $P = 4.5/12$ for $N = 12$ although their Z-values are the same. The second Z-value is given more weight in the calculations because it is based on a larger sample (see Section 6.2).

The linear transformation was introduced here to illustrate the concept of scaling. In practice, it is better to use the normal distribution as in RASC. This is because a linear transformation would imply that the

TABLE 6.1

Example of Z-values for selected relative frequencies P. The Z^*-values in last column are linearly related to the frequencies and are approximate Z-values.

P	Z	Z^*
0.00	$-q_c$	-2.930
0.05	-1.645	-1.319
0.10	-1.282	-1.172
0.20	-0.842	-0.879
0.30	-0.524	-0.586
0.40	-0.253	-0.293
0.50	0.000	0.000
0.60	0.253	0.293
0.70	0.524	0.586
0.80	0.842	0.879
0.90	1.282	1.172
0.95	1.645	1.319
1.00	q_c	2.930

frequency density function of the interval between two events along the RASC scale is uniform. This, in turn, would mean that frequency density functions of individual events along the RASC scale would have different shapes depending on the value of Z^*; e.g. for $Z^*_{AB} = 0$, A and B would have U-shaped density functions with local minima at their mean locations. It is more realistic to assume that the individual species have density functions with maxima at or near their mean values. The mode and mean coincide for the normal (Gaussian) curve model used in RASC. This model is not satisfactory for small densities in the tails where artificial truncation is applied when the cumulative frequency of the sample is observed to be either 0 or 1 (see before). It is good to keep in mind that decrease in density away from the mode could be different for different taxa. Also, for the same species it could be different in the stratigraphically upward and downward directions (cf. Chapters 2 and 9).

The scaling algorithms presented in this chapter form the second part of the RASC program for ranking and scaling of biostratigraphic events and other events which can be uniquely identified. An optimum sequence constructed by means of a ranking algorithm provides the starting point for estimating average "distances" between successive events. The frequency of cross-over (mismatch) of the events in the sections is used for this purpose. These distances are clustered by constructing a dendrogram which can be used as a standard and permits definition of average interval zones (cf. Fig. 2.2). This chapter will include artificial examples in which the theory of scaling is illustrated and tested by applying it to sets of random normal numbers in computer simulation experiments.

6.2 Scaling versus ranking

The techniques described in this chapter have in common that distances are estimated between successive events in the optimum sequence obtained by the ranking algorithms described in the previous chapter. In a ranking, the successive events follow each other and no allowance can be made for the situation that some events should be closer together than others along a relative time scale. It can be useful to position the events along a scale with variable intervals between them. For example, suppose that two microfossils have observed extinction points (A and B) in 10 sections with A occurring 5 times above B, and 5 times below B. If a fence diagram were constructed, in which each event is connected to itself in other sections, the lines connecting event A would cross those connecting the event B in a number of places. It could be said that the relative cross-over (mismatch) frequency is $P_{AB} = 0.5$ because the number of matches is equal to the number of mismatches. This analogy generally does not hold true if P is a positive number not equal to 0.5 because, in general, the frequency of cross-overs is partly determined by the spatial pattern of the geographic locations of the sections. However, if the number of sections is not too small, the frequency P_{AB} always can be regarded as an estimate of the probability that A occurs above B. The interval between A and B along the relative time scale used for scaling should be nearly zero if P_{AB} is close to 0.5, and greater if P_{AB} tends to zero or one. Suppose that A occurs, for example, 9 times above B and only once below B. Then A and B should be separated by a longer distance along the relative time scale, corresponding to $P_{AB} = 0.9$.

The purpose of the scaling techniques is to estimate distances in time between successive events, not only from the cross-over frequencies between successive events, but also by using the cross-over frequencies between all events with mismatch in location in the observed sequences for segments of the optimum sequence.

Figure 6.2 from Agterberg and Gradstein (1988) provides an example of output from a scaling algorithm. The number codes of the events (exits of microfossils) and the microfossil names are shown on the right side. Each code is followed by the estimated distance from its event to the event below it. These distances have been plotted in the horizontal direction toward the left. They were clustered during a sequence of linking steps. The two successive events (32 and 29) in the scaled optimum sequence with the shortest distance (0.0067) between them were linked first. After scanning the set of unused interfossil distances, single events or clusters of events were linked pairwise, at each linking step, by using the shortest distances between them until the longest interfossil distance (between 20 and 24) was reached. The resulting clusters based on interfossil distances in time resemble assemblage zones (cf. Section 2.2).

The solution of Figure 6.2 for 54 taxon exits in 21 wells on the Labrador Shelf and northern Grand Banks shows a number of distinct and progressively younger clusters. A shading pattern was used to enhance the stratigraphically most useful parts of individual clusters. In total, 10 preferred RASC zones are shown. These are separated by relatively long interfossil distances. Several of such intervals between clusters represent stratigraphic hiatuses (Gradstein et al., 1985). In order to construct Figure 6.2, the output of the RASC program listed in Agterberg and Nel (1982) was combined with a DISSPLA graphics package (copyrighted in 1975 by Integrated Software System Corporation). A version of this DISSPLA program called DENO was published by Jackson et al. (1984). DENO was used to construct the optimum sequences and dendrograms of nine data bases in Gradstein et al. (1985, Appendix I). The input data for Figure 6.2 were processed by using the modified Hay method with threshold parameters $k_c = 7$, $m_{c1} = 2$ and $m_{c2} = 4$. The optimum sequence resulting from ranking was used as a starting point for scaling. It was slightly reordered during the application of the scaling algorithm (see later). The distances between successive events shown in Figure 6.2 can be added in order to obtain distance of each event from a common origin coinciding with the first event (No. 4 in Fig. 6.2). The resulting RASC distances can be related to geological time (in Ma) on the basis of those events for which the age is relatively well known (see Chapter 9).

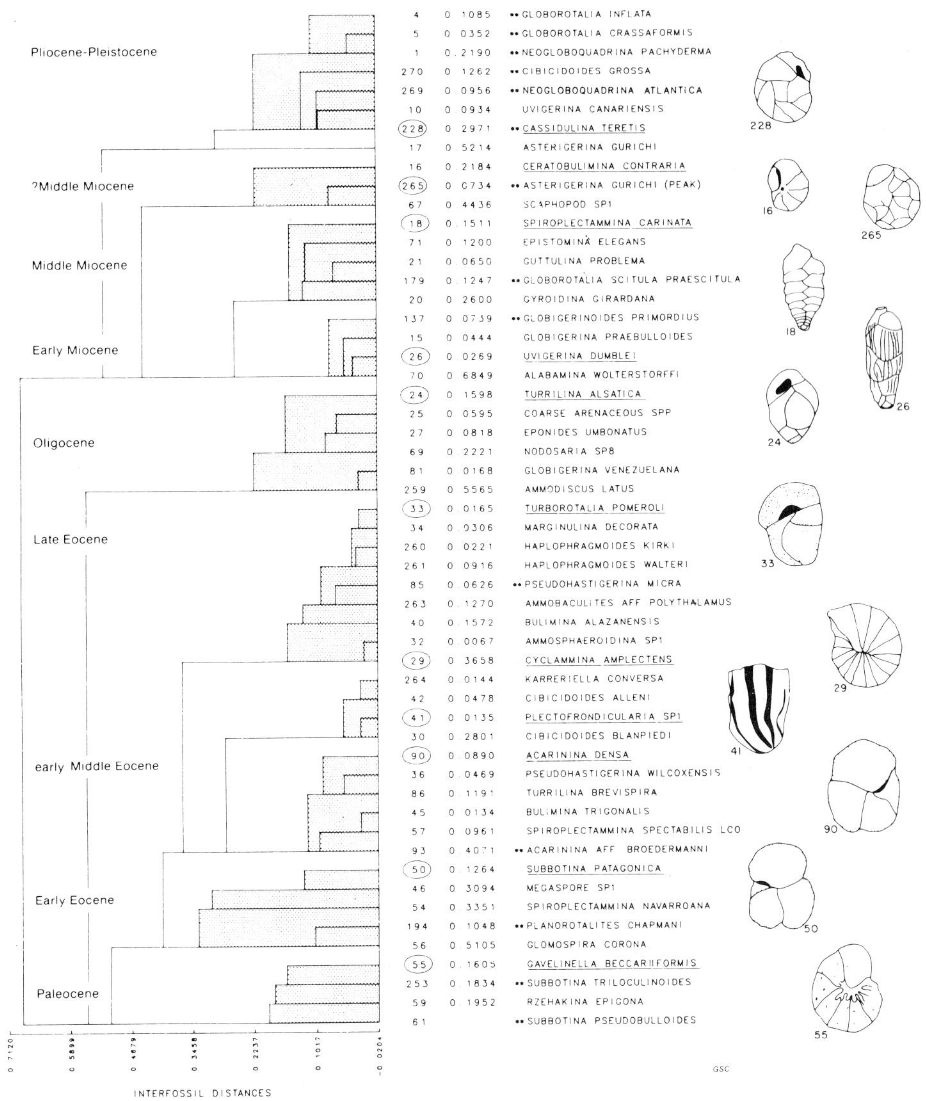

Fig. 6.2 Scaled optimum sequence for 21 wells on Labrador Shelf and Grand Banks ($k_c=7$, $m_{c1}=2$, $m_{c2}=4$). Dendrogram values along horizontal axis are interfossil distances (=intervals between successive exits) also given in numerical form in the vertical direction. Each distance represents distance between an event and its successor of which the dictionary code number and name are printed on the next line. The tenfold zonation is representative for the regional Cenozoic stratigraphy. There are eleven unique events, shown with double asterisks. These unique events occurred in fewer than $k_c=7$ sections so that they were not used for scaling. Their interfossil distances were estimated later, by re-inserting them into the scaled optimum sequence on the basis of their relative stratigraphic positions (with respect to events that were used) in the one or more sections containing them. A shading pattern was used to enhance the stratigraphically most useful parts of the dendrogram. The large distances on either side of the Eocene, Oligocene and Miocene assemblages are sedimentary cycle boundaries (cf. Gradstein et al., 1985, pp. 146-151).

Figure 6.3 shows DENO output for the Hay example (cf. Fig. 4.2, Table 5.5). All 10 events were used and the threshold parameters m_{c1} and m_{c2} were set equal to 2. The relatively short intervals between events 1 to 7 in Figure 6.3b reflect the fact that these events tend to be coeval on the average in the lower parts of the sections (see Fig. 4.2). On the other hand, events 8, 9 and 10 tend to occur above the others. Clearly, the dendrogram (scaled optimum sequence (Fig. 6.3b)) contains more information than the optimum sequence (Fig. 6.3a). As another example of this, it may be considered that events 9 and 10 are coeval on the average according to Figure 6.3a. This would imply that there is 50 percent probability that event 9 occurs above 10. However, in Figure 6.3b, event 9 occurs above 10 with distance of $D=0.4354$. It will be shown in the next section that the estimated probability P_e corresponding to D satisfies $P_e = \Phi(D)$. Consequently, event 9 would occur above 10 with probability $P_e = \Phi(0.4354) = 0.67$ or 67 percent which is slightly greater than 50 percent. Although W (event 9) occurs three times above Δ (event 10), and Δ three times above W in Figure 4.2, it also can be seen that if W occurs above Δ, the latter event is coeval to six (Section B), one (Section G) and two (Section H) other events, respectively. On the other hand, if Δ occurs above W, the latter event is not coeval to any other events. Because all possible pairwise comparisons are considered simultaneously in scaling, event 9 (W) is placed above 10 (Δ) in the scaled optimum sequence instead of at the same position.

6.3 Statistical model for scaling of stratigraphic events

The existence of events which interchange places with one another in different sections can be explained by assuming that each event is described by a different probability distribution. As pointed out before, the exact probability distributions of the events are not known. However, it can be assumed that the distributions of the direct and indirect distance estimates are approximately normal because these are averages of two and three event distances, respectively, and averages tend to be normally distributed (cf. Fig. 2.18). It will be shown that this allows estimation of the parameters of the model. An advantage of this statistical approach is that, later, the fitted model can be tested against the observed data. This

OPTIMUM FOSSIL SEQUENCE

a

SEQUENCE POSITION	FOSSIL NUMBER	RANGE	FOSSIL NAME
1	9	0 — 3	HO DISCOASTER TRIBRACHIATUS
2	10	0 — 3	LO DISCOLITHUS DISTINCTUS
3	8	2 — 5	LO DISCOASTER CRUCIFORMIS
4	6	1 — 5	LO RHABDOSPHAERA SCABROSA
5	4	4 — 6	LO COCCOLITHUS SOLITUS
6	7	5 — 8	LO DISCOASTER MINIMUS
7	5	5 — 8	LO COCCOLITHUS GAMMATION
8	1	7 — 10	LO DISCOASTER DISTINCTUS
9	3	7 — 11	LO DISCOASTER GERMANICUS
10	2	8 — 11	LO COCCOLITHUS CRIBELLUM

b

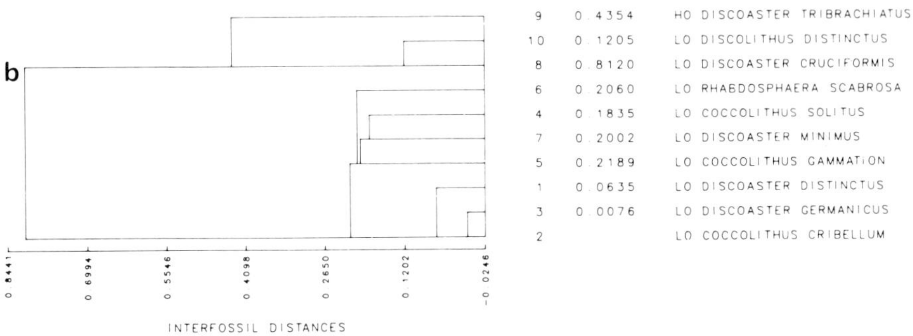

Fig. 6.3 DENO output for the Hay example (from Agterberg and Gradstein, 1998). The clustering of events 1 to 6 in the dendrogram (b) reflects the relatively large number of cross-overs and many coeval events near the base of most sections used (cf. Fig. 4.2).

final testing either verifies or negates the results obtained by means of the statistical model.

Figure 6.4 shows the basic model initially adopted for the scaling algorithms. Each event (e.g. A) would assume a position x_{Ai} in section i where x_{Ai} is the distance to A from an origin with arbitrary location along the relative time scale (x-axis in Fig. 6.4). The distance x_{Ai} is assumed to be the realization of a random variable X_A whose probability distribution is shown in Figure 6.4. Similar random variables are defined for the other events B, C,...

The random variable X_A satisfies the normal (Gaussian) probability distribution $N(EX_A, \sigma^2)$ with expected (or mean) value EX_A and variance σ^2. The mean values of the events differ from one another but the standard deviations of all events are assumed to be equal to σ in the model of Figure 6.4.

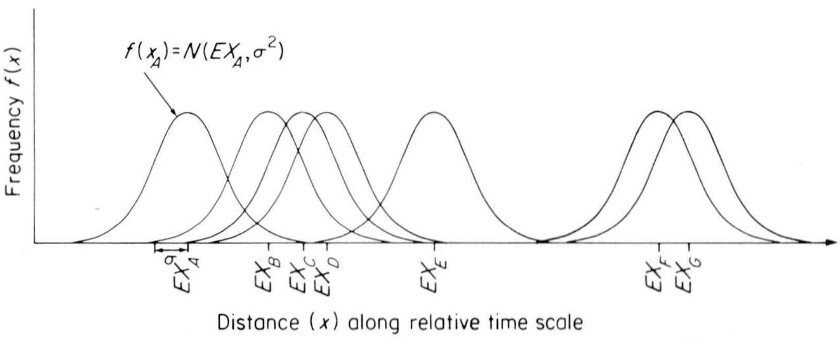

Fig. 6.4 Probabilistic model for clustering of biostratigraphic events (A, B, C, ...) along relative time scale (x-axis). Relative position of event (for example, A) in section or well is random variable (X_A) which is distributed normally around average location (EX_A) with standard deviation σ.

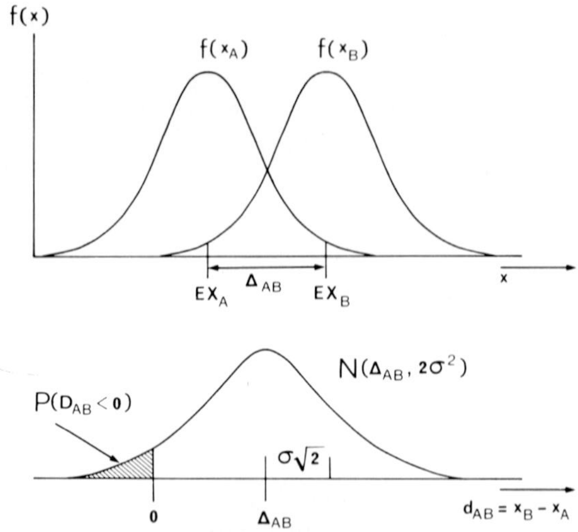

Fig. 6.5 Direct estimation of distance Δ_{AB} between events A and B from cross-over frequency $P(D_{AB}<0)$. Random variable $D_{AB}(=X_B-X_A)$ is negative only when order of A and B in section is reverse of order of EX_A and EX_B. Variance of D_{AB} is twice as large as variance σ^2 of individual events A and B.

The normal distribution curves for events A and B are shown separately in Figure 6.5. Because the time scale is relative, it will not be possible to estimate σ which determines the scale along the x-axis. (In the RASC program σ^2 is set equal to 0.5, see later) However, it is possible to estimate the ratio $\Delta_{AB}/(\sigma\sqrt{2})$ for the distance between the population means $\Delta_{AB} = EX_B - EX_A$ from the relative cross-over frequency P_{AB}.

For this purpose, P_{AB} is considered to provide a good estimate of the probability $P(X_B - X_A > 0) = P(D_{AB} > 0)$ which satisfies

$$P(D_{AB}>0) = \frac{1}{\sqrt{(2\pi)}} \cdot \int_{-\Delta_{AB}/\sigma\sqrt{2}}^{\infty} e^{-z^2/2}\, dz \tag{6.1}$$

This formula follows from the fact that the difference $D_{AB} = X_B - X_A$ has a normal distribution $N(\Delta_{AB}, 2\sigma^2)$ which is shown in the bottom part of Figure 6.5. The distance between events A and B for a specific section can be written as $d_{AB} = x_B - x_A$. The hatched area in Figure 6.5 is for $P(D_{AB}<0) = 1 - P(D_{AB}>0)$. If Φ represents fractile of the normal distribution in standard form, it follows that

$$\Phi(\Delta_{AB}/\sigma\sqrt{2}) = \frac{1}{\sqrt{(2\pi)}} \cdot \int_{-\infty}^{\Delta_{AB}/\sigma\sqrt{2}} e^{-z^2/2}\, dz \tag{6.2}$$

Consequently,

$$P(D_{AB}>0) = \Phi(\Delta_{AB}/\sigma\sqrt{2}) \tag{6.3}$$

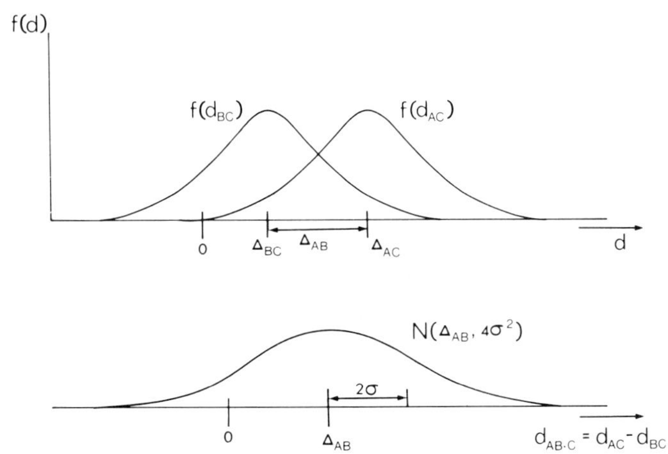

Fig. 6.6 Indirect estimation of distance Δ_{AB} between events A and B from cross-over frequencies with event C. Indirect distance $D_{AB \cdot C} = D_{AB} - D_{BC}$ has variance which is four times as large as variance of individual events A, B and C.

A precise estimate of P_{AB} which would allow the determination of Δ_{AB} is seldom available in practical applications because this would require a very large number of sections containing both A and B. However, it generally is possible to estimate Δ_{AB} indirectly by using pairs of cross-over frequencies linking A and B to other events; for example, by using the pair P_{AC} and P_{BC}. A distance of this type will be written as $D_{AB \cdot C}$. As illustrated in Figure 6.6, $D_{AB \cdot C} = D_{AC} - D_{BC}$ is normally distributed with $N(\Delta_{AB}, 4\sigma^2)$. Because σ^2 is arbitrary (σ determines scale along x-axis), the variance of the normal distribution was set equal to the constant $\sigma^2 = 0.5$. As a result of this simplification, it follows that

$$P(D_{AB} > 0) = P(D_{AB.C} > 0) = \Phi(\Delta_{AB}) \tag{6.4}$$

In the middle term of Equation (6.4), the event C can be replaced by any other event from which an indirect estimate of Δ_{AB} can be obtained. In practice, it usually turns out that there are many events showing inconsistencies with both events for which the interval Δ along the x-axis is being estimated. Averaging of many indirect distance estimates yields a more precise estimate of Δ. Once Δ_{AB} in Equation (6.4) has been estimated, it can be used to estimate $P(D_{AB} > 0)$. The resulting "theoretical" probability should be close to P_{AB}. Although, for model verification, it is not meaningful to make separate comparisons of this type, it can be useful to compare many observed and theoretical probabilities simultaneously by means of a chi-squared test (see Section 6.11).

It should be kept in mind that the model of Figure 6.4 is not necessarily realistic because it is unlikely that all events would have the same normal curve with variance equal to σ^2 for their exit location distributions. However, in practice, an estimate of indirect distance such as $D_{AB \cdot C}$ is based on two separate distances (D_{AC} and D_{BC}) and, each of these two random variables, in turn, is based on two separate distances (X_A, X_C and X_B, X_C) although X_C is used twice. Hence $D_{AB \cdot C}$ is based on three random variables (X_A, X_B, and X_C) that cannot be estimated separately. Because of the central-limit theorem of statistical theory, $D_{AB \cdot C}$ tends to be normally distributed even if the frequency curves of events A, B and C are not normal and have unequal variances (cf. Fig. 2.18).

Even if random variables for indirect distances such as $D_{AB \cdot C}$ are not normally distributed with equal variances, then the computation of an unweighted or weighted average of a number of indirect distance estimates, almost certainly, will yield a final estimate of Δ with a normal distribution because the central limit theorem applies to this new averaging process as well. However, although the final distance estimates may be precise estimates of the expected values (EX_A, EX_B, EX_C, etc. in Fig. 6.4) of the exit distributions, the corresponding variances σ^2_A, σ^2_B, σ^2_C, ... are not necessarily all equal to 0.5. Neither are all exit distributions necessarily normal. To assume normality with $\sigma^2 = 0.5$ for all distributions usually provides a crude approximation of the exit distributions only (see Chapter 8 for further discussion).

Unweighted distances for Hay example

Table 6.2A shows the relative cross-over frequencies $P_{ij} = S_{ij}/R_{ij}$ for the Hay example. The order of the events is that of the optimum sequence shown previously in Table 5.5. The elements in Table 6.2A are identical to those in Table 5.3A except that two pairs with $R_{ij} = 2$ were set equal to zero because the threshold parameter $m_{c2} = 3$ was used. Each of the frequencies of Table 6.2A was changed into a fractile of the standard normal distribution or Z-value (see Table 6.2B).

Table 6.1 shows Z-values for selected relative frequencies. Because $P_{ji} = 1 - P_{ij}$, it follows that $Z_{ji} = -Z_{ij}$. When the optimum sequence is used as a starting point, all or most of the Z-values in the upper triangle of the Z-matrix are positive. Negative values occur in the upper triangle only for elements with $P_{ij} < 0.5$ corresponding to events whose scores were ignored in order to break a cycle in which these events were participating during ranking by means of the modified Hay method. It is noted that scores temporarily ignored for constructing the optimum sequence are restored to their positions before use of the scaling algorithms of RASC is initiated.

Clearly, a relative frequency P_{ij} for a small sample will be subject to considerable uncertainty and this error is propagated into the Z_{ij}-value derived from it. This is the reason for defining the minimum sample size m_{c2} (=3 for Table 6.2). It means that Z_{ij}-values based on fewer than m_{c2} pairs of occurrences will not be used. In the original RASC program (Agterberg and Nel, 1982a, b) no distinction was made between m_{c1} and

m_{c2}. However, later work has shown that better results can be obtained by setting $m_{c2} > m_{c1}$. For the example of Table 5.3, $m_{c2}=3$ and $m_{c1}=1$.

When an average distance between two events is estimated from Z-values for 10 events, it could be based on as many as nine seperate estimates of the distance. The direct estimate of the distance between events i and j follows from Z_{ij} and the indirect estimates involving other events k follow from the differences $Z_{ik}-Z_{jk}$ ($k \neq i,j$) where i and $j = i+1$ are successive rows. However, because $Z_{ij} = -Z_{ji}$, the differences $Z_{kj}-Z_{ki}$ ($k \neq i,j$), where i and $j = i+1$ are successive columns, also can be used. For example, the direct estimate of distance between events 4 and 7 which occur in columns 5 and 6, respectively, satisfies $D(4\text{-}7) = Z_{56}=0.210$. The corresponding indirect estimates are $Z_{16}-Z_{15} = 1.645—1.068 = 0.577$, $Z_{26}-Z_{25} = 1.282—0.524 = 0.758$, and six other, similar differences between Z-values in adjacent columns. The differences for all pairs of events are shown in Table 6.2C.

In the RASC program, Z-values in the upper triangle are used only. The lower triangle is used to retain information on sample sizes. Addition of indirect and direct estimates yields the sum of the N^* separate estimates. For events 4 and 7, Sum=1.56 (see Table 6.2C). The average of all $N^*=9$ estimates of the interval between events 4 and 7 amounts to Sum/9 = 0.174. This is called an unweighted estimate of distance between successive events in the output of the RASC program. The complete set of 9 intervals is shown in Table 6.3. The cumulative RASC distance or distance from the first event (No. 9) is shown in the last column of Table 6.3. Because of missing values (see Table 6.2) or pairs of cross-over frequencies which both are equal to one (see later), distance estimates may be based on fewer than N^* (=9 for the example) pairs of events. Theoretically, the direct estimate of distance (cf. Fig. 6.5) has half the variance of the indirect estimates (cf. Fig. 6.6). Thus it should be weighted twice as heavily. This will be done in weighted distance estimation in which errors in P_{ij} due to small sample sizes also will be considered.

Weighted distance estimates

The relative cross-over frequencies P_{ij} are calculated from scores (S_{ij}) on samples of different sizes (R_{ij}). For this reason, it is preferable to compute weighted mean distances Δ_{eij} in which the weights assigned to the direct and indirect estimates of distance are primarily determined by

TABLE 6.2

Unweighted distance estimation to obtain intervals between successive events along RASC distance scale for Hay example. **A.** P-matrix of relative frequencies for the 10 events in order of optimum sequence. Values excluded because of threshold $m_{2c} = 3$ are shown as 000. **B.** Z-values corresponding to P-values. Note that threshold q_c is equal to 1.645. **C.** Values are differences between values in successive columns of Table 6.2B. Zero differences for pairs of q_c-values are shown as 000 and were not used. Bottom row shows sums for columns with number of values (N^*) used for obtaining sum.

A	9	10	8	6	4	7	5	1	3	2
9	x	3.0/6	5.0/5	4.0/4	6.0/7	7.0/7	9.0/9	8.0/8	6.0/6	8.0/8
10	3.0/6	x	2.5/3	000	3.5/5	4.5/5	5.0/6	4.5/5	3.5/4	4.5/5
8	0.0/5	0.5/3	x	000	3.0/4	4.5/5	5.0/5	5.0/5	4.0/4	5.0/5
6	0.0/4	000	000	x	3.0/4	1.5/3	3.0/4	2.5/3	3.0/4	3.0/4
4	1.0/7	1.5/5	1.0/4	1.0/4	x	3.5/6	4.5/7	4.5/6	4.5/6	3.0/6
7	0.0/7	0.5/5	0.5/5	1.5/3	2.5/6	x	3.5/7	4.0/6	3.5/5	4.5/6
5	0.0/9	1.0/6	0.0/5	1.0/4	2.5/7	3.5/7	x	4.5/8	4.0/6	5.0/8
1	0.0/8	0.5/5	0.0/5	0.5/3	1.5/6	2.0/6	3.5/8	x	2.5/5	5.0/7
3	0.0/6	0.5/4	0.0/4	1.0/4	1.5/6	1.5/5	2.0/6	2.5/5	x	3.0/6
2	0.0/8	0.5/5	0.0/5	1.0/4	3.0/6	1.5/6	3.0/8	2.0/7	3.0/6	x

B	9	10	8	6	4	7	5	1	3	2
9	x	0.000	1.645	1.645	1.068	1.645	1.645	1.645	1.645	1.645
10	0.000	x	0.967	000	0.524	1.282	0.967	1.282	1.150	1.282
8	-1.645	-0.967	x	000	0.674	1.282	1.645	1.645	1.645	1.645
6	-1.645	000	000	x	0.674	0.000	0.674	0.967	0.674	0.674
4	-1.068	-0.524	-0.674	-0.674	x	0.210	0.366	0.674	0.674	0.000
7	-1.645	-1.282	1.282	0.000	-0.210	x	0.000	0.430	0.524	0.674
5	-1.645	-0.967	-1.645	-0.674	-0.366	0.000	x	0.157	0.430	0.318
1	-1.645	-1.282	-1.645	-0.967	-0.674	-0.430	-0.157	x	0.000	0.566
3	-1.645	-1.150	-1.645	-0.674	-0.674	-0.524	-0.430	0.000	x	0.000
2	-1.645	-1.282	-1.645	-0.674	0.000	-0.674	-0.318	-0.566	0.000	x

C		10	8	6	4	7	5	1	3	2
9		0.000	1.645	000	-0.577	0.577	000	000	000	000
10		x	0.967	000	000	0.758	-0.315	0.315	0.132	0.132
8		0.678	x	000	000	0.608	0.363	0.000	0.000	0.000
6		000	000	x	0.674	-0.674	0.674	0.293	-0.293	0.000
4		0.544	-0.150	0.000	x	0.210	0.156	0.308	0.000	-0.674
7		0.363	0.000	1.282	-0.210	x	0.000	0.430	0.094	0.150
5		0.678	-0.678	0.971	0.308	0.366	x	0.157	0.273	-0.112
1		0.363	-0.363	0.678	0.293	0.244	0.273	x	0.000	0.566
3		0.495	-0.495	0.971	0.000	0.150	0.094	0.430	x	0.000
2		-0.363	-0.363	0.971	0.674	-0.674	0.356	-0.248	0.566	x
	Sum/N^*	3.98/8	0.56/8	4.87/6	1.16/7	1.56/9	1.60/8	1.69/8	0.51/8	0.06/8

TABLE 6.3

Unweighted distance analysis of values shown in Table 6.2 continued to obtain RASC distances of events. The origin of the scale is set at the first event. Consequently, the distance for event 9 is equal to zero. Event 10 has distance of 0.435. Event 2 has the largest cumulative RASC distance (=2.140).

	Events	N*	Sum	Interval	Distance
1	9-10	8	3.98	0.935	0.435
2	10-8	8	0.56	0.070	0.506
3	8-6	6	4.87	0.812	1.318
4	6-4	7	1.16	0.166	1.484
5	4-7	9	1.56	0.174	1.658
6	7-5	8	1.60	0.200	1.858
7	5-1	8	1.69	0.211	2.069
8	1-3	8	0.51	0.064	2.132
9	3-2	8	0.06	0.008	2.140

the sizes of the samples used to obtain the Z-values. The weight-corrected equation for estimating the distance between events i and j is:

$$\Delta_{eij} = \frac{w_{ij} Z_{ij} + \sum_{k \neq i,j} w_{ij.k} (Z_{ik} - Z_{ik})}{w_{ij} + \sum_{k \neq i,j} w_{ij.k}} \tag{6.5}$$

where the weights w_{ij} and $w_{ij \cdot k}$ are

$$w_{ij} = \frac{R_{ij} e^{z^2_{ij}}}{2\pi P_{ij}(1-P_{ij})}; \quad w_{ij.k} = \frac{(w_{ik})(w_{ik})}{w_{ik} + w_{jk}}; \; k \neq i,j \tag{6.6}$$

In order to derive these equations, use was made of theory of weighting coefficients (cf. Bliss, 1935; Fisher and Yates, 1964; Finney, 1971. The weights were derived in the following manner. The observed proportion P_{ij} is assumed to be the realization of a random variable P which is related to a standard normal variable Z such that

$$P = \frac{1}{(2\pi)^{\frac{1}{2}}} \int_{-\infty}^{Z} e^{-\frac{1}{2}s^2} \, ds \tag{6.7}$$

where s denotes position along the linear scale used.

The proportion P can be assumed to originate from a binomial random variable with expected value $E(P) = P_{ij}$ and variance

$$\sigma^2(P) = \{P_{ij}(1-P_{ij})\}/R_{ij} \tag{6.8}$$

where R_{ij}, as before, is the number of times that events i and j occurred in the same section.

It is known that, approximately,

$$\sigma^2(P) = \left(\frac{\partial p}{\partial z}\right)^2 \sigma^2(Z) \tag{6.9}$$

where p and z represent the density functions of P and Z, *respectively.* These equations can be combined into

$$\sigma^2(P) = \left(\frac{1}{2\pi}\right) e^{(-Z_{ij}^2)} \sigma^2(Z) \tag{6.10}$$

Each weight w_{ij} is obtained as

$$w_{ij} = \frac{1}{\sigma^2(Z)} = \frac{R_{ij} e^{-Z_{ij}^2}}{2\pi P_{ij}(1-P_{ij})} \tag{6.11}$$

Weights $w_{ij \cdot k}$ are obtained by addition of similar variances $\sigma^2(Z)$ of the values Z_{ik} and Z_{jk}. If $Z_{ij} = q_c$, the P_{ij} value corresponding to q_c is used together with the original R_{ij} value in Equation (6.11).

Table 6.4 shows intervals which are weighted distances $\Delta_{ei,i+1}$ (i = 1, ..., N-1) estimated for successive events in the optimum sequence. For example, the weighted distance between events 4 and 7 is calculated as follows. From Table 6.2 it follows, for events 4 and 7, that $R_{56} = 6$, $P_{56} = 3.5/6$ and $Z_{56} = 0.210$. Consequently, $w_{56} = 3.76$ (Eq. 6.11 or 6.6). Likewise, for the same example, $w_{15} = 2.91$ and $w_{16} = 1.57$. Hence, $w_{56.1} = 1.02$ (Eq. 6.6). The sum of 9 weights is $W = 3.76 + 1.02 + 0.8 =$ 15.0 (see Table 6.4). The corresponding sum (numerator, right side of Eq. 6.5) is 2.34. The weighted distance between events 4 and 7 therefore is

TABLE 6.4

Weighted distance analysis of values shown in Table 6.2. The Z-values were weighted according to sample size (see Eq. 6.5 and 6.6 in text). Standard deviations were computed by using Eq. 6.13. Note that the interval between events 3 and 2 (on bottom row) is negative. As a result, event 9 has RASC distance (=2.149) which is less than that of event 8 (=2.155).

	Events	W	Sum	Interval	$s(\bar{x})$	Distance
1	9-10	10.3	3.27	0.317	0.100	0.317
2	10-8	7.0	1.24	0.176	0.289	0.493
3	8-6	4.7	3.62	0.770	0.203	1.262
4	6-4	9.2	2.44	0.266	0.163	1.529
5	4-7	15.0	2.34	0.157	0.153	1.686
6	7-5	14.8	2.32	0.157	0.085	1.893
7	5-1	15.2	2.96	0.195	0.082	2.038
8	1-3	12.6	1.47	0.117	0.090	2.155
9	3-2	13.3	-0.08	-0.006	0.124	2.149

$\Delta_e = 2.34/15.0 = 0.157$. This value is among the intervals listed in Table 6.4.

For simplification, Equation (6.5) can be rewritten as:

$$x = W^{-1} \sum_{i=1}^{N^*} w_i x_i \qquad (6.12)$$

with

$$x = \Delta_{AB}; \quad W = \sum_{i=1}^{N^*} w_i$$

and

$$x_1 = Z_{AB}, \quad w_1 = w_{AB}$$

$$x_2 = Z_{AC} - Z_{BC}, \quad w_2 = w_{AB.C}$$

$$x_3 = Z_{AD} - Z_{BD}, \quad w_3 = w_{AB.D}$$

with similar expressions for x_i ($i=4, 5, ...$). In these expressions, A and B denote two successive events, and other events are written as C, D, ... The weight W and sum $\Sigma w_i x_i$ for the Hay example were given in Table 6.4. The corresponding standard deviation $s(\bar{x})$ shown in the last column of Table 6.4 is the positive square root of

$$s^2(x) = W^{-1}(N^*-1)^{-1} \sum_{i=1}^{N^*} w_i (x_i - x)^2 \tag{6.13}$$

As before, the number of pairs of Z-values used for estimation is written as N^*. This includes the Z-value for the direct estimate. The standard deviation for the distance between events 4 and 7 amounts to 0.153 (see Table 6.4). This is nearly equal to the value of the interval itself (= 0.157). It would indicate that the latter is not significantly different from zero. A rapid test of this hypothesis (approximate t-test) consists of multiplying the standard deviation by 2 and subtracting the result from the estimated distance. If the difference is negative, the distance could well be zero. Application of this test to the values listed in Table 6.4 shows that only 3 of the intervals computed for the Hay example would be greater than zero with probability greater than 95 percent. Equation (6.13) is based on the assumption that the x_i-values are realizations of stochastically independent random variables. This condition may not be satisfied in practice and the estimated standard deviations may be too small.

When all possible comparisons can be made as for the pair of events 4 and 7, $N^* = N$-1 where N denotes total number of events. However, in the RASC computer program, N^* may be less than N-1 for the following two reasons: (1) The total number of comparisons is reduced by one for each value x_i that cannot be computed because one of the Z-values needed is missing (this includes the case that both Z-values are missing); (2) if $S_{ij} = R_{ij}$, $P_{ij} = 1$ and the corresponding Z-value is set equal to the threshold value q_c (= 1.645 in Table 6.2). Pairs of Z-values both equal to q_c, and with zero-difference, are not used for estimating the average distance Δ_{eij} unless a pair of this type is contained within a cluster of mutually inconsistent events. For this reason, pairs of values (Z_{ik}, Z_{jk}) in successive columns (i, $j=i+1$) are tested by letting k decrease from $k = i+1$. Suppose that, for a given value of k, $Z_{ik} = Z_{jk} = q_c$. This pair is not used for the distance estimation unless a pair of Z-values, which are not both equal to q_c, is found for a smaller value of k. In the RASC program, it is assumed that this situation is encountered as soon as five pairs of Z-values equal to q_c have been identified for decreasing k.

Likewise, pairs of values (Z_{ik}, Z_{jk}) in successive rows can be tested by letting k increase from $k = i+2$.

Both preceding situations occur in the Hay example for estimation of the distance between events 8 and 6. Because the Z-values for these events combined with event 9 both are equal to $q_c = 1.645$ (see first row of Table 6.2B), and because the pair (8, 6) also has two non-determined values, $N^* = 9 - 3 = 6$. The corresponding weight (W) in Table 6.4 is only 4.7. The standard deviation (= 0.203) for the corresponding interval (= 0.770) is relatively large. Nevertheless, application of the preceding approximate t-test suggests that the latter value is statistically significant.

When a large number of events for a long time interval is used, N^* is likely to be much smaller than N-1 in all distance calculations, because events belonging to relatively young assemblages (e.g. Late Miocene in Fig. 6.2) normally all occur above events in older assemblages (e.g. Early Eocene in Fig. 6.2). Distance estimates based on few pairs of Z-values are relatively imprecise. In the RASC program there is an option that distances based on N^* less than m_{c2} are replaced by zeros.

The choice of a value for q_c usually is not critical, because most pairs of q_c-values will not be used for distance estimation. D'Iorio (1990) has performed a study of the effect of systematically changing q_c for his database (cf. Section 8.2). The average distance between successive events increases when q_c becomes larger but, in general, the relative order of the events is not changed significantly. As a "default", q_c is set equal to 1.645 in the RASC program. This corresponds to a cross-over frequency of $P = 0.95$ (see Table 6.1). The user can replace the default value by any other value. In general, q_c should be greater than 1 and less than 2. It should be kept in mind that the value of q_c is selected because, theoretically, a cross-over frequency of 1 corresponds to an infinitely large Z-value and distance estimation would not be possible. It can be assumed that the scores from which cross-over frequencies are calculated satisfy binomial frequency distributions. For small samples, the probability that a cross-over frequency is equal to 1 (or 0), then is relatively large even when a minimum sample size (m_{c2}) has been defined. This problem is restricted to the tails of the normal (Gaussian) frequency curve and can be solved by choosing a q-value which, effectively, changes the range of the normal curve from $(-\infty, \infty)$ to $(-q_c, q_c)$.

Reordering of events in the scaled optimum sequence

The last interval estimated in Table 6.4 is negative. For this reason, it is desirable to reorder the events before a dendrogram of successive interfossil distances is constructed. The cumulative distance from the first event (No. 9) in the original optimum sequence obtained by ranking can be calculated for each event in weighted as well as unweighted distance analysis. In Table 6.4, the distance between events 9 and 2 (2.149) is less than that between 9 and 3 (2.155). If distances from event 9 are used, it follows that event 2 should lie above 3 in the scaled optimum sequence. The events always can be reordered on the basis of this cumulative distance. This allows the clustering of successive distances as shown, for example, in Figure 6.2.

The standard deviations of the distances between successive events cannot be recalculated readily after a reordering which removes negative distances. This is because successive distance estimates are not stochastically independent. In order to obtain the new standard deviations, it is necessary to repeat all calculations taking the reordered optimum sequence as the starting point. Because different Z-values then are used for estimation, the distance estimates will change as is illustrated in Table 6.5 for the Hay example. New negative distances may be computed at this stage and the procedure would have to be repeated again. These new calculations can be performed by using the final reordering option of the RASC program. The objective of final reordering is to obtain a set of distances between successive events which are all positive so that the corresponding standard deviatons also are known. This result readily could be achieved for the Hay example. However, when the data base is large, and when k_c and m_{c2} are small, it may not be possible to obtain a single set of consecutive distances which are all positive. This is because the iterative process does not necessarily converge to a single solution. As a default, at most four complete reorderings are allowed in the RASC program. If convergence to a situation of positive distances is not obtained in four or more steps, either the result without final reordering can be accepted, or the result obtained after four or more reorderings. In the latter solutions, the number of negative distances probably will have been reduced considerably.

Figure 6.7 illustrates that the preceding iterative process for final reordering does not necessarily converge to a single solution. Suppose that the numbers in Figure 6.7 represent estimated distances between pairs of

TABLE 6.5

Example of weighted distance analysis after reordering. The optimum sequence used as input for scaling was not the ranking result used for Tables 6.2 to 6.4 but the scaled optimum sequence in the ranking of events in last column of Table 6.4. Differences between Tables 6.4 and 6.5 are restricted to values in two rows at the bottom only.

	Events	Interval	$s(\bar{x})$	Distance
1	9-10	0.317	0.100	0.317
2	10-8	0.176	0.289	0.493
3	8-6	0.770	0.203	1.263
4	6-4	0.266	0.163	1.530
5	4-7	0.157	0.153	1.686
6	7-5	0.157	0.085	1.843
7	5-1	0.195	0.082	2.038
8	1-2	0.118	0.147	2.156
9	2-3	0.006	0.124	2.162

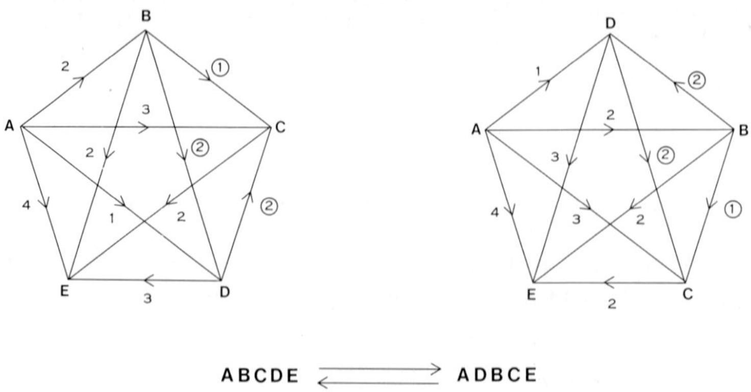

Fig. 6.7 Artificial example for demonstrating that the final reordering option of the RASC computer program does not necessarily converge to unique solution. See text for further explanation.

events A, B, C, D and E. A positive distance from one event to another is indicated by an arrow pointing from the one event to the other. For example, the distance from A to B is +2 and that from C to D is –2. Let the optimum sequence ABCDE have only one negative distance (between C and D). Because this distance is greater than that between B and C (= +1), the reordered sequence becomes ADBCE. The distances for this artificial example have been chosen in such a way that this new sequence again has only one negative distance (between D and B) and reordering ADBCE gives the original sequence ABCDE. Consequently, a unique solution with positive distances between successive events does not exist. Situations similar to the one illustrated in Figure 6.7 do occur in practice, especially in situations where the estimated distances are not very precise.

6.4 Artificial example

The purpose of this section is to illustrate the theory of scaling as developed in the previous section by using the artificial example of Table 4.12 based on random normal numbers. Although the theory leads to valid results for large samples, small-sample fluctuations may be considerable. This aspect will be evaluated here. In general, the understanding of statistical models applied to observed data can be helped considerably by simulation experiments. Nevertheless, it should be kept in mind that numbers are used, of which it may be known beforehand that they should fit well because all expected values were determined by the scientist conducting the experiment. In practical applications to real data, the conditions artificially created for a simulation experiment may not be satisfied. The artificial example of Table 4.12 clearly demonstrates some features of the theory outlined in the previous section. However, it differs from natural situations by (1) small number of events, (2) large number of sequences, (3) all events are observed in all sequences, and (4) the positions of the events satisfy normal distributions with equal variances.

By counting, it was determined that A is followed by B in 116 of the 150 sequences of Table 4.12. (This implies that B precedes A in 34 sequences.) Likewise, A precedes C in 130 sequences and B comes before C in 85 sequences. These three numbers (n) are shown in the first column of Table 6.6. They were transformed into relative frequencies (f) by dividing them by 150 (see column 2 of Table 6.6). By consulting a table of cumulative frequencies for the normal distribution in standard form, the f-values were converted into Z-values. Multiplication by $\sqrt{2}$ then yields direct estimates of the distances between the events. For example, $D_{AB} = 0.750\sqrt{2} = 1.061$. Only one indirect estimate of the distance D_{AB} can be obtained. It is equal to 1.335 which represents the difference between 1.571 (direct estimate of D_{AC}) and 0.236 (direct estimate of D_{BC}). The arithmetic average of the direct and indirect estimates of distance is

TABLE 6.6

RASC method of scaling applied to data of artificial example. For meaning of column headings, see text.

	n	f	z	D (direct)	D (indirect)	Do (Ave)	D (Ave)	E (D)
AB	116	0.7733	0.750	1.061	1.335	1.198	1.152	1.000
AC	130	0.8667	1.111	1.571	1.297	1.434	1.480	1.500
BC	85	0.5667	0.167	0.236	0.510	0.373	0.327	0.500
SSD				0.079	0.152	0.060	0.053	

shown in column 6 of Table 6.6. This is followed by a weighted distance estimate which satisfies $D(\text{Ave}) = (1.061 + 1.335/2)/1.5 = 1.152$. Finally, the expected value E(D) is shown in the last column.

Comparison of the three estimates of distance is facilitated by computing the sum of squared deviations (SSD) from the expected value of each estimate. For example, for the direct estimate of D in Table 6.6,

$$\text{SSD} = (1.061 - 1.000)^2 + (1.571 - 1.500)^2 + (0.236 - 0.500)^2 = 0.079.$$

The SSD values also are shown in Table 6.6. The results suggest that the variance of the indirect estimate which is proportional to its SSD value is about twice as large as that of the direct estimate. The weighted average distances are most precise because they have the smallest variance.

The analysis shown in Table 6.6 was repeated for the 5 smaller subsamples. In all instances, the weighted mean distance provided the best estimate (see Table 6.7). It also can be seen, however, that in small samples, the estimated distance may differ considerably from its expected value.

In the preceding statistical analysis of which the results are shown in Tables 6.6 and 6.7, the weighted distance estimate D(Ave) was obtained by assigning twice as much weight to the direct estimate as to the indirect estimate. Because weights are inversely proportional to variances, this simply reflects the fact that, on the average, the variance of indirect estimates is twice as large as that of direct estimates (see Fig. 6.6).

Suppose, however, that the equation for estimating weighted distances $\Delta_{ei,\,i+1}$ as in the RASC program is used. From the values in the

TABLE 6.7

Statistical analysis of Table 6.6 repeated for the five subsamples.

Subsample		D (direct)	D (indirect)	Do (Ave)	D (Ave)
1 ...	AB	1.030	1.089	1.060	1.050
	AC	1.571	1.512	1.542	1.551
	BC	0.482	0.541	0.512	0.502
	SSD	0.006	0.010	0.006	0.006
2 ...	AB	0.741	0.911	0.826	0.798
	AC	1.030	0.860	0.945	0.973
	BC	0.119	0.289	0.289	0.176
	SSD	0.459	0.462	0.426	0.424
3 ...	AB	1.191	1.809	1.500	1.112
	AC	1.571	0.953	1.262	1.365
	BC	-0.238	0.380	0.071	-0.032
	SSD	0.586	0.968	0.491	0.314
4 ...	AB	0.881	2.236	1.559	1.333
	AC	2.594	1.239	1.917	2.142
	BC	0.358	1.713	1.036	0.810
	SSD	1.231	3.067	0.774	0.619
5 ...	AB	1.571	1.089	1.330	1.410
	AC	1.571	2.053	1.812	1.732
	BC	0.482	0.000	0.241	0.321
	SSD	0.331	0.564	0.273	0.254

columns for f and Z in Table 6.6, it readily is computed that $W_{AB} = 77.593$, $W_{AC} = 60.139$, and $W_{BC} = 94.549$. On the other hand, $W_{AB \cdot C} = 36.758$, $W_{AC \cdot B} = 42.618$, and $W_{BC \cdot A} = 33.880$. The latter three weights are not exactly half as large as the first three weights. The reason for this discrepancy is that the values of f and Z are approximations only. They were estimated from samples and used instead of the population values in the RASC method. The RASC weighted distances become 1.149, 1.457 and 0.308, instead of 1.152, 1.480 and 0.327 shown in Table 6.6 for D(Ave). Their correponding SSD value becomes 0.061 indicating that the D(Ave) values (with SSD = 0.053) of Table 6.6 are better in this artificial example.

6.5 Computer simulation experiments

As already pointed out in Section 4.9, the type of experiment described in the previous section can be performed on a computer using a pseudo-random number generator. Computer simulation experiments on biostratigraphic events also have been performed by Edwards (1982) and Harper (1984). Edwards (1982) dealt with the problem previously illustrated in Figure 2.19. She assumed that a taxon has a true extinction point in time which was randomly displaced upwards or downwards in a section due to sediment mixing. The aim of experiments of this type is to model the distribution of exits (or entries).

Harper (1984) performed computer simulation to create artificial successions of taxa in sections. Three types of optimum sequences were obtained for exits and entries of these taxa by means of the RASC program: (a) presorting option only; (b) modified Hay method only; and (c) scaling sequence as derived from (b). Harper demonstrated that, for his successions, (a) consistently gave results that were better than those of (b) and recommended that (a) instead of (b) be used as the input sequence for (c). However, results of (b) could be improved by making m_{c1} as small as possible (see Section 7.4). A revision made in the RASC program on the basis of Harper's results was to allow the usage of different threshold values for number of pairs of events (m_{c1} and m_{c2}; also see before). These examples illustrate that computer simulation experiments can provide results which are useful because they are complementary to results obtained from the analysis of natural data sets.

The computer simulation experiments used to test RASC were as follows. Twenty artificial stratigraphic events were generated for each of 50 sections whereby the interval between the expected positions of the events was kept constant in each run. For comparison, in the experiment of Table 4.12, three events were studied in each of 150 sections and, in total, 3 x 150 = 450 random normal numbers were used. In addition, twenty artificial stratigraphic events were generated for each of 25 sections using only 3 spacings as described in Section 4.9. The latter abbreviated database was shown previously in Tables 4.13 to 4.15. For each full experiment to be described here, $20 \times 50 = 1000$ random normal numbers with $\sigma = 1$ were used. Every experiment was performed twice using a different set of 1000 random normal numbers but the same set of 1000 random normal numbers was used in each of two sets of experiments where the interval between expected positions E(D) was set equal to 1.0, 0.5, 0.3, 0.2, 0.1 and 0.0, respectively. For scaling the threshold parameter q_c was set equal to 2.326 corresponding to $P=0.99$ which is midway between 49/50 and 50/50. The latter two values are the largest frequencies P which are possible in the database for 50 sections (without ties). For the abbreviated database for 25 sections, q_c was set equal to 2.054 corresponding to $P=0.98$.

Table 6.8 shows the artificial sequences of events created for the first two sections for all 6 intervals in one of the two sets of experiments. Table 6.8 illustrates that for E(D) = 1.0 (σ = 1.0), there are relatively few inconsistencies in the observed sequences. On the other hand, for E(D) = 0.1 it is difficult to recognize from the sequences that the expected sequence is 1, 2, ..., 20. The sequences for E(D) = 0.0 are, of course, completely random.

Table 6.9 shows optimum sequences obtained by (a) presorting option; (b) ditto, followed by modified Hay method; (c) scaling (unweighted differences); (d) scaling (weighted differences); and (e) ditto, after final reordering.

TABLE 6.8

First two artificial sequences used in complete set of computer simulation experiments (20 events in 50 sections) with E(D) equal to 1.0, 0.5, 0.3, 0.2, 0.1, and 0.0, respectively.

Series No.	E(D)	Expected Sequence Numbers																			
		1	2	3	4	5	6	7	8	9	10	11	12	13	14	15	16	17	18	19	20
1	1.0	1	2	4	5	3	6	8	10	9	7	11	12	13	14	15	17	16	18	19	20
2	1.0	1	4	3	2	7	6	5	8	9	11	10	12	13	15	14	16	18	17	19	20
1	0.5	1	2	5	4	3	6	10	8	9	11	13	14	12	15	7	17	16	18	19	20
2	0.5	1	4	3	2	7	8	9	6	11	5	12	13	10	15	18	19	16	14	17	20
1	0.3	5	1	4	2	10	3	6	8	11	9	15	13	14	17	12	16	7	19	18	20
2	0.3	1	4	3	7	2	8	9	11	6	12	13	18	15	5	10	19	16	20	17	14
1	0.2	5	4	1	2	10	11	3	6	8	9	15	17	14	13	12	16	19	18	20	7
2	0.2	1	4	7	3	11	8	9	13	18	12	2	15	6	19	20	10	16	5	17	14
1	0.1	5	10	4	2	1	11	17	15	14	8	9	13	6	3	16	12	19	20	18	7
2	0.1	1	4	7	18	11	19	9	13	8	12	3	15	20	2	6	16	17	10	14	5
1	0.0	5	10	17	15	11	4	14	13	16	19	2	9	20	8	1	18	6	12	3	7
2	0.0	1	18	19	4	20	11	13	15	7	12	9	8	17	16	3	10	6	14	2	5

TABLE 6.9

Optimum sequences obtained by 5 methods (a to e) of ranking applied to 5 artificial sequences (I to V) of events. Mean intervals were 1.0 (I), 0.5 (II), 0.3 (III), 0.2 (IV) and 0.1 (V), respectively. A and B denote results obtained for two different datasets (1 and 2).

A:Set 1																				
Ia e	1	2	3	4	5	6	7	8	9	10	.1	12	13	14	15	16	17	18	19	20
IIa e	1	3	2	4	5	6	7	8	9	11	11	12	13	14	15	16	17	18	19	20
IIIa e	1	3	2	4	5	6	7	8	9	9	10	12	13	14	15	16	17	18	19	20
IVa b	1	3	2	4	5	6	7	8	11	9	10	12	15	13	14	16	17	18	19	20
IVc	1	3	2	4	6	5	7	8	11	9	10	12	13	14	15	16	18	17	19	20
IVd e	1	3	2	4	5	6	7	8	11	9	10	12	13	14	15	16	18	17	19	20
Va b	3	4	5	1	2	6	8	11	7	9	10	15	13	12	14	16	18	17	19	20
Vc	1	3	4	2	5	6	8	7	11	9	10	15	14	12	13	16	18	17	19	20
Vd e	1	3	4	5	2	6	8	7	11	9	10	15	14	12	13	16	18	17	19	20
B:Set 2																				
Ia IIIb	1	2	3	4	5	6	7	8	9	10	11	12	13	14	15	16	17	18	19	20
IIIc e	1	2	3	4	5	6	7	8	9	10	12	11	13	14	15	16	17	18	19	20
IVa b	1	2	3	4	5	6	7	8	10	9	11	12	14	13	15	16	17	18	19	20
IVc	1	2	3	4	5	6	7	8	10	9	12	11	14	13	15	17	16	19	18	20
IVd e	1	2	3	4	5	6	7	8	10	9	12	11	14	13	15	16	17	19	18	20
Va	1	2	3	4	5	6	8	7	10	9	12	14	11	16	13	17	19	15	18	20
Vb	1	2	3	4	5	6	8	7	10	9	12	14	11	16	13	19	17	15	18	20
Vc	1	2	5	3	4	7	6	9	8	10	12	14	16	11	13	17	19	15	18	20
Vd e	1	2	5	3	4	7	6	8	9	10	12	14	16	11	13	17	19	15	20	18

From the results in Table 6.9, it can be concluded that all optimum sequences are either equal to or close to the expected sequence. However, it cannot be decided which optimum sequence is best (also see Section 7.2). Results obtained from these computers simulation experiments will be further discussed in Chapter 8. In the remainder of this section, the smaller database originally presented in Section 4.9 will be used. It is sufficiently large to show that (1) the scaling method presented earlier in this chapter provides unbiased estimates of intervals between successive events, and (2) the standard deviations computed by means of Equation (6.13) may be too small because the direct and indirect distance estimates that are being averaged are not stochastically independent.

Scaling of abbreviated database.

The original SEQ files for this data base were shown in Tables 4.13 to 4.15. All three sets were subjected to probabilistic ranking (presorting) followed by modified Hay method. The resulting optimum sequences and dendrograms (obtained after final reordering) are shown in Figures 6.8 to 6.10 (from Gradstein et al., 1985). As pointed out before, an identical

SEQUENCE POSITION	FOSSIL NUMBER	RANGE
1	1	0 – 2
2	2	1 – 3
3	3	2 – 4
4	4	3 – 5
5	5	4 – 6
6	6	5 – 7
7	7	6 – 8
8	8	7 – 9
9	9	8 – 10
10	10	9 – 11
11	11	10 – 12
12	12	11 – 13
13	13	12 – 14
14	14	13 – 15
15	15	14 – 16
16	16	15 – 17
17	17	16 – 18
18	18	17 – 19
19	19	18 – 20
20	20	19 – 21

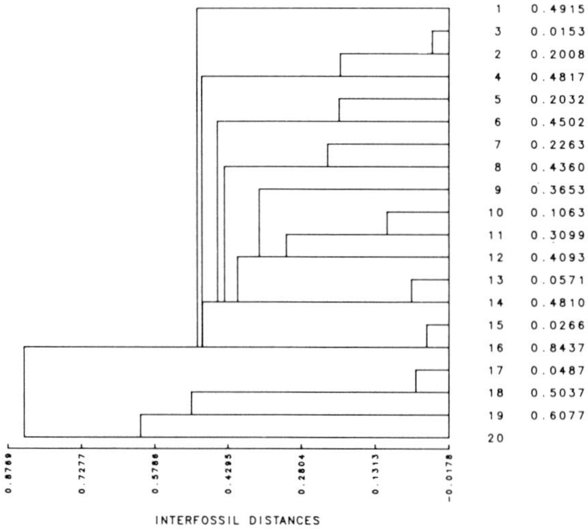

Fig. 6.8 Optimum sequence and dendrogram for sample drawn at random from population (theoretical model of equally spaced events labelled 1 to 20 along RASC scaled with E(D) = 0.5). Original SEQ file is shown in Table 4.13.

sequence of random normal numbers was used in all three experiments. Consequently, there is a general resemblance of input and output for the three SEQ files. The deviation from the theoretical model, consisting of equally spaced events ordered 1 to 20, increases when E(D), representing spacing between successive events in the theoretical model, decreases.

The modified Hay method did not change the probabilistic ranking result in the experiments with E(D)=0.5 or 0.3. Probabilistic ranking results were changed somewhat by subsequent application of the modified Hay method for E(D)=0.1. Eight 3-event cycles occurred and were broken by temporarily zeroing the (first) pair of elements in the cycle with the smallest difference as shown in Table 6.10. In total, seven (12, 13) pairs and one (11, 14) pair were ignored in order to obtain the optimum sequence of Figure 6.10. However, a detailed comparison of the probabilistic ranking result (Table 6.11A) with the final optimum sequence (Table 6.11B) shows that the probabilistic ranking is closer to the true sequence than the modified Hay ranking. The numbers in the bottom rows of Tables 6.11A and B are absolute values of differences between ranking results and true order numbers. Their sum is 22 for Table 6.11A and 33 for Table 6.11B suggesting that the probabilistic ranking result is slightly better in this type of application (also see Section 7.4). For further comparison, Table 6.11C shows differences for the scaled optimum sequence of Figure

SEQUENCE POSITION	FOSSIL NUMBER	RANGE
1	1	0 – 2
2	3	1 – 3
3	4	2 – 4
4	2	3 – 5
5	5	4 – 6
6	6	5 – 7
7	7	6 – 8
8	8	7 – 9
9	9	8 – 10
10	10	9 – 11
11	11	10 – 12
12	12	11 – 13
13	14	12 – 14
14	13	13 – 15
15	16	14 – 16
16	15	15 – 17
17	17	16 – 18
18	18	17 – 19
19	19	18 – 20
20	20	19 – 21

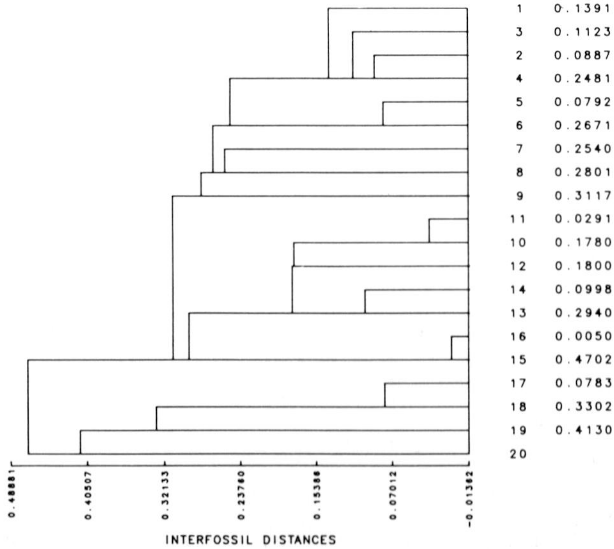

Fig. 6.9 Optimum sequence and dendrogram for computer simulation experiment of Fig. 6.8 repeated with E(D) = 0.3. Original SEQ file is shown in Table 4.14.

SEQUENCE POSITION	FOSSIL NUMBER	RANGE
1	4	0 – 2
2	5	1 – 3
3	3	2 – 4
4	1	3 – 5
5	2	4 – 6
6	6	5 – 7
7	8	6 – 8
8	7	7 – 9
9	11	8 – 10
10	9	9 – 11
11	14	10 – 12
12	16	11 – 13
13	10	12 – 14
14	12	13 – 15
15	15	14 – 16
16	13	15 – 17
17	17	16 – 18
18	18	17 – 19
19	19	18 – 20
20	20	19 – 21

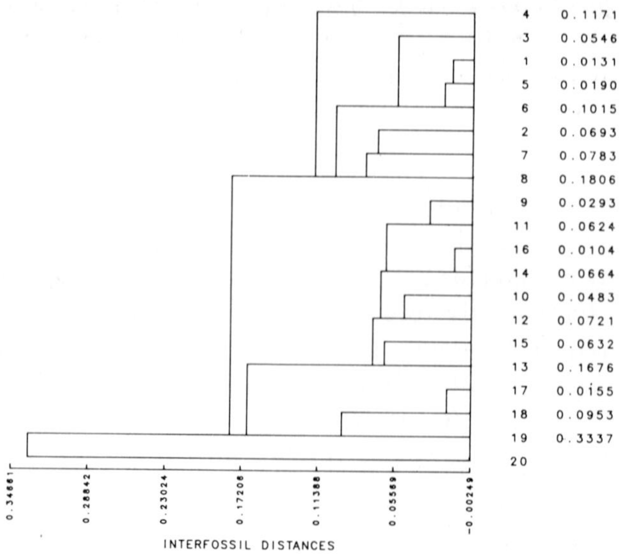

Fig. 6.10 Optimum sequence and dendrogram for computer simulation experiment of Fig. 6.8 repeated with E(D) = 0.1. Original SEQ file is shown in Table 4.15.

6.10. These add to 28 which is about midway between the preceding two sums.

TABLE 6.10

Eight 3-event cycles detected during application of modified Hay method (after probabilistic ranking) to SEQ file shown in Table 4.15.

A

4	3	5
x	**11**	14
14	x	10
11	15	x

B

2	1	6
x	11	14
14	x	**12**
11	**13**	x

C

7	6	2
x	11	**13**
14	x	11
12	14	x

D

10	9	11
x	9	**13**
16	x	9
12	16	x

E

14	9	11
x	11	**13**
14	x	9
12	16	x

F

16	14	12
x	10	16
15	x	**12**
9	**13**	x

G

10	16	14
x	10	**13**
15	x	10
12	15	x

H

16	13	12
x	**12**	16
13	x	12
9	13	x

TABLE 6.11

Comparison of true optimum sequence with optimum sequences resulting from (A) probabilistic ranking, (B) modified Hay method after probabilistic ranking, and (C) scaling after probabilistic ranking and modified Hay method. Absolute values of differences between estimated and true ranks can be regarded as penalty points. In the RASC step model (Chapter 7) these penalty points will be added to obtain a statistic from which Kendall's rank correlation can be computed.

A	1	2	3	4	5	6	7	8	9	10	11	12	13	14	15	16	17	18	19	20
	3	4	5	1	6	2	7	8	9	10	11	14	12	16	15	13	17	18	19	20
	2	2	2	3	1	4	0	0	0	0	0	2	1	2	0	3	0	0	0	0
B	1	2	3	4	5	6	7	8	9	10	11	12	13	14	15	16	17	18	19	20
	4	5	3	1	2	6	8	7	11	9	14	16	10	12	15	13	17	18	19	20
	3	3	0	3	3	0	1	1	2	1	3	4	3	2	0	3	0	0	0	0
C	1	2	3	4	5	6	7	8	9	10	11	12	13	14	15	16	17	18	19	20
	4	3	1	5	6	2	7	8	9	11	16	14	10	12	15	13	17	18	19	20
	3	1	2	1	1	4	0	0	0	1	5	2	3	2	0	3	0	0	0	0

Estimates of intervals between successive events in the optimum sequence using both unweighted ($\bar{x}_1$) and weighted ($\bar{x}_2$) scaling analysis are shown in Tables 6.12 to 6.14 for the three experiments, respectively. It can be seen that N^* tends to decrease towards the top and the base of the composite section if E(D) increases. This is because of reduced probability of inconsistencies between events near top and base, respectively, which leads to pairs of q_c-values not used for distance estimation (see before).

Standard deviations obtained by Equation (6.13) for the weighted scaling results are shown in the last columns of these three tables. Because the standard deviation of the normal random numbers used in the computer simulation experiments was equal to unity (instead of equal to $(2)^{-\frac{1}{2}}$ as in RASC), E(D) must be divided by $\sqrt{2}$ in order to obtain expected values for $\bar{x}_2$. For example, the expected value for all numbers in the fourth ($\bar{x}_1$) and fifth ($\bar{x}_2$) column of Table 6.12 is $0.5/\sqrt{2}=0.354$. The corresponding expected values for Tables 6.13 and 6.14 are 0.212 and 0.071, respectively. However, in these two tables the order of the twenty events is not 1 to 20 as in Table 6.12. For example, in Table 6.13 event 1 is follwed by event 3 instead of event 2. Consequently, the expected value for this interval is 2×0.212 instead of 0.212. Mean intervals computed from 19 separate distances between successive events are shown at the bottom of each table. In all instances, the mean is close to the expected value of the interval between events with consecutive numbers, although the scatter of the individual $\bar{x}_1$ and $\bar{x}_2$ values is considerable. When the

TABLE 6.12

Unweighted ($\bar{x}_1$) and weighted ($\bar{x}_2$) estimates of intervals in scaled optimum sequences for computer simulation experiment with E(D)=0.5. Standard deviations $s(\bar{x}_2)$ of weighted estimates are shown in last column.

	Events	N^*	$\bar{x}_1$	$\bar{x}_2$	$s(\bar{x}_2)$
1	1-2	9	0.455	0.427	0.083
2	2-3	9	-0.038	-0.015	0.093
3	3-4	7	0.192	0.242	0.066
4	4-5	10	0.541	0.482	0.079
5	5-6	10	0.181	0.203	0.084
6	6-7	14	0.466	0.450	0.063
7	7-8	13	0.169	0.226	0.071
8	8-9	14	0.391	0.436	0.081
9	9-10	14	0.288	0.346	0.082
10	10-11	14	0.096	0.101	0.034
11	11-12	13	0.297	0.310	0.035
12	12-13	11	0.907	0.409	0.051
13	13-14	13	0.049	0.057	0.047
14	14-15	13	0.420	0.481	0.074
15	15-16	13	0.030	0.027	0.039
16	16-17	11	0.766	0.844	0.115
17	17-18	8	-0.083	0.099	0.161
18	18-19	8	0.470	0.504	0.065
19	19-20	6	0.536	0.608	0.113
		Mean Interval:	0.296	0.326	

TABLE 6.13

Same as Table 6.12 for E(D)=0.3.

	Events	N^*	$\bar{x}_1$	$\bar{x}_2$	$s(\bar{x}_2)$
1	1-3	12	0.110	0.139	0.072
2	3-4	15	0.207	0.209	0.062
3	4-2	15	-0.122	-0.089	0.078
4	2-5	17	0.335	0.346	0.056
5	5-6	17	0.012	0.079	0.075
6	6-7	16	0.274	0.267	0.046
7	7-8	17	0.281	0.254	0.064
8	8-9	17	0.259	0.280	0.068
9	9-10	19	0.329	0.338	0.062
10	10-11	19	-0.042	-0.029	0.068
11	11-12	19	0.194	0.222	0.077
12	12-14	18	0.183	0.180	0.059
13	14-13	18	0.061	0.100	0.065
14	13-16	17	0.308	0.283	0.068
15	16-15	17	-0.053	0.005	0.063
16	15-17	17	0.444	0.470	0.059
17	17-18	15	0.017	0.078	0.066
18	18-19	15	0.320	0.330	0.073
19	19-20	10	0.309	0.413	0.088
		Mean Interval:	0.180	0.204	

reordering of events in Tables 6.13 and 6.14 is considered, the mean expected values for these two tables are increased to 0.302 (instead of 0.212) and 0.112 (instead of 0.071), respectively.

Table 6.15 contains summary statistics for the three data sets. Separate standard deviations (s and $\hat{\sigma}$) were computed from the samples of 19 values with respect to the sample mean (unbiased estimate, 18 degrees of freedom) and the population mean (unbiased estimate, 19 degrees of freedom). For E(D)=0.5, the pairs of estimates (s and $\hat{\sigma}$) are nearly equal to one another. For E(D)=0.3 and 0.1, $\hat{\sigma}(\bar{x}_1)$ and $\hat{\sigma}(\bar{x}_2)$ are larger than $s(\bar{x}_1)$ and $s(\bar{x}_2)$ because the order of the events in the corresponding optimum sequences differs from the expected order (1, 2, ..., 20). Ordinary product-moment correlation coefficients $r(\bar{x}_1, \bar{x}_2)$ are shown at the bottom of Table 6.15. They indicate that the unweighted and weighted analysis results are strongly correlated.

TABLE 6.14

Same as Table 6.12 for $E(D) = 0.1$.

	Events	N^*	$\bar{x}_1$	$\bar{x}_2$	$s(\bar{x}_2)$
1	4-5	19	0.207	0.182	0.063
2	5-3	19	-0.103	-0.064	0.066
3	3-1	19	0.035	0.055	0.060
4	1-2	19	0.146	0.133	0.044
5	2-6	19	-0.126	-0.102	0.055
6	6-8	19	0.266	0.250	0.053
7	8-7	19	-0.081	-0.078	0.040
8	7-11	19	0.361	0.286	0.067
9	11-9	19	-0.061	-0.029	0.064
10	9-14	19	0.100	0.103	0.063
11	14-16	19	-0.019	-0.011	0.059
12	16-10	19	0.064	0.077	0.062
13	10-12	19	0.038	0.048	0.048
14	12-15	19	0.058	0.072	0.039
15	15-13	19	0.055	0.063	0.053
16	13-17	19	0.196	0.168	0.071
17	17-18	19	-0.008	0.016	0.048
18	18-19	19	0.117	0.095	0.055
19	19-20	19	0.285	0.334	0.069
		Mean Interval:	0.078	0.084	

The standard deviations $s(\bar{x}_2)$ and $\hat{\sigma}(\bar{x}_2)$ as computed from the $\bar{x}_2$ values are considerably greater than individual $s(\bar{x}_2)$ values listed in the last column of each table. The explanation of this discrepancy is that the N^* separate x_2-values, on which $\bar{x}_2$ and $s(\bar{x}_2)$ are based, are not stochastically independent. (Two random variables are stochastically independent if the expected value of the cross-product for deviations from their means is equal to zero. The expected value of the product-moment correlation coefficient of two stochastically independent variables also is zero). The computer simulation experiments show that the effect of mutual interdependence obviously can be strong for smaller as well as for larger expected intervals between successive events. It does not lead to noticeable bias in the $\bar{x}$-values but may result in standard deviations which are 2 or 3 times too small.

TABLE 6.15

Comparison of estimates of Tables 6.12 to 6.14 to population parameters. See text for explanations of expressions in first column.

	$E(D)=0.5$	$E(D)=0.3$	$E(D)=0.1$
$E(X)$	0.354	0.212	0.071
Mean ($\bar{x}_1$)	0.296	0.180	0.078
Mean ($\bar{x}_2$)	0.326	0.204	0.084
$s(\bar{x}_1)$	0.227	0.160	0.132
$\hat{\sigma}(\bar{x}_1)$	0.228	0.194	0.193
$s(\bar{x}_2)$	0.225	0.152	0.121
$\hat{\sigma}(\bar{x}_2)$	0.221	0.203	0.194
$r(\bar{x}_1, \bar{x}_2)$	0.981	0.980	0.987

Mutual interdependence of x_i

The following considerations are helpful for understanding the mutual interdependence of separate estimates x_i of the interval between two events A and B. Let C and D be two other events which can be used for indirect estimation of the distance between A and B.

Three events A, B and C are related by the six probabilities P_{ABC}, P_{ACB}, P_{CAB}, P_{BAC}, P_{BCA} and P_{CBA}. It follows immediately that

$$P_{AB} = P_{ABC} + P_{ACB} + P_{CAB}$$

$$P_{AC} = P_{ABC} + P_{ACB} + P_{BAC}$$

$$P_{BC} = P_{ABC} + P_{BAC} + P_{BCA} \tag{6.14}$$

Similar expressions can be written out for P_{BA}, P_{CA} and P_{BC} but it is simpler to regard these probabilities as complementary to P_{AB}, P_{AC} and P_{BC}, or

$$P_{BA} = 1-P_{AB};\ P_{CA} = 1-P_{AC};\ P_{CB} = 1-P_{BC} \tag{6.15}$$

In Section 6.1 it was pointed out that the Z-value of a probability P can be approximated by a Z^*-value which simply is a linear function of P, or $Z^*=2.93(P\text{-}0.5)$. This relation can be used to prove that the indirect estimate $Z_{AB\cdot C}=Z_{AC}\text{-}Z_{BC}$ is approximately equal to

$$Z^*_{AB.C} = Z^*_{ACB} - Z^*_{BCA} \tag{6.16}$$

where Z^*_{ACB} and Z^*_{BCA} are linear functions of P_{ACB} and P_{BCA}, respectively. Consequently, for the two events C and D, it follows that

$$Z_{AB.C} \simeq 2.93\,(P_{ACB} - P_{BCA})$$

$$Z_{AB.D} \simeq 2.93\,(P_{ADB} - P_{BDA}) \tag{6.17}$$

The latter two values cannot be regarded as stochastically independent estimates of the interval between A and B. Even if C and D are independent of one another, as well as from A and B, the estimates of Equation (6.17) both depend on the positions occupied by A and B in the sections.

In the computer simulation experiments, all events are present in all sections. The preceding argument explains why Equation (6.13) provides poor results in that situation. In most practical applications of ranking and scaling, the events do not all occur in all sections. As pointed out before (cf. Fig. 1.1), frequency distributions of events usually are positively skew with most events occurring in relatively few sections. If the two events C and D in the preceding argument occur in different sections, the two estimates of Equation (6.17) would be stochastically independent.

Equation (6.13) only would provide an approximately unbiased estimate in the unlikely situation that all events used occur in different sections. In practice, it can be expected that the estimated standard deviations will be too small, although bias will not be as severe as in the computer simulation experiments. This type of bias also must be considered in results of the normality test (see next section). In that situation, the computer simulation experiments initially helped to point out the problem. This led to a revised normality test in which the bias could be estimated and eliminated in practical applications (see Chapter 8).

6.6 Normality test

When an optimum sequence has been obtained, individual sequences for sections can be compared to it. In practice, this is important as a tool to spot anomalous events which occur either much higher or much lower than expected in a section. The RASC computer program provides three optional outputs for ranking or scaling results immediately preceding the normality test which applies to scaled optimum sequences only. These are:

(1) An occurrence table can be constructed with the final ranking plotted in the vertical direction. Sections in this table are represented by columns. If an event occurs in a section its presence is indicated by an X.

(2) Each section can be compared to the optimum sequence by using a system of scoring penalty points when an event is out of place. This procedure is called the Step Model (cf. Section 7.3). The relative order of every pair of events is checked against their order in the optimum sequence. If the order is different, one penalty point is scored. Coeval events each receive half a point. Obviously, an event with many penalty points is likely to be either too high or too low in a section. A drawback of the step model is that events which belong to clusters of events with many internal inconsistencies are likely to accumulate high total scores even if they occur in normal positions. Thus, it may not be easy to distinguish between anomalous events which are out of place and events which are part of a cluster.

(3) Scattergrams: the order of events in the sections is plotted against the optimum sequence.

The preceding three outputs can be obtained before or after scaling. The normality test can only be performed after scaling. Two outputs for the normality test are shown in Tables 6.16 and 6.17 for the Hay example. After scaling, each event occurs at a fixed distance from the origin which was set at the position of the first event in the scaled optimum sequence. Hence, the score of each event, with the exception of the first and last events, can be compared with the scores of its two neighbors in every section. The amount by which it is out of place can be evaluated statistically. For this purpose, the second-order difference will be used

TABLE 6.16

RASC normality test output for 9 sections of Hay example. E=event code number; L=event level number in stratigraphically downward direction; X=cumulative RASC distance; and U=second order difference. Single asterisk indicates that event is out of place with probability greater than 95%. Two asterisks indicate that event is out of place with probability greater than 99%.

Section A – Vaca Valley		*E*	*L*	*X*	*U*
HI	*Discoaster tribrachiatus*	9	1	0.000	
LO	*Discoaster cruciformis*	8	2	0.493	0.700
LO	*Discoaster minimus*	7	3	1.686	-1.616
LO	*Rhabdosphaera scabrosa*	6	4	1.263	1.683
LO	*Coccolithus gammation*	5	4	1.843	-0.894
LO	*Coccolithus solitus*	4	4	1.529	0.940
LO	*Discoaster germanicus*	3	4	2.155	-0.632
LO	*Coccolithus cribellum*	2	4	2.149	-0.105
LO	*Discoaster distinctus*	1	4	2.038	
Section B – Pacheto Syncline		*E*	*L*	*X*	*U*
HI	*Discoaster tribrachiatus*	9	1	0.000	
LO	*Discoaster distinctus*	10	2	0.317	2.778*
LO	*Rhabdosphaera scabrosa*	6	2	1.263	-0.366
LO	*Caccolithus gammation*	5	2	1.843	-0.894
LO	*Coccolithus solitus*	4	2	1.529	0.470
LO	*Discoaster minimus*	7	2	1.686	0.313
LO	*Discoaster germanicus*	3	2	2.155	-0.475
LO	*Coccolithus cribellum*	2	2	2.149	
Section C – Tres Pinos		*E*	*L*	*X*	*U*
HI	*Discoaster tribrachiatus*	9	1	0.000	
LO	*Discoaster distinctus*	1	2	2.038	-2.234
LO	*Coccolithus gammation*	5	3	1.843	0.502
LO	*Coccolithus cribellum*	2	4	2.149	
Section D – Upper Reliz Creek		*E*	*L*	*X*	*U*
LO	*Discolithus distinctus*	10	1	0.317	
HI	*Discoaster tribrachiatus*	9	2	0.000	0.810
LO	*Discoaster cruciformis*	8	3	0.493	0.857
LO	*Coccolithus gammation*	5	4	1.843	-1.507
LO	*Discoaster minimus*	7	5	1.686	0.509
LO	*Discoaster distinctus*	1	6	2.038	-0.241
LO	*Coccolithus cribellum*	2	7	2.149	

which is approximately equal to the sum of the scores of the neighbors minus twice the score of the event itself.

If the second-order difference value of an event in a section receives two asterisks in the output, it is out of place with a probability of 99 percent. One asterisk signifies an event that occurs too high or too low with a probability of 95 percent. It is to be expected that, on the average, one and five percent of all events tested will be assigned two asterisks and

TABLE 6.16 (continued)

Section E – New Idria		*E*	*L*	*X*	*U*
HI	*Discoaster tribrachiatus*	9	1	0.000	
LO	*Rhabdosphaera scabrosa*	6	2	1.263	-0.997
LO	*Coccolithus solitus*	4	3	1.529	-1.303
LO	*Discoaster cruciformis*	8	4	0.493	2.229
LO	*Discoaster minimus*	7	5	1.686	-0.724
LO	*Discoaster germanicus*	3	6	2.155	-0.586
LO	*Discoaster distinctus*	1	7	2.038	-0.078
LO	*Coccolithus gammation*	5	8	1.843	0.809
LO	*Coccolithus cribellum*	2	8	2.149	
Section F – Media Aqua Creek		*E*	*L*	*X*	*U*
LO	*Discolithus distinctus*	10	1	0.317	
HI	*Discoaster tribrachiatus*	9	2	0.000	0.810
LO	*Discoaster cruciformis*	8	3	0.493	1.044
LO	*Discoaster minimus*	7	3	1.686	-1.074
LO	*Coccolithus cribellum*	2	4	2.149	-0.770
LO	*Coccolithus gammation*	5	5	1.843	0.337
LO	*Coccolithus solitus*	4	5	1.529	0.595
LO	*Discoaster germanicus*	3	6	2.155	-0.399
LO	*Discoaster distinctus*	1	6	2.038	
Section G – Upper Canada		*E*	*L*	*X*	*U*
HI	*Discoaster tribrachiatus*	9	1	0.000	
LO	*Discoaster cruciformis*	8	2	0.493	-0.248
LO	*Discolithus distinctus*	10	2	0.317	1.280
LO	*Coccolithus gammation*	5	3	1.843	-0.798
LO	*Coccolithus cribellum*	2	3	2.149	-0.418
LO	*Discoaster distinctus*	1	3	2.038	-0.819
LO	*Coccolithus solitus*	4	4	1.529	1.556
LO	*Discoaster germanicus*	3	4	2.155	-1.517
LO	*Discoaster minimus*	7	5	1.686	
Section H – Las Cruces		*E*	*L*	*X*	*U*
LO	*Coccolithus solitus*	4	1	1.529	
HI	*Discoaster tribrachiatus*	9	2	0.000	3.372**
LO	*Coccolithus gammation*	5	3	1.843	-1.595
LO	*Discoaster distinctus*	1	3	2.038	-1.917
LO	*Discolithus distinctus*	10	3	0.317	3.038*
LO	*Discoaster minimus*	7	4	1.686	
Section I – Lodo Gulch		*E*	*L*	*X*	*U*
LO	*Discolithus distinctus*	10	1	0.317	
HI	*Discoaster tribrachiatus*	9	2	0.000	1.580
LO	*Rhabdosphaera scabrosa*	6	3	1.263	-0.997
LO	*Coccolithus solitus*	4	4	1.529	0.047
LO	*Coccolithus gammation*	5	5	1.843	-0.118
LO	*Discoaster distinctus*	1	6	2.038	0.227
LO	*Discoaster germanicus*	3	6	2.155	-0.428
LO	*Coccolithus cribellum*	2	7	2.149	

TABLE 6.17

Normality test applied to all (=51) second-order differences of Table 6.16. The expected frequencies (E) of the ten classes are all equal to 5.1. The chi-squared test is used for comparing observed (O) and expected (E) frequencies with one another. Abnormally large values in the last column would be indicated by asterisks (one and two asterisks for lack of fit with probability greater than 95% and 99%, respectively).

Class	O	E	O-E	$(O\text{-}E)^2/E$
1	3	5.1	-2.1	0.39
2	5	5.1	-0.1	0.00
3	8	5.1	2.9	0.74
4	7	5.1	1.9	0.32
5	5	5.1	-0.1	0.00
6	3	5.1	-2.1	0.39
7	5	5.1	-0.1	0.00
8	7	5.1	1.9	0.32
9	3	5.1	-2.1	0.39
10	5	5.1	-0.1	0.00
			Chi-squared = 2.56	

one asterisk, respectively, when there are no anomalies as in the computer simulation experiments described in the preceding section. The 51 second-order differences listed in Table 6.16 for the Hay example show one double asterisk (*HI Discoaster tribrachiatus*) in Section H and two single asterisks. This is approximately as expected for a normal dataset of this size and no clear anomalies are indicated.

The model which assumes that each event has a normal probability curve with the same variance, can be used to estimate a set of expected frequencies for all observed event locations. The second-order differences then are also normally distributed. For convenience, this theoretical distribution is divided into 10 classes with equal expected frequencies. In total, 10 expected frequencies for 51 values are compared to the corresponding observed frequencies in the bottom part of Table 6.17 for the Hay example.

In the last column of Table 6.17, the squared difference, divided by expected value corrected for autocorrelation (see Chapter 8) is shown. Each of these values is approximately distributed as chi-squared with a single degree of freedom, and the total for 10 classes has seven degrees of freedom. Probable chi-squared departures from normality are marked by asterisks. A more detailed explanation of the procedures followed in the normality test will be given in Chapter 8.

It should be kept in mind that the normality test is based on the assumption that all second-order differences have the same variance. This condition is only approximately satisfied if individual events have different variances. In the latter situation, the averaging process helps to stabilize the variance. It also is noted that the number of second-order differences considered is two less than the number of events per section. It is possible to consider slightly different types of second-order differences by considering coeval events.

6.7 Marker horizon option of the RASC method

This section deals with an option of the RASC computer program in which the location of chronostratigraphic (marker) horizons such as seismic markers or bentonite beds resulting from volcanic ash fall can be considered. If events of this type can be correlated with certainty between sections, they should be assigned zero variance along the relative time scale used for estimating RASC distances, when they are considered in conjunction with other stratigraphic events. In practice, this means that marker horizons will be given more weight than other events in the calculations. Marker horizons are entered like other stratigraphic events in the SEQ file for the RASC program. However, they are identified as special events in the input specifications for RASC.

The underlying statistical theory for marker horizons is explained in Figure 6.11 which can be compared to Figure 6.5 earlier in this chapter. The event A has been replaced by a marker horizon with zero variance in Figure 6.11. This means that D_{AB} is normally distributed with mean Δ_{AB} and variance σ^2. The Z-value for relative cross-over frequency P_{AB} has to be divided by $\sqrt{2}$ before it can be used as a direct estimate of the distance between events A and B which is compatible with the direct estimate of Figure 6.5 and the indirect estimates of Figure 6.6. Its weight also has to be adjusted accordingly.

When indirect estimates such as $\Delta_{AB \cdot C}$ involving a marker horizon are obtained, there are two possibilities: (1) the event used for the indirect comparison (C) is a marker horizon; and (2) either A or B is a marker horizon. In the first situation, D_{AC} and D_{BC} are normal with variance σ^2

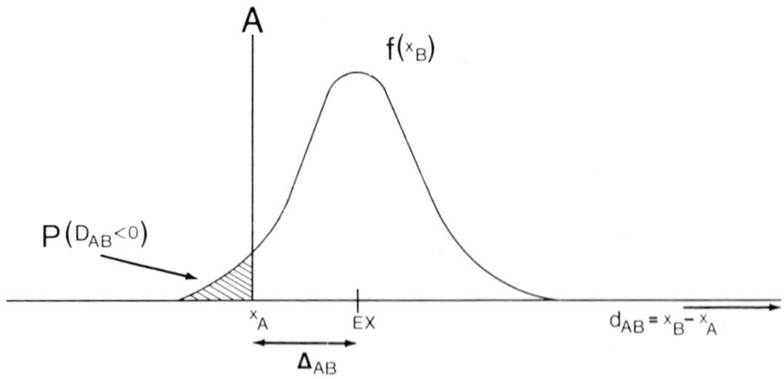

Fig. 6.11 Direct estimation of distance Δ_{AB} between events A and B from relative cross-over frequency when A is a marker horizon with zero variance. The variance of D_{AB} is equal to the variance of event B.

(as D_{AB} in Figure 6.11). Consequently, their difference can be assigned variance equal to $2\sigma^2$. The second case results in difference $D_{AB \cdot C}$ with variance $3\sigma^2$.

The preceding theory of marker horizons is further illustrated by using a modified version of the artificial example of Table 6.6. Table 6.18 shows results comparable to those of Table 6.6 under the assumption that B is a marker horizon. As before, a table of sequences for A, B and C was derived (cf. Table 4.12). However, this time all random normal numbers for B were replaced by 2.000. For example, the first three event "distances" of Table 4.11 previously were 1.422, 0.130, 2.732 and, therefore, gave BAC. This time they result in 1.422, 2.000, 2.732 and, therefore, ABC. (For one sequence, the locations of A and B became both equal to 2.000 and 0.5 was added to the tallies of both AB and BA.)

TABLE 6.18

RASC method of scaling applied to data of artificial example. Event B is marker horizon. These results should be compared to those shown in Table 6.6 where B, like A and C, had unit variance.

	n	f	z	D(direct)	D(indirect)	D_0(Ave)	D(Ave)	E(D)
AB	129.5	0.8633	1.095	1.095(1)	1.283(3)	1.189	1.142	1.000
AC	130	0.8667	1.111	1.571(2)	1.383(2)	1.477	1.477	1.500
BC	92	0.6133	0.288	0.288(1)	0.476(3)	0.382	0.335	0.500
SSD				0.059	0.094	0.050	0.048	

Comparison of Table 6.18 to Table 6.6 shows the following differences. The cross-over frequencies for AB and BC have increased significantly. This reflects the fact that the variance of B was set equal to zero. The direct estimates of distance for AB and BC were not multiplied by $\sqrt{2}$ as in Table 6.6. However, D (indirect) was estimated from D (direct) by simple addition or subtraction as before. The relative variances of D (direct) and D (indirect) are shown in brackets in Table 6.18. As before D_0 (Ave) is the arithmetic average and D(Ave) represents the weighted average. The weights are inversely proportional to the variances. For example, D(Ave) of AB is equal to $(1.095 + 1.283/3)/(1 + 1/3) = 1.142$. The SSD values in Table 6.18 are less than those in Table 6.6.

6.8 Unique event option of RASC program

The purpose of this option is that a rare (unique) event can be entered into a regional standard by comparing its position in a single or few sections to those of the more abundant taxa used for constructing this standard. The unique event option is useful when an index fossil is observed in one or a few sections only. Because of its rarity, the index fossil cannot be used for construction of the optimum sequence. However, it can be fitted in with the other events afterwards and this may help to define assemblage (average interval) zones from the dendrogram. Like marker horizons, the unique events have to be identified in the input specifications of the RASC program.

The unique event option can be used to solve the following hypothetical problem. A feature is observed in two sections and it is of interest to determine whether this feature represents a single stratigraphic event. The feature, then, can be entered by codes as a different unique event for each section. The resulting positions of these two unique events in the standard can be compared to each other and this may be helpful for deciding whether the same event is present in both sections.

Figure 6.12 shows the technique used for treating unique events. The event A with distance x_a from the origin is observed immediately above the unique event in a section. Event S (with x_s) is coeval to the unique event and event B (with x_b) occurs below it. Also shown on the left of Figure 6.12 are the locations x'_b and x''_b for two other events B' and B'' observed below B. As a first approximation, the unique event is assigned

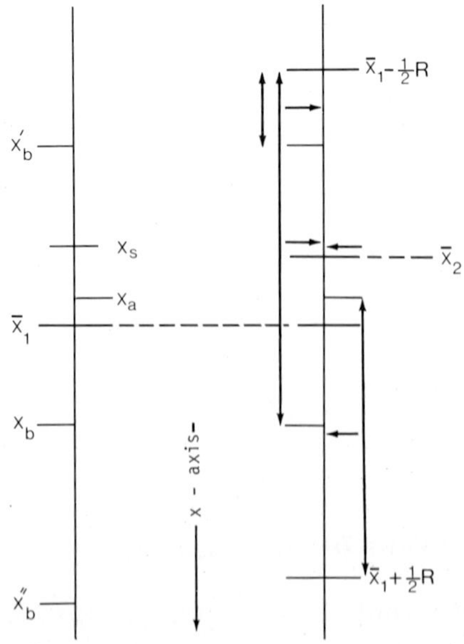

Fig. 6.12 Simple example to illustrate application of unique event option. A unique event was observed in a single section simultaneous to the event S, stratigraphically below the event A and above the events B, B' and B". The cumulative RASC distances of the latter five events are shown along the scale on the left. The positions of S, A and B were averaged to obtain first approximation $\bar{x}_1$ for the unique event. The second approximation was based on the RASC distances of all events within the range R.

the position $\bar{x}_1$ representing the arithmetic average of x_a, x_s, and x_b. In practical applications, S may be missing. (The special situation that A or B is missing would occur only if the unique event were to occupy the first or last position in a section.)

More than a single event may be observed in the positions immediately above, simultaneous to, or below the unique event. Then, x_a, x_s, or x_b will be computed as averages for these events which, in turn, will be averaged to estimate $\bar{x}_1$.

A range of $\bar{x}_1 \pm (1/2)R$ can be defined for all events encountered within the vicinity of $\bar{x}_1$ with a probability greater than 5 percent. Because $\sigma^2 = 0.5$, $(1/2)R = 1.96\sigma = 1.386$. The events in the scaled optimum sequence with locations in the interval $\bar{x}_1 \pm (1/2)R$ can be identified. For the simplified example of Figure 6.12, these are the events A, S, B, and B' (but not B"). For each event above the unique event in the section (A), a value is computed which is the average of its location (x_a) and the value $\bar{x}_1 + (1/2)R$. Similarly, for each event below the unique event (B or B'), a value is computed which is the average of its location (x_b or x'_b) and the value $\bar{x}_1$ - (1/2)R. These average values which are shown as arrows in the diagram on the right of Figure 6.12 are averaged together with the values (x_s) for events observed to be simultaneous with the unique event. This gives the second approximation $\bar{x}_2$. If the unique event occurs in more than one section, the preceding calculation is performed for each section and the resulting values of $\bar{x}_2$ are averaged.

The choice of a range R in the method of Figure 6.12 is somewhat arbitrary. However, the location of the second approximation $\bar{x}_2$ is independent of R when the number of events within the interval $\bar{x}_1 \pm (1/2)R$ remains constant. Although the unique event option generally is used for the construction of biozonations from the scaled optimum sequence, it also can be used in association with an optimum sequence obtained by ranking. In that situation, the sequence numbers for events in the optimum sequence are used as x-values and R is set equal to a larger value (e.g. $R = 3.0$).

Examples of using the unique event option to include index fossils in biozonations are given elsewhere (see e.g. Fig. 6.2). The following example illustrates the concept of re-including an event that initially was excluded. Event 6 in the Hay example occurs in 4 sections only. By setting the threshold parameter k_c equal to 5, it can be excluded from the computations required for ranking and scaling. Table 6.19 shows optimum sequences obtained from 8 events with later re-insertion of event 6. In both sequences, event 6 is positioned between events 8 and 4 (but closer to 4 than to 8) as in the results previously obtained for the Hay example.

6.9 Binomial and trinomial models for scaling

As already pointed out in Section 3.4, the RASC model for scaling can be evaluated in terms of observed probabilities by which the events succeed one another in the wells or were observed to be coeval. The relation between binomial and trinomial models will be considered in the final sections of this chapter.

Suppose that two stratigraphic events (either entries or exits or one of each) for different taxa are expressed as E_i and E_j. If E_i and E_j occurred relatively close in geological time, it may be that E_i is observed to occur

TABLE 6.19

Test of unique event option applied to Hay example. Event 6 which occurs in 4 sections only was excluded by setting $k_c=5$. Later it was re-inserted in the optimum sequence derived by ranking as well as in the scaled optimum sequence.

Event	Ranking	Scaling
9	0.00	0.000
10	1.00	0.317
8	2.00	0.493
6*	2.87	1.017
4	3.00	1.198
7	4.00	1.415
5	5.00	1.534
1	6.00	1.724
2	7.00	1.868
3	8.00	1.875

above E_j in some outcrop sections or wells, and E_j above E_i in others. It also may be that E_i and E_j are locally coeval. In order to avoid confusion, events for pairwise comparison will be denoted by using the letter A instead of E. Binomial models provide estimates of the probability of A_1 or that E_i occurs above E_j in a section. If this probability is written as P_1, and the probability of A_2 (E_j occurs above E_i) as P_2, then $P_2 = 1\text{-}P_1$. As originally pointed out by Edwards and Beaver (1978), only a trinomial model can result in an estimate of the probability of occurrence of A_3 or that E_i and E_j are observed to be coeval in a section in addition to the probabilities that A_1 and A_2 occur. The development of trinomial models is of importance because biostratigraphic events are frequently observed to be coeval and this possibility should be considered in the statistical models.

Statistical theory of the binomial and trinomial distributions can be found in standard reference volumes such as Johnson and Kotz (1969, Chapters 3 and 11). Consider a series of independent trials, in each of which just one of 3 mutually independent events A_1, A_2 and A_3 must be observed, and in which the probability of occurrence of event A_k (k = 1, 2, or 3) is equal to P_k for each trial with the sum of the three probabilities equal to one. The trinomial distribution then is the joint distribution of the random variables N_1, N_2 and N_3 representing the numbers of occurrences of the events A_1, A_2 and A_3, respectively, in N trials. It is defined by

$$P(N_1, N_2, N_3) = N!. \prod_{k=1}^{3} (P_k^{N_k} / N_k!) \tag{6.18}$$

The distribution of N_1, N_2 or N_3 considered separately is binomial with $P(N_k)$ satisfying Equation (3.4). Also, if one of the events, say A_3, is ignored, then the other two satisfy Equation (3.4) provided that N is replaced by N-N_3 and P_k by $P_k/(1-N_3/N)$. The maximum likelihood estimator of P_k ($k = 1, 2,$ or 3) is N_k/N.

The preceding theory can be illustrated by means of the following simple example. For a set of 18 wells on the Canadian Atlantic Margin, *Uvigerina canariensis* (Fossil no. 10) and *Asterigerina gurichi* (Fossil no. 17) were both observed in $N=5$ wells. The exit of no. 10 occurred twice ($N_1=2$) above and once ($N_2=1$) below that of no. 17, respectively. In the remaining two wells ($N_3=2$), the events were observed to be coeval. Consequently, it can be estimated that no. 10 occurs above, below or coeval to no. 17 with probabilities of 40, 20, and 40 percent, respectively. If coeval events are ignored, then the estimates of the probabilities that no. 10 occurs above and below no. 17 are 67 and 33 percent, respectively.

Of course, the uncertainties of the preceding estimated percentages are considerable. For the observed relative frequency 2/5, the 95 percent confidence limits for P_1 are 5.3 and 85.3 percent, respectively. For 2/3, the 95 percent confidence limits are 9.4 and 99.2 percent. These confidence limits were looked up in Hald's (1952) statistical tables. They also can be computed by using various approximation formulas (see Johnson and Kotz, 1969, Chapter 3; Southam et al., 1975).

The preceding practical example, clearly illustrates that simple binomial or trinomial theory results in imprecise estimates of the probabilities if the sample size N is small. For large samples, however, the theory is satisfactory.

Binomial model based on multiple pairwise comparisons

Gradstein and Agterberg (1982) originally developed the scaling technique mainly to cope with the problem that nearly all of their sample sizes (N) were small. Each stratigraphic event E_i was assumed to occupy a position along a linear scale L. The positions assumed by E_i in individual sections fluctuate at random about an average value along L. The "distance" D_{ij} between average positions of two events E_i and E_j then can be converted into the probability that E_i occurs above E_j. Alternately, the probability can be converted into the distance. The advantage of this method is that the distance need not only be estimated from the relative positions of E_i and E_j in the sections but (N^*) double pairs (E_i, E_k) and (E_j, E_k) with $k \neq i,j$ can also be used. Even if all sample sizes are small, N^* may be large, and precise estimates of the average positions of the events along the linear scale L can be obtained.

Ties were treated as follows. If $F\,(=N_1)$ represents the observed frequency of A_1 (event E_i occurs above E_j), and $T\,(=N_3)$ the number of ties for these two events, then the score $S = N_1 + N_3/2$ was used for estimating the probability of occurrence of A_1 with $P_1 = S/N$. This implies that A_2 is observed to occur $N_2 + N_3/2 = N\text{-}S$ times, and its probability of occurrence is $P_2 = 1\text{-}S/N$. In this approach, an observed tie receives the same weight as either one of the direct observations of the events A_1 or A_2. However, it is recognized that no preference can be given to A_1 or A_2 if A_3 is observed. Although the average positions of E_i and E_j along the linear scale could coincide (with $P_1 = P_2 = 0.5$), and observed ties will tend to decrease the distance between the average positions, the scaling model does not allow for explicit estimation that two events are observed to be coeval in a section. Instead of this, a tie is interpreted as a coincidence due to sampling method (e.g. use of well cuttings) or due to occurrence of sudden events at the time of deposition which favored fossilization of several taxa in "patches" (cf. Fig. 2.7).

A distinction should be made between the frequency curve for relative abundance of occurrence of a fossil taxon through time at a given place and the probability curves for its entry and exit (cf. Chapter 2). Methods by which frequencies are averaged may give range zones which are shorter than those resulting from "conservative" methods in which more weight is assigned to places where events occur relatively high or low in the stratigraphic column in relation to other events. For example, if exit E_1 is observed above exits E_2 and E_3 in one section but below E_2 and E_3 in many other sections, then a conservative method would place the upper limit of the range for taxon 1 above those of taxa 2 and 3. On the other hand, this point would fall below those of taxa 2 and 3 when the average location of E_1 is determined. From a statistical point of view, the estimation of an average exit is more satisfactory because the position of the endpoint is more susceptible to random fluctuations. Moreover, the average value is more robust if events are locally out of place due to anomalous circumstances such as sediment mixing or misidentification. In the RASC computer program, individual sections can be compared to the "standard" which consists of a set of average distance values along the linear scale L (normality test; also see Gradstein, 1984).

6.10 Application of Glenn and David's trinomial model

As outlined in Section 3.4, Glenn and David's (1960) model is an extension of the Thurstone-Mosteller model which uses Gaussian curves for the distribution of positions of events along a linear scale L as is done in

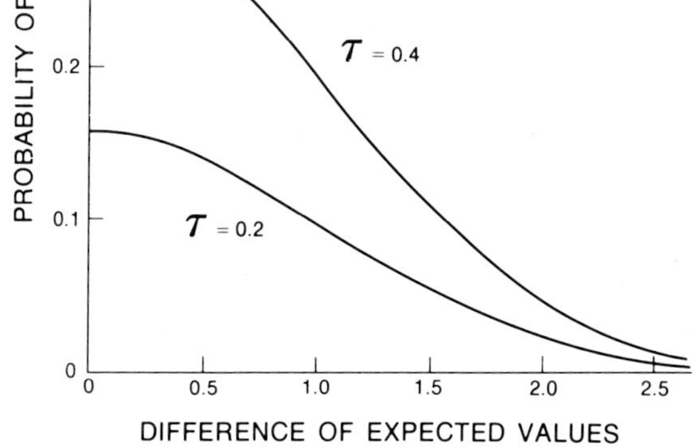

Fig. 6.13 Probability of a tie as a function of "distance" (δ) between mean positions of events along linear distance scale (after Glenn and David, 1960).

the RASC model. As a first step for calculating average distances between events along this scale, the observed "cross-over" frequencies (P) are converted to Z-values according to the transformation $\Phi^{-1}(P) = Z$. This is the inverse of $P = \Phi(Z)$ where Φ denotes the fractile (cumulative frequency) of a normal distribution in standard form. The model without ties can be extended to the model with ties as follows.

Suppose that the random variable D represents "distance" along the linear scale L between two events in a single section. D is assumed to have unit variance and its average value is δ. Glenn and David (1960) have introduced a threshold parameter τ. A tie of the two events is assumed to occur when D is less than τ and greater than $-\tau$. The probability of a tie (P_3) then depends on both τ and the mean distance (δ) between the two events considered. This relationship is illustrated in Figure 6.13 for $\tau=0.2$ and $\tau=0.4$.

It is readily shown that Glenn and David's model results in the following three probabilities for A_1, A_2 and A_3:

$$P_1 = \Phi(\delta - \tau)$$

$$P_2 = 1 - \Phi(\delta + \tau)$$

$$P_3 = \Phi(\delta + \tau) - \Phi(\delta - \tau) \tag{6.19}$$

Consequently,

$$P_1 + P_3 = \Phi(\delta + \tau)$$

$$P_2 + P_3 = \Phi(-\delta + \tau) \tag{6.20}$$

This indicates that δ and τ can be estimated from P_1, P_2 and P_3. A set of observed frequencies using the format (F,T/R) are shown in Table 6.20A. This is the Hay example as used in Agterberg and Nel (1982b) and Agterberg (1984). It is convenient to define

TABLE 6.20

Example of 10 biostratigraphic events forming optimum sequence as in Agterberg and Nel (1982b, Table 7, p. 74). **A.** Numbers F, T/R are for pairwise comparison using a trinomial model. If rows are labelled by the index i and columns by j, then F denotes the number of times E_j follows E_i in the sequences, T represents number of ties, and R is number of times E_i and E_j were observed in the same section. Example: the first entry of the second column (4, 2/7) indicates that event 1 follows event 2 four times while the two events were observed to be coeval in two sections. Because $R=7$, this implies that event 1 precedes event 2 in SEQ file for one section. **B.** Matrix consisting of elements $A=(F+T)/R$ corresponding to Table 6.20A.

A

	2	1	3	5	7	4	6	8	9	10
2	x	4,2/7	2,2/6	3,4/8	4,1/6	2,2/6	2,2/4	5,0/5	8,0/8	4,1/5
1	1,2/7	x	1,3/5	3,3/8	4,0/6	4,1/6	2,1/3	5,0/5	8,0/8	4,1/5
3	2,2/6	1,3/5	x	3,2/6	3,1/5	3,3/6	2,2/4	4,0/4	6,0/6	3,1/4
5	1,4/8	2,3/8	1,2/6	x	3,1/7	3,3/7	2,2/4	5,0/5	9,0/9	4,2/6
7	1,1/6	2,0/6	1,1/5	3,1/7	x	3,1/6	1,1/3	4,1/5	7,0/7	4,1/5
4	2,2/6	1,1/6	0,3/6	1,3/7	2,1/6	x	2,2/4	3,0/4	6,0/7	3,1/5
6	0,2/4	0,1/3	0,2/4	0,2/4	1,1/3	0,2/4	x	1,0/2	4,0/4	1,1/2
8	0,0/5	0,0/5	0,0/4	0,0/5	0,1/5	1,0/4	1,0/2	x	5,0/5	2,1/3
9	0,0/8	0,0/8	0,0/6	0,0/9	0,0/7	1,0/7	0,0/4	0,0/5	x	3,0/6
10	0,1/5	0,1/5	0,1/4	0,2/6	0,1/5	1,1/5	0,1/2	0,1/3	3,0/6	x

B

	2	1	3	5	7	4	6	8	9	10
2	x	6/7	4/6	7/8	5/6	4/6	4/4	5/5	8/8	5/5
1	3/7	x	4/5	6/8	4/6	5/6	3/3	5/5	8/8	5/5
3	4/7	4/5	x	5/6	4/5	6/6	4/4	4/4	6/6	4/4
5	5/8	5/8	3/6	x	4/7	6/7	4/4	5/5	9/9	6/6
7	2/6	2/6	2/5	4/7	x	4/6	2/3	5/5	7/7	5/5
4	4/6	2/6	3/6	4/7	3/6	x	4/4	3/4	6/7	4/5
6	2/4	1/3	2/4	2/4	2/3	2/4	x	1/2	4/4	2/2
8	0/5	0/5	0/4	0/5	1/5	1/4	1/2	x	5/5	3/3
9	0/8	0/8	0/6	0/9	0/7	1/7	0/4	0/5	x	3/6
10	1/5	1/5	1/4	2/6	1/5	2/5	1/2	1/3	3/6	x

$$A_{ij} = (F_{ij} + T_{ij})/R_{ij}$$

$$A_{ji} = (F_{ji} + T_{ij})/R_{ij} \tag{6.21}$$

These values are shown in matrix form in Table 6.20B, with A_{ij} in the upper triangle and A_{ji} in the lower triangle. The transformation $\Phi^{-1}(A_{ij})$

was made with the result shown in Table 6.21A. Finally, separate estimates (d and t) of δ and τ were obtained as

$$d = [\phi^{-1}(A_{ij}) - \phi^{-1}(A_{ji})]/2$$

$$t = [\Phi^{-1}(A_{ij}) - \Phi^{-1}(A_{ji})]/2 \qquad (6.22)$$

The d-values computed from the values of Table 6.21A are shown in the upper triangle of Table 6.21B and the t-values in its lower triangle. The d-values can be treated in exactly the same way as the Z-values were treated in scaling for obtaining average distances between events along the linear scale L. Each of the t-values can be regarded as an estimate of τ. A frequency distribution of the 32 observed t-values of Table 6.21B is shown in Table 6.22. Their average amounts to $\bar{t} = 0.4520$ which seems to be a fairly precise estimate of τ. (The standard deviation of $\bar{t}$ is 0.046.)

Glenn and David (1960) have shown that the preceding simple averaging method does not result in a least squares solution of τ and the distances between events. They proposed a modified model replacing the Gaussian curves along the distance scale L by cosine curves. Then the preceding expressions for d and t represent the least squares solution when $\Phi^{-1}(A_{ij})$ is replaced by $\arcsin(2A_{ij} - 1)$. Application of the arcsin transformation to the values of Table 6.20B yields Table 6.21C instead of Table 6.21A. Table 6.21D was derived from Table 6.21C in the same way as Table 6.21B from Table 6.21A and also can be used for estimating τ and the distances. The modified average value now amounts to $\bar{t}=0.4080$ as shown in Table 6.22.

A more elaborate test of the preceding version of Glenn and David's model consisted of its application to 48 events each occurring at least 5 times in the set of 18 wells used by Gradstein (1984). First an optimum sequence was obtained (probabilistic ranking followed by modified Hay method with $m_{c1}=1$). This sequence was split into two segments consisting of 21 and 27 events, respectively. τ was estimated separately by the two methods (Gaussian Model and Cosine Model) for these two groups which contain 75 (Group 1 in Table 6.23) and 173 (Group 3) individual t-values, respectively. Group 2 in Table 6.23 is for 39 t-values arising from comparison of events in Group 1 to events in Group 3. The average values of t (Gaussian Model) are 0.2419, 0.1914 and 0.2179, respectively. These

TABLE 6.21

A. Values $\Phi^{-1}(A)$ corresponding to Table 6.20B. Values for samples with $R=2$ were not used and are written as x. Values corresponding to 1 and 0 are written as a and -a, respectively. For some subsequent calculations a was set equal to $q_c=1.645$. **B.** Values d (in upper triangle) and t (in lower triangle) obtained by Eq. (6.22). The values aa and aaa are undetermined. **C.** Same as Table 6.21A except that the transformation arcsin $(2A\text{-}1)$ was used. For some subsequent calculations a was set equal to $q_c=1.571$ (instead of 1.645). **D.** Same as Table 6.21B except that the transformation arcsin $(2A\text{-}1)$ was used.

		2	1	3	5	7	4	6	8	9	10
A	2	x	1.068	0.430	1.150	0.967	0.430	a	a	a	a
	1	-0.180	x	0.841	0.674	0.430	0.967	a	a	a	a
	3	0.430	0.891	x	0.967	0.841	a	a	a	a	a
	5	0.318	0.318	0.000	x	0.180	1.068	a	a	a	a
	7	-0.430	-0.430	-0.253	0.180	x	0.430	0.430	a	a	a
	4	0.430	-0.430	0.000	0.180	0.000	x	a	0.674	1.068	0.841
	6	0.000	-0.430	0.000	0.000	0.430	0.000	x	x	a	x
	8	-a	-a	-a	-a	-0.841	-0.674	x	x	a	a
	9	-a	-a	-a	-a	-a	-1.068	-a	-a	x	0.000
	10	-0.841	-0.841	-0.674	-0.930	-0.841	-0.253	a	-0.430	0.000	x
B	2	x	0.624	0.000	0.416	0.699	0.000	0.823	aaa	aaa	1.243
	1	0.444	x	0.000	0.178	0.430	0.699	1.038	aaa	aaa	1.243
	3	0.430	0.841	x	0.484	0.547	0.823	0.823	aaa	aaa	1.160
	5	0.734	0.496	0.484	x	0.000	0.444	0.823	aaa	aaa	1.038
	7	0.269	0.000	0.294	0.180	x	0.215	0.000	1.243	aaa	1.243
	9	0.430	0.269	0.823	0.624	0.215	x	0.823	0.674	1.068	0.547
	6	0.823	0.607	0.823	0.823	0.430	0.823	x	x	aaa	x
	8	aa	aa	aa	aa	0.402	0.000	x	x	aaa	1.038
	9	aa	aa	aa	aa	aa	0.000	aa	aa	x	0.000
	10	0.402	0.402	0.486	0.607	0.402	0.294	x	0.607	0.000	x
C	2	x	0.796	0.340	0.848	0.730	0.340	a	a	a	a
	1	-0.143	x	0.644	0.524	0.340	0.730	a	a	a	a
	3	0.340	0.644	x	0.730	0.644	a	a	a	a	a
	5	0.253	0.253	0.000	x	0.143	0.796	a	a	a	a
	7	-0.340	-0.340	-0.201	0.143	x	0.340	0.340	a	a	a
	4	0.340	-0.340	0.000	0.143	0.000	x	1.571	0.524	0.796	0.644
	6	0.000	-0.340	0.000	0.000	0.340	0.000	x	x	a	x
	8	-a	-a	-a	-a	-0.644	-0.524	x	x	a	a
	9	-a	-a	-a	-a	-a	-0.796	-a	-a	x	0.000
	10	-0.644	-0.644	-0.524	-0.340	-0.644	-0.201	x	-0.340	0.000	x
D	2	x	0.469	0.000	0.298	0.535	0.000	0.785	aaa	aaa	1.107
	1	0.326	x	0.000	0.135	0.340	0.535	0.955	aaa	aaa	1.107
	3	0.340	0.644	x	0.365	0.422	0.785	0.785	aaa	aaa	1.047
	5	0.550	0.388	0.365	x	0.000	0.326	0.785	aaa	aaa	0.955
	7	0.195	0.000	0.221	0.143	x	0.170	0.000	1.107	aaa	1.107
	4	0.340	0.195	0.785	0.469	0.170	x	0.785	0.524	0.796	0.422
	6	0.785	0.615	0.785	0.785	0.340	0.785	x	x	aaa	x
	8	aa	aa	aa	aa	0.464	0.000	x	x	aaa	0.955
	9	aa	aa	aa	aa	aa	0.000	aa	aa	x	0.000
	10	0.464	0.464	0.524	0.615	0.464	0.221	x	0.615	0.000	x

values are not significantly different from each other at the 5 percent level of significance when analysis of variance is applied. This demonstrates

TABLE 6.22

Frequency distribution of t-values shown in lower triangles of Tables 6.21B and 6.21D. G.M. denotes Gaussian Model; C.M. - Cosine Model; N - sample size; S.D. - Standard Deviation.

Class Limits	G.M.	C.M.
0.000	4	4
0.001 - 0.200	1	4
0.201 - 0.400	5	8
0.101 - 0.600	11	7
0.601 - 0.800	5	9
0.801 - 1.000	6	0
N	32	32
Mean	0.4520	0.4080
S.D.	0.2603	0.2489
S.D./$N^{1/2}$	0.0460	0.0440

TABLE 6.23

Glenn and David's trinomial model applied to 48 exits of Cenozoic Foraminifera observed in 18 wells on northwestern Atlantic Margin. Abbreviations as in Table 6.22. Groups resulted from splitting the optimum sequence after 21 events. Group 1 (see Table 6.24 for original data) is for pairwise comparisons of events belonging to first 21 events, Group 3 is same for last 27 events, and Group 2 is for comparison of events of Group 1 to events of Group 3.

	Group 1		Group 2		Group 3	
Class Limits	G.M.	C.M.	G.M.	C.M.	G.M.	C.M.
0.000	28	28	20	20	73	73
0.001 - 0.200	5	14	1	2	13	21
0.201 - 0.400	22	14	14	8	47	43
0.401 - 0.600	11	11	8	6	27	25
0.601 - 0.800	6	6	3	3	10	10
0.801 - 1.000	1	1	0	0	2	2
1.001 - 1.200	1	0	0	0	1	0
N	75	75	39	39	173	173
Mean	0.2419	0.2242	0.1914	0.1854	0.2179	0.2008
S.D.	0.2402	0.2321	0.2196	0.2249	0.2288	0.2204
S.D./$N^{1/2}$	0.0277	0.0268	0.0352	0.0360	0.0174	0.0168

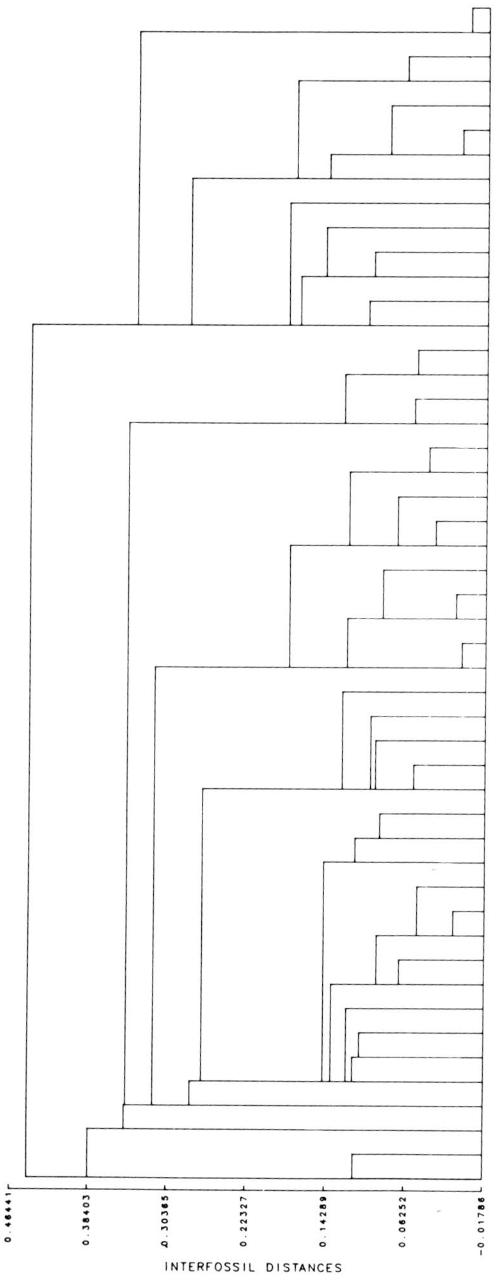

77	0.0000	ELPHIDIUM SP
228	0.3377	CASSIDULINA TERETIS
10	0.0643	UVIGERINA CANARIENSIS
65	0.1760	COSCINODISCUS SP1
22	0.0814	COSCINODISCUS SPP
17	0.0082	ASTERIGERINA GURICHI
67	0.1427	SCAPHOPOD SP1
16	0.2832	CERATOBULIMINA CONTRARIA
71	0.1832	EPISTOMINA ELEGANS
18	0.1457	SPIROPLECTAMMINA CARINATA
21	0.0971	GUTTULINA PROBLEMA
20	0.1714	GYROIDINA GIRARDANA
15	0.1025	GLOBIGERINA PRAEBULLOIDES
26	0.4465	UVIGERINA DUMBLEI
70	0.0525	ALABAMINA WOLTERSTORFFI
24	0.1264	TURRILINA ALSATICA
25	0.0552	COARSE ARENACEOUS SPP
27	0.3457	EPONIDES UMBONATUS
81	0.0403	GLOBIGERINA VENEZUELANA
69	0.1212	NODOSARIA SP8
33	0.0718	TURBOROTALIA POMEROLI
31	0.0334	PTEROPOD SP1
82	0.1809	GLOBIGERINA LINAPERTA
29	0.0862	CYCLAMMINA AMPLECTENS
34	0.0123	MARGINULINA DECORATA
85	0.1226	PSEUDOHASTIGERINA MICRA
40	0.0062	BULIMINA ALAZANENSIS
118	0.3178	EPISTOMINA SP5
41	0.1270	PLECTOFRONDICULARIA SP1
30	0.0980	CIBICIDOIDES BLANPIEDI
35	0.0930	SPIROPLECTAMMINA DENTATA
42	0.0544	CIBICIDOIDES ALLENI
32	0.2685	QUADRIMORPHINELLA INCAUTA
86	0.0884	TURRILINA BREVISPIRA
49	0.1135	OSANGULARIA EXPANSA
53	0.1452	UVIGERINA BATJESI
57	0.0505	SPIROPLECTAMMINA SPECTABILIS
90	0.0138	ACARININA DENSA
36	0.0912	PSEUDOHASTIGERINA WILCOXENSIS
93	0.0684	ACARININA AFF. BROEDERMANNI
45	0.1372	BULIMINA TRIGONALIS
43	0.1215	BULIMINA MIDWAYENSIS
50	0.1081	SUBBOTINA PATAGONICA
46	0.1150	MEGASPORE SP1
54	0.2800	TEXTULARIA PLUMMERAE
52	0.3469	ACARININA SOLDADOENSIS
56	0.3837	GLOMOSPIRA CORONA
55	0.1139	GAVELINELLA BECCARIIFORMIS
59		RZEHAKINA EPIGONA

Fig. 6.14 Dendrogram for distances between successive events estimated by Glenn and David's trinomial model assuming Gaussian probability curves for events. Each event (except the last one) is followed by estimate of distance connecting it to the event immediately below it. These distances were plotted toward the left and clustered.

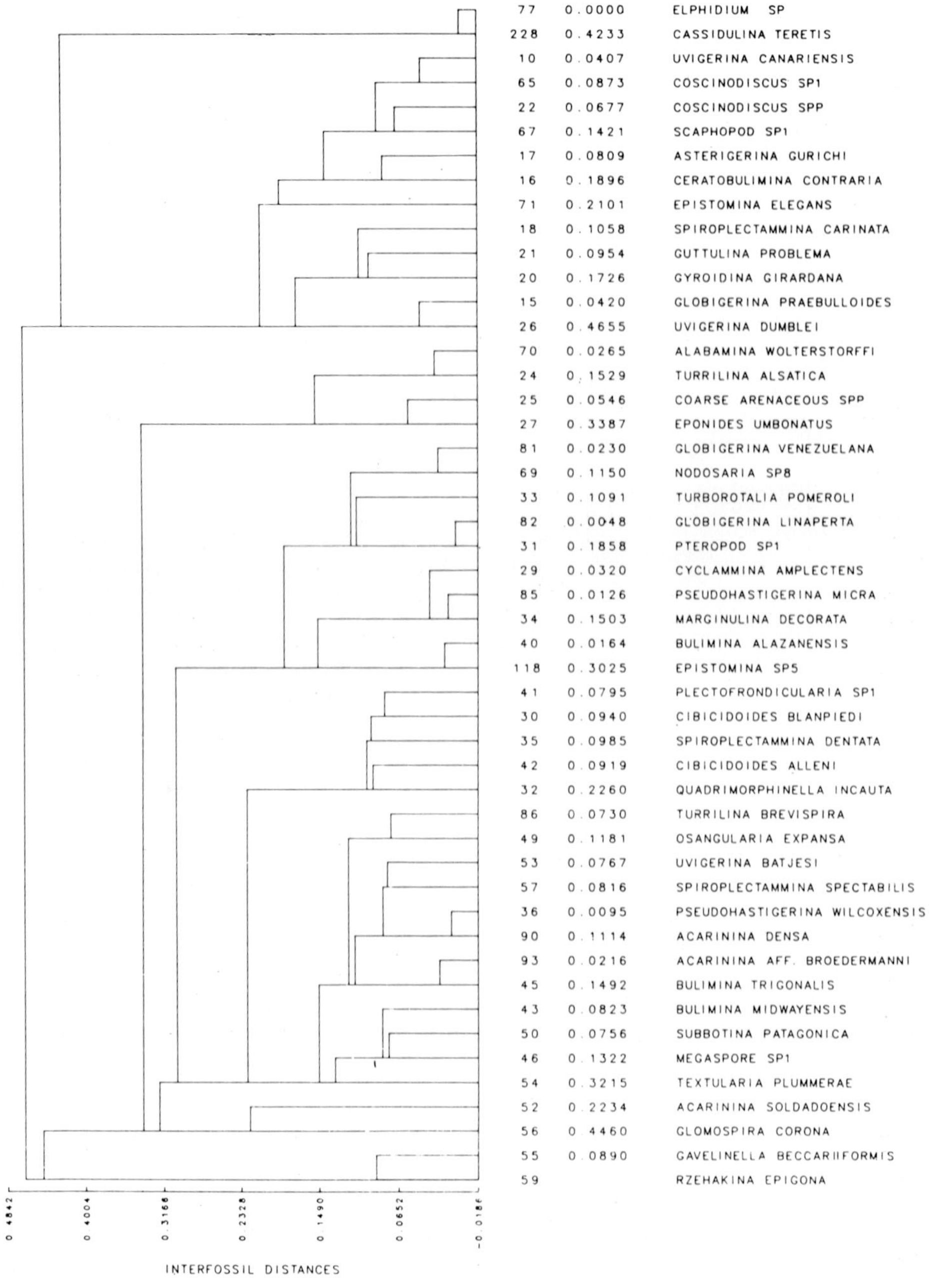

Fig. 6.15 Same as Fig. 6.14 except that cosine-shaped probability curves (instead of Gaussian curves) were assumed for events. Note that differences between patterns of Figs. 6.14 and 6.15 are small, indicting that choice of shape of probability curves for events probably is not of critical importance.

TABLE 6.24

Estimation of probabilities (P_f and P_t) and frequencies (f_e and t_e) corresponding to observed successions (f) and ties (t). Trinomial model was applied to first 21 events (Group 1) of optimum sequence for 48 exits of Cenozoic Foraminifera also used in Table 6.23 and Fig. 6.14. Last columns show estimated values for scores (s) based on modified binomial model using RASC weighted distance analysis. See text for explanations of other column headings. Event numbers of column 1 are explained in Fig. 6.14 (from Agterberg, 1984).

E_i-E_j	f,t/r	d_f	P_f	f_e	P_t	t_e	s	d_s	P_s	s_e
10-17	2,2/5	0.32	0.53	2.7	0.18	0.9	3.0	0.39	0.65	3.3
10-16	5,1/6	0.47	0.59	3.5	0.17	1.0	5.5	0.94	0.83	5.0
10-17	2,1/3	0.76	0.70	2.1	0.15	0.4	2.5	1.43	0.92	2.8
17-16	3,3/6	0.15	0.46	2.8	0.19	1.1	4.5	0.54	0.71	4.2
17-71	2,0/3	0.43	0.58	1.7	0.18	0.5	2.0	1.03	0.85	2.5
17-18	6,1/7	0.62	0.65	4.5	0.16	1.1	6.5	1.19	0.88	6.2
17-20	6,1/7	0.86	0.73	5.1	0.13	0.9	6.5	1.49	0.93	6.5
17-15	6,1/7	1.03	0.79	5.5	0.11	0.8	6.5	1.73	0.96	6.7
65-16	4,1/5	-0.09	0.37	1.9	0.19	1.0	4.5	0.84	0.80	4.0
77-228	1,2/4	0.00	0.40	1.6	0.19	0.8	2.0	0.00	0.50	2.0
228-22	2,0/3	0.58	0.63	1.9	0.16	0.5	2.0	1.21	0.89	2.7
16-22	3,0/5	0.23	0.50	2.5	0.19	0.9	3.0	0.01	0.51	2.5
16-67	3,2/6	0.14	0.46	2.8	0.19	1.1	2.0	-0.32	0.49	2.9
16-71	3,1/6	0.28	0.52	3.1	0.18	1.1	3.5	0.49	0.69	4.1
16-18	11,3/16	0.47	0.59	9.4	0.17	2.8	12.5	0.64	0.74	11.8
16-20	11,2/14	0.71	0.68	9.5	0.15	2.1	12.0	0.95	0.83	11.6
16-15	11,4/15	0.88	0.74	11.1	0.13	2.0	13.0	1.18	0.88	13.2
16-26	7,0/8	0.98	0.77	6.2	0.12	1.0	7.0	1.33	0.91	7.3
22-71	2,1/3	0.52	0.61	1.8	0.17	0.5	2.5	0.48	0.68	2.0
22-21	2,1/3	0.84	0.73	2.2	0.13	0.4	2.5	0.80	0.79	2.4
22-18	3,0/5	0.70	0.68	3.4	0.15	0.8	3.0	0.63	0.74	3.7
22-20	4,1/5	0.94	0.76	3.8	0.12	0.6	4.5	0.93	0.82	4.1
22-15	4,0/5	1.11	0.81	4.0	0.10	0.5	4.0	1.17	0.88	4.4
67-21	4,1/5	0.75	0.70	3.5	0.15	0.7	4.5	0.84	0.80	4.0
67-18	5,1/6	0.61	0.64	3.9	0.16	1.0	5.5	0.68	0.75	4.5
71-21	1,1/3	0.33	0.54	1.6	0.18	0.5	1.5	0.32	0.63	1.9
71-18	2,2/6	0.18	0.48	2.9	0.19	1.1	3.0	0.15	0.56	3.4
71-20	4,1/6	0.43	0.57	3.4	0.18	1.1	4.5	0.46	0.68	4.1
71-15	4,2/6	0.60	0.64	3.8	0.16	1.0	5.0	0.70	0.76	4.5
71-26	4,0/5	0.70	0.68	3.4	0.15	0.8	4.0	0.84	0.80	4.0
71-27	2,1/3	1.38	0.87	2.6	0.08	0.2	2.5	1.58	0.94	2.8
21-18	3,5/11	-0.15	0.35	3.8	0.19	2.1	5.5	-0.17	0.43	4.8
21-20	3,6/9	0.10	0.44	4.0	0.19	1.7	6.0	0.14	0.55	5.0
21-15	4,2/10	0.27	0.51	5.1	0.18	1.8	5.0	0.38	0.65	6.5
21-26	2,0/4	0.37	0.55	2.2	0.18	0.7	2.0	0.52	0.70	2.8
21-70	5,1/6	0.82	0.72	4.3	0.14	0.8	5.5	0.85	0.80	4.8
21-24	7,1/8	0.87	0.74	5.9	0.13	1.1	7.5	1.00	0.84	6.7
21-27	5,0/6	1.05	0.79	4.7	0.11	0.7	5.0	1.26	0.90	5.4
18-20	7,5/15	0.24	0.50	7.5	0.19	2.8	9.5	0.30	0.62	9.3
18-15	10,3/16	0.41	0.57	9.1	0.18	2.8	11.5	0.54	0.71	11.3
18-26	7,0/8	0.52	0.01	4.9	0.16	1.3	7.0	0.68	0.75	6.0
18-70	6,1/8	0.96	0.77	6.1	0.12	1.0	6.5	1.01	0.84	6.8
18-24	10,1/12	1.02	0.78	9.4	0.12	1.4	10.5	1.16	0.88	10.5
18-25	11,0/12	1.14	0.82	9.8	0.10	1.2	11.0	1.43	0.92	11.1
20-15	8,2/14	0.17	0.47	6.6	0.19	2.6	9.0	0.24	0.60	8.3
20-26	5,1/8	0.27	0.51	4.1	0.18	1.5	5.5	0.38	0.65	5.2
20-70	6,0/7	0.72	0.68	4.8	0.15	1.0	6.0	0.71	0.76	5.3
20-24	9,1/11	0.77	0.70	7.7	0.14	1.6	9.5	0.86	0.81	8.9
20-25	9,1/10	0.90	0.74	7.4	0.13	1.3	9.5	1.13	0.87	8.7
20-27	5,0/7	0.95	0.76	5.3	0.12	0.9	5.0	1.13	0.87	6.1
15-26	4,0/8	0.10	0.45	3.6	0.19	1.5	4.0	0.14	0.56	4.4
15-70	4,3/8	0.55	0.62	5.0	0.16	1.3	5.5	0.47	0.68	5.5
15-24	9,1/12	0.60	0.64	7.7	0.18	2.1	9.5	0.62	0.73	8.8
15-25	10,1/12	0.73	0.69	8.2	0.15	1.8	10.5	0.89	0.81	9.8
15-27	5,0/7	0.78	0.71	4.9	0.14	1.0	5.0	0.89	0.81	5.7
15-81	4,1/6	1.13	0.81	4.9	0.10	0.6	4.5	1.30	0.90	5.4
26-24	5,1/7	0.50	0.60	4.2	0.17	1.2	5.5	0.48	0.69	4.8
26-25	1,0/4	0.63	0.65	2.6	0.16	0.6	1.0	0.75	0.77	3.1
26-27	3,1/5	0.68	0.67	3.4	0.15	0.8	3.5	0.75	0.77	3.9
70-24	4,0/6	0.05	0.43	2.6	0.19	1.1	4.0	0.15	0.56	3.4
70-25	6,1/8	0.18	0.48	3.8	0.19	1.5	6.5	0.42	0.66	5.3
70-27	2,0/3	0.23	0.50	1.5	0.19	0.6	2.0	0.42	0.66	2.0
70-81	2,0/3	0.58	0.63	1.9	0.16	0.5	2.0	0.83	0.80	2.4
70-31	3,0/4	0.81	0.72	2.9	0.14	0.6	3.0	1.02	0.85	3.4
24-25	6,1/9	0.12	0.45	4.1	0.19	1.7	6.5	0.26	0.66	5.4
24-27	4,0/6	0.18	0.48	2.9	0.19	1.1	4.0	0.27	0.61	3.6
24-81	3,0/4	0.53	0.61	2.4	0.17	0.7	3.0	0.83	0.80	3.2
25-27	3,0/5	0.06	0.43	2.1	0.19	1.0	3.0	0.00	0.50	2.5
25-31	7,1/9	0.63	0.65	5.9	0.16	1.4	7.5	0.60	0.73	6.5
27-81	2,0/3	0.35	0.54	1.6	0.18	0.5	2.0	0.41	0.66	2.0
27-31	5,0/7	0.58	0.63	4.4	0.16	1.1	5.0	0.60	0.73	5.1
27-82	2,0/3	0.61	0.65	1.9	0.16	0.5	2.0	0.68	0.75	2.3
81-31	3,0/5	0.23	0.50	2.5	0.19	0.9	3.0	0.19	0.58	2.9
81-82	3,1/4	0.27	0.51	2.0	0.19	0.7	3.5	0.27	0.61	2.4
31-82	4,0/5	0.03	0.42	2.1	0.19	1.0	4.0	0.08	0.53	2.7

that Glenn and David's trinomial model indeed can be used for describing the frequencies of observed ties.

The d-values were treated as Z-values in the RASC computer program (now setting $m_{c2} = 3$ and using the unweighted method for scaling). The resulting dendrograms are shown in Figure 6.14 (Gaussian Model) and Figure 6.15 (Cosine Model). It may be concluded that the differences between results obtained by these two models are minimal. On average, successive distances in Figures 6.14 and 6.15 are shorter than those in dendrograms resulting from runs with the RASC program. All successive distances in Figure 6.14 are less than 0.5. Because τ is

TABLE 6.25

Comparison of observed and estimated frequencies for 75 pairwise comparisons of Table 6.24. First six columns are for trinomial model and last three columns for binomial (RASC weighted scaling) model. If model provides good fit, the *U*-values are approximately distributed as chi-squared with single degree of freedom. Totals are shown in bottom line.

T_e	T_o	U_t	F_e	F_o	U_f	S_e	S_o	U_s
9.09	13	1.69	33.31	39	0.97	45.85	45.5	0.00
10.91	12	0.11	44.53	49	0.45	53.49	53	0.00
9.14	12	0.89	40.30	41	0.01	45.77	47	0.03
8.90	15	4.18	30.07	29	0.04	35.93	36.5	0.01
10.52	10	0.02	46.72	51	0.39	54.95	56	0.02
8.86	5	1.68	36.00	42	1.00	42.51	44.5	0.09
8.35	6	0.66	34.28	36	0.09	39.58	39	0.01
10.34	4	3.88	32.14	39	1.46	40.39	41	0.01
7.15	2	3.71	22.60	29	1.81	26.29	30	0.52
83.25	79	16.83	319.95	355	6.22	384.76	392.5	0.70

approximately equal to 0.2, most probabilities of a tie between successive events are about 15 percent (cf. Fig. 6.13).

6.11 Comparison of observed and estimated probabilities

A detailed comparison of estimated trinomial and binomial probabilities with observed frequencies is shown in Table 6.24 for Group 1 in Table 6.23 only. A temporary change in notation restricted to this section is that *f*, *t*, *r* and *s* are used instead of *F*, *T*, *R* and *S* for pairwise comparison; *F*, *T* and *S* will be used instead to denote sums of *f*-, *t*- and *s*- values (see Table 6.25). The distances d_f in Table 6.24 are as in Figure 6.14. For example, the distance 0.32 between events 10 and 17 is equal to the sum of three successive differences (0.0643, 0.1760 and 0.0814) in Figure 6.14. According to the original equations for the Glenn-David model, the estimate of τ (= 0.2419) should be subtracted from these distances and the fractile of the normal distribution in standard form determined for estimation of P_1. In order to distinguish it from another estimate of P_1 (see later), this estimate is written as P_f. For example, the distance $d_f = 0.3217$ gave $d_f\text{-}\bar{t} = 0.3217\text{-}0.2419 = 0.0798$ from which $P_f = 0.53$ was derived. Multiplication of the estimated probability P_f by sample size $r=5$ resulted in the estimated frequency $f_e = 2.7$ for number of times event 10 occurs above event 17. This estimated frequency can now be compared to the observed frequency $f=2$ in the second column of Table 6.24.

It is also possible to estimate P_2 and P_3. Because $P_2 = 1\text{-}P_1\text{-}P_3$, the probability of a tie, written as P_t, is shown only, followed by the corresponding estimated frequency t_e. For the previous example, $P_t = 0.18$ and $t_e = 2.8$ (to be compared to $t=2$).

The 75 pairs of events are divided into 9 groups in Table 6.24. The estimated frequencies t_e and f_e were added for these groups, with the totals shown as T_e and F_e in Table 6.25 for comparison to corresponding sums of observed frequencies written as T_o and F_o. The quantities $U_t = (T_o - T_e)^2/T_e$ and $U_f = (F_o - F_e)^2/F_e$ are also given in Table 6.25.

If the model provides a good fit to the observations, each of the quantities U_t and U_f is approximately distributed as chi-squared with a single degree of freedom. The totals ΣU_t and ΣU_f would be distributed as chi-squared with approximately 9 degrees of freedom. The 95 percent confidence limit for this distribution amounts to 16.9. This suggests that the observed frequencies are well described by Glenn and David's model. On the other hand, the discrepancy that the T_e-values are less than the T_o-values in the upper part of Table 6.25 and greater in its lower part may be significant. The number of degrees of freedom is not known exactly for this test. It is, however, probably less than 9 and this would increase the 95 percent confidence limit from 16.9 to below $\Sigma U_t = 16.8$.

In this chapter the method of scaling was presented and initially illustrated by using the two examples of the previous chapter on ranking (Lower Tertiary nannoplankton from the California Coast Range and Cenozoic Foraminifera from the northwestern Atlantic continental margin). The basic assumptions of this approach were tested by using artificial data sets consisting of ranking normal numbers and computer simulation experiments. Important options of the RASC computer program introduced in this chapter were the normality test, the marker horizon option and the unique event option. By using the same two examples it also was shown that a modified version of the trinomial model of Glenn and David (1960) can be used for description of observed frequencies of coeval biostratigraphic events. The stratigraphic significance of the threshold parameter τ is not immediately obvious. It can be said that a new distribution for ties (see Fig. 6.13) has been introduced in addition to the probability distributions for events along the linear scale L. The height of the new distribution for ties is roughly proportional to the value of τ. In general, τ therefore expresses the likelihood that events are coeval.

In the RASC model, observed ties are not ignored but each tie of two events E_i and E_j is scored as a 50 percent probability that E_i occurs above E_j and a 50 percent probability that E_j occurs above E_i. The last four columns of Table 6.24 show observed scores in comparison with estimated frequencies. The estimated probabilities P_s (for E_i occurring above E_j) satisfy $P_s = \Phi(d_s)$ where d_s was estimated by means of the weighted scaling option of the RASC computer program in which variations of sample size are considered. The agreement between observed and estimated scores is excellent (also see Table 6.25, for comparisons of group totals, S_e and S_o for estimated and observed scores, respectively). Because the origin of the RASC scale is set at the location of the first event in a scaled optimum sequence, N events obtain N^* (=N-1) cumulative RASC distances after scaling. In general, these N^* values can be used to estimate the $N(N\text{-}1)/2$ probabilities that one event occurs above (or below) another event. These expected probabilities for pairwise comparison are close to the observed probabilities, because the former were computed from the latter. This conclusion is supported by application of the chi-squared test for goodness of fit after grouping pairs of events (cf. Table 6.25, last column). The number of degrees of freedom to be used in this chi-squared test, however, remains unknown, because of autocorrelation of the estimated RASC distances. The latter topic will be discussed in more detail in Chapter 8 in relation to the normality test.

In addition to providing a good fit, the RASC method has several options (normality test; marker horizon, unique event and weighted scaling options) which are not available for the modified Glenn-David model. For these reasons, this trinomial model should only be used when it is necessary to model observed frequencies of coeval events.

CHAPTER 7

RANK CORRELATION AND PRECISION OF SCALED OPTIMUM SEQUENCE

7.1 Introduction

Suppose that a number of objects has been ranked in two different ways, e.g. by using different characteristics. One then may be interested in the mutual agreement or disagreement of the two rankings. Rank correlation methods are described in detail by Kendall (1975). Many authors have applied these methods in biostratigraphy for comparing sequences of events, e.g. as obtained by different methods, with one another (see, for example, Brower, 1985, 1989; Harper, 1984). In the first part of this chapter, rank correlation will be discussed in connection with the RASC step model. Examples of application will be given. A method for estimating the precision of the cumulative RASC distances of the scaled optimum sequence will be presented in the second part of this chapter.

7.2 Rank correlation coefficients

The two measures of rank correlation discussed by Kendall and widely available in systems of statistical software (e.g. SAS) are Kendall's tau (τ) and Spearman's rho (ρ). They are estimated by using the following equations:

$$\hat{\tau} = \frac{2S}{n(n-1)} \tag{7.1}$$

$$\hat{\rho} = 1 - \frac{6\,SSD}{n(n^2-1)} \tag{7.2}$$

where S is a total score of +1 for pairs of elements having the same order in both series and -1 otherwise. The total number of elements is written as *n*. Spearman's rho is based on the sum of squared differences (SSD) of rankings of the elements in the two series compared to one another.

Both rank correlation coefficients emulate Pearson's product-moment correlation coefficient for a bivariate relationship in that they vary between 0 for lack of correlation and +1 or -1 for maximum positive or negative correlation. Unless there is complete agreement or disagreement, tau and rho are not the same for any given pair of rankings. Rho tends to give more weight to inversions of ranks which are farther apart. In practice, it is often found that, when neither coefficient is close to unity, rho is about 50 per cent greater than tau in absolute value (Kendall, 1975, p. 12). Although rho is easier to calculate than tau, Kendall has shown that from practical as well as theoretical points of view, tau is preferable to rho. For example, after completing two rankings of the same set of objects, it may be that some new objects become available for ranking. In that situation, rho must be completely recalculated, whereas the addition of new members does not require a complete recalculation of tau. For the latter reason, it is also easier to evaluate the influence of addition of individual objects on tau than on rho.

Kendall's (1975, p. 3) first example consists of the following two rankings of ten objects A, ..., J:

	A	B	C	D	E	F	G	H	I	J
Ranking 1:	7	4	3	10	6	2	9	8	1	5
Ranking 2:	5	7	3	10	1	9	6	2	8	4

Then $n=10$ objects have $n(n\text{-}1)/2=45$ possible pairs. Table 7.1 is a complete list of scores being +1 if two elements forming a pair have the same order in both rankings; and -1, otherwise. In total, there are $P=21$ positive and $Q=24$ negative scores in this table. The sum of all elements is equal to -3. Hence, according to Equation (7.1), $\hat{\tau}$ = -0.07.

In order to estimate Spearman's rho, the sum of squared differences (SSD) is needed. Individual squared differences for the examples are shown in the following tabulation:

TABLE 7.1

Listing of all 45 pairs and their scores for Kendall's (1975) first example with 10 rank members A-J.

Pair	Score	Pair	Score
AB	-1	CJ	+1
AC	+1	DE	+1
AD	+1	DF	+1
AE	+1	DG	+1
AF	-1	DH	+1
AG	+1	DI	+1
AH	-1	DJ	+1
AI	-1	EF	-1
AJ	+1	EG	+1
BC	+1	EH	+1
BD	+1	EI	-1
BE	-1	EJ	-1
BF	-1	FG	-1
BG	-1	FH	-1
BH	-1	FI	+1
BI	-1	FJ	-1
BJ	-1	GH	+1
CD	+1	GI	-1
CE	-1	GJ	+1
CF	-1	HI	-1
CG	+1	HJ	-1
CH	-1	IJ	-1
CI	-1		

	A	B	C	D	E	F	G	H	I	J
Ranking 1:	7	4	3	10	6	2	9	8	1	5
Ranking 2:	5	7	3	10	1	9	6	2	8	4

Differences d	2	-3	0	0	5	-7	3	6	-7	1
Differences^2 d^2	4	9	0	0	25	49	9	36	49	1

By summing the entries in the bottom row, we find SSD = 182. Consequently, according to Equation (7.2) $\hat{\rho}$ = -0.103 which is somewhat smaller than $\hat{\tau}$ = -0.07.

Kendall's tau and Spearman's rho have been calculated for the optimum sequences of Table 6.9 obtained by running RASC on 50 artificial sequences in computer simulation experiments. Table 7.2 shows the two ranking correlation coefficients between every optimum sequence and the underlying true sequence consisting of integer numbers from 1 to 20. All ranking statistics of Table 7.2 are rather large, indicating relatively strong positive correlation. As expected, there is a general decrease in strength of correlation when the spacing between expected values along the real line decreases from 1.0 to 0.1. For set 1, scaled optimum sequences are somewhat better than optimum sequences obtained by ranking but the opposite holds true for set 2. From these computer simulation experiments, it cannot be decided which type of optimum sequence is best. It only can be concluded that these optimum sequences are approximately equally good. A similar conclusion will be drawn from the results of Harper's (1984) computer simulation experiments to be discussed in Section 7.4. It does not follow from this conclusion that ranking of stratigraphic events is to be preferred to scaling because the latter technique requires more computing. In practical applications, the advantage of scaling with respect to ranking is that clusters of events separated by hiatuses can be identified so that a regional biozonation can be constructed. It is desirable that the optimum sequence obtained by ranking which forms the input for scaling is as good as possible because estimates of intervals between successive events are less precise if the events subjected to scaling are out of order (cf. Section 7.5).

7.3 RASC step model

In RASC, stratigraphic events are assigned numbers in the dictionary and these numbers are used in the rankings. Suppose that the 10 objects (A, ..., J) of Kendall's first example are numbered 1 to 10:

TABLE 7.2

Kendall's tau and Spearman's rho for optimum sequences of Table 6.9 correlated to underlying true sequence consisting of integer numbers from 1 to 20.

A(Set 1)	Tau	Rho	B(Set 2)	Tau	Rho
IIa-e	0.990	0.999	IIIc-e	0.979	0.997
IIIa-e	0.979	0.997	IVa-b	0.979	0.997
IVa-b	0.947	0.990	IVc	0.958	0.994
IVc	0.947	0.991	IVd-e	0.968	0.996
IVd-e	0.968	0.994	Va	0.895	0.979
Va-b	0.853	0.955	Vb	0.884	0.970
Vc	0.884	0.974	Vc	0.863	0.961
Vd-e	0.874	0.970	Vd-e	0.863	0.959

	1	2	3	4	5	6	7	8	9	10
Ranking 1:	7	4	3	10	6	2	9	8	1	5
Ranking 2:	5	7	3	10	1	9	6	2	8	4

Then the rankings rewritten as RASC input sequences become:

Sequence 1:	9	6	3	2	10	5	1	8	7	4
Sequence 2:	5	8	3	10	1	7	2	9	6	4

In the RASC step model, which can be applied after computation of an optimum sequence, the observed sequences for all stratigraphic sections are compared with this optimum sequence. The latter represents an average ranking based on the observed sequences for all sections. Suppose that, in Kendall's first example, sequence 1 is the optimum sequence and sequence 2 is one of many section sequences on which sequence 1 is based. In the step model, the position of each event is compared to its position in the optimum sequence. A penalty point is scored each time the event is out of place with respect to another event in comparison with their order in the optimum sequence. Table 7.3 shows the penalty points scored for the example.

Table 7.3 has separate columns for the number of times an event occurs "too high" or "too low" in the section. For example, event no. 9, with position no. 1 in the optimum sequence, occurs three places from the bottom in the section. It occurs "too high" with respect to all other events in the optimum sequence except events 6 and 4. Its total number of penalty points is equal to 7. Another example is as follows. In the section, event no. 1 occurs above nos. 2, 9 and 6, instead of above these events as in the optimum sequence. Consequently, it has penalty score 3 for occurring "too high". Its other penalty point arises because, in the section, event 1 is observed below event 8. Event 1's total score, therefore, is 4 penalty points. The column totals for "too high" and "too low" must be equal to one another. It also can be seen that these totals are equal to Q (=24 for the example), representing the total number of -1 scores used previously for estimating S, which is needed to compute tau (see Table 7.1). P can be

TABLE 7.3

Comparison of assignment of penalty points in RASC method with computation of tau on basis of Kendall's first example. Sums of columns for events that are "too high" and "too low" are both equal to $Q=24$. Total number of penalty points is $2Q=48$. Tau is fully determined by Q and total number of events ($n=10$).

Optimum Sequence Position	Event in Optimum Sequence	Event in Section (Sequence 2)	"too high"	"too low"	Penalty Points
1	9	5	0	7	7
2	6	8	0	7	7
3	3	3	2	2	4
4	2	10	2	5	7
5	10	1	3	2	5
6	5	7	5	0	5
7	1	2	3	1	4
8	8	9	6	0	6
9	7	6	3	0	3
10	4	4	0	0	0
		Sum=	24	24	48

obtained from Q because P + Q = n(n-1)/2, representing the total number of ordered pairs of events.

Suppose that the total number of penalty points is written as T (= $2Q$). Then the relation between τ and T can be written as:

$$\hat{\tau} = 1 - \frac{T}{P+Q} = 1 - \frac{T}{n(n-1)/2} \qquad (7.3)$$

This equation, for example, can be used to evaluate the relative strength of correlation of each of the three series in the previous example of Table 6.11. It already was pointed out that the total numbers of penalty points amount to 22, 33 and 28 for the situations of Tables 6.11A, B and C, respectively. Because $n = 20$, it follows from Equation (7.3) that the corresponding tau-values are 0.884, 0.826 and 0.853.

Table 7.4 shows another example of application. The 25 original input sequences of Table 4.15 (cf. Sections 4.9 and 6.5) were correlated to the scaled optimum sequence extracted from this dataset after final reordering (see Fig. 6.10). All tau-values for rank correlation in Table 7.4

TABLE 7.4

Kendall's tau for 25 sequences of Table 4.15 correlated to scaled optimum sequence of Fig. 6.10. Values probably different from zero are marked by one ($\alpha = 0.05$) and two ($\alpha = 0.01$) asterisks, respectively.

Seq.	Tau	Seq.	Tau
1	0.31*	14	0.39**
2	0.07	15	0.54**
3	0.61**	16	0.26
4	0.44**	17	0.27
5	0.33*	18	0.09
6	0.32*	19	0.37*
7	0.17	20	0.45**
8	0.49**	21	0.57**
9	0.34*	22	0.56**
10	0.42**	23	0.48**
11	0.93**	24	0.03
12	0.40**	25	0.41**
13	0.49**		

are positive but the differences between values are relatively large. The smallest tau-value is 0.03 and the largest one is 0.61. Values that differ significantly from 0 are marked by asterisks in Table 7.4. A single asterisk indicates that a value exceeds the threshold value for level of significance equal to $\alpha=0.05$; two asterisks mean that the significance level for $\alpha=0.01$ is exceeded as well. Most computer programs for rank correlation provide statistics for testing the significance of Kendall's tau and Spearman's rho (also see Kendall, 1975, Chapter 4). It can be shown that S in Equation (7.1) has variance equal to

$$var S = n(n-1)(2n+5)/18 \tag{7.4}$$

In the example of Table 7.4, $n = 25$. Consequently, var $S = 1833.3$ with corresponding standard deviation $\hat{\sigma}(S)=42.82$. For $n>13$, S is approximately normally distributed. If there is no rank correlation, $E(S)=0$. Then it is possible to estimate X, representing the smallest value of S which is significantly different from zero. After application of a continuity correction (cf. Kendall, 1975, p. 54) which simply consists of subtracting 1 from X, it follows that

$$X-1 = Z_c\hat{\sigma}(S); \; P_c = \Phi(Z_c) \tag{7.5}$$

If the absolute value of S is tested, $\alpha=2(1-P_c)$. If $\alpha=0.05$, $P_c=0.975$ and $Z_c=1.96$. For the example, $\hat{\sigma}(S)=42.82$ and Equation (7.5) gives $X=1.96\times42.82+1=84.93$.

From Equation (7.1) it follows that, for $\alpha=0.05$, the critical value of tau is 0.283. If $\alpha=0.01$, this threshold value becomes 0.372. For this reason, values in Table 7.4 which are greater than 0.283 and 0.372 are followed by one and two asterisks, respectively.

7.4 Presorting and ranking by Harper

In a study evaluating various ranking techniques, Harper (1984) found that probabilistic ranking (presorting option) provided slightly better rankings than the modified Hay method. Harper was interested in comparing competing ranking algorithms in stratigraphic paleontology on the basis of computer-simulated sections. By means of a computer program he (1) generated a hypothetical, and thus known, succession of taxa in time, and (2) simulated their succession in strata at several local sample sites.

If desired, steps (1) and (2) may be repeated for many (50 or 100, for example) iterations and the local site data for each iteration sent to user routines for inferred rankings (inferred succession of events in time). First, data for first and last occurrences (entries and exits) taken together, then data for exits-only, then data for entries- only were sent. For each simulated data set, Kendall and Spearman rank correlation coefficients were computed, and the inferred rankings compared with the known succession of events in time. The performance of two competing ranking algorithms may be compared by

(1) obtaining for each submitted dataset the differences between corresponding Kendall and Spearman rank correlation coefficients computed for the two algorithms, and

(2) testing the observed differences for statistical significance.

Harper (1984) used his computer program to compare three ranking algorithms (presorting, ranking and scaling) provided by Agterberg and Nel (1982a, b) as well as to determine whether the algoithms work as well for datasets combining exits and entries versus datasets for exits-only or entries-only. He concluded from a series of experiments that Agterberg and Nel's presorting algorithm (= probabilistic ranking) performed somewhat better than the modified Hay and scaling algorithms. All three methods performed slightly but significantly better on data for exits-only or entries-only as opposed to combined data. The reader is referred to Harper (1984) for a full discussion of his approach and complete results for all experiments performed. Only a few examples will be given here with emphasis on how Harper's approach can be used in practice; e.g. for choosing the threshold parameters k_c and m_{c_1}.

The computer program begins by generating ranges for 50 taxa over 80 time intervals. A random number generator is used for determining "true" entries and exits of each taxon in a range chart. Next stratigraphic succession data for n_s sample sites are generated by random sampling of the range chart. This sampling is controlled by choosing a value for

(1) the probability (P_1) that a given taxon is present at a local site;

(2) the probability (P_2) that a taxon is sampled at a given horizon at a sample site given that it occurs in the time interval represented by the horizon; and

(3) the probability (P_3) that two adjacent horizons correspond to the same time interval.

Harper conducted 3 experiments (A, B and C) of which the parameters are shown in Table 7.5. For each sample site, n_t sets of stratigraphic succession data were obtained, with n_t representing the number of iterations. Run, sample site, and sequence data were sent to the RASC computer program in order to obtain three types of optimum sequences (a) probabilistic ranking (presorting only); (b) modified Hay method only; and (c) scaled optimum sequence as derived from (b). The threshold parameters employed are shown in Table 7.5. Harper (1984, Fig. 4-6) compared experimentally-obtained optimum sequences with the "true" optimum sequence on the range chart by using Kendall's rank correlation coefficients. In total, 1950 tau-values were calculated, one for each comparison; all turned out to be relatively close to +1, and significantly greater than zero. This signifies that all rankings were good. However, by comparing methods with one another, and looking at small differences between average tau-values, it can be determined which one of a pair of techniques is better. Average differences between tau-values for comparing presorting with the modified Hay method are shown in the bottom four rows of Table 7.5. Each of the values shown is the average of 50 differences between tau-values, except the two values in the last column which were based on 100 differences; n.o. indicates that an average for 100 runs was not obtained for Run C. A negative value signifies that the modified Hay method gave poorer rankings than presorting. Except for Run B (first run), the negative values are significantly different from zero as determined by Student's t-test (Harper, 1984, Tables 2-7). The results for exits and entries are similar as can be expected, and the first two values in the last two columns also duplicate one another.

It may be concluded that, for the experiments performed, probabilistic ranking gave better results than use of the modified Hay method only, when k_c is relatively small. When k_c is large, the two methods probably give rankings that are equally good. The results of the experiments also suggest the possibility that, by increasing the ratio k_c/m_c, the performance of the modified Hay method can be improved. The presorting option (renamed probabilistic ranking in Section 5.5) was introduced in

TABLE 7.5

Results for three computer simulation experiments (A, B and C) conducted by Harper (1984) (for explanation see text).

		A	B	C
Number of sites:	n_s	22	16	6
Probability of presence:	P_1	0.20	0.20	0.10
Sampling probability:	P_2	0.55	0.80	0.85
Adjacency probability:	P_3	0.10	0.10	0.20
Number of datasets:	n_t	50(or 100)	50(or 100)	50
Minimum number of sites:	k_c	5	7	3
Minimum number of pairs:	m_c	4	5	3
Ratio:	k_c/m_c	1.25	1.40	1.00
	exits	−0.013	−0.003	−0.022
Average difference	entries	−0.014	−0.003	−0.020
between tau-values:	both	−0.004	−0.001	−0.007
	both(100)	−0.005	−0.000	n.o.

Agterberg and Nel (1983a) and routinely has been used in RASC runs after 1980. The results of presorting are independent of the choice of the threshold parameters m_{c1} and m_{c2} which apply to the modified Hay method and scaling, respectively. As a result of Harper's experiments, the RASC program was modified in 1983 to allow the choice of separate threshold parameters for these two techniques. Before then, all runs including those performed by Harper had $m_{c1} = m_{c2}$.

Application of the modified Hay method after probabilistic ranking can be regarded as a fine-tuning operation in situations when there are many missing data. The presorting could yield poor results when many frequencies are undetermined. Then it should be useful to compare the ranking of each event with all others in order to find the optimum permutation as is done in the modified Hay method. Ideally, the threshold parameter m_{c1} should be set equal to 1 so that all frequencies are considered. However, a decrease in m_{c1} frequently corresponds to an increase in number of cycles (inconsistencies involving 3 or more events). It then is necessary to use a value greater than 1 in order to reduce the number of iterations.

Harper (1984) also found negative differences between tau-values when the modified optimum sequence resulting from scaling was compared to the optimum sequence resulting from the modified Hay method only. However, the lower tau-values in this instance may have been caused by the fact that Harper (1984, p. 16) regarded as tied successive events which were less than 0.5 apart along the RASC scale. A modified formula for estimating Kendall's rank correlation coefficient was used to accommodate tied events. On average, events preceding other events along the RASC scale, occur before those other events on the range chart as well, even when distances between successive events are small. Scoring them as tied, therefore, results in a somewhat smaller tau-value. This may explain why the optimum sequence from the modified Hay method, in which no ties were allowed, yielded somewhat higher tau-values.

Finally, Harper (1984) showed that exits and entries, run separately, gave somewhat higher tau-values than when both were mixed together. This was to be expected (also see Edwards and Beaver, 1978) because, on the average, exits will be moved downward, and entries upward, with respect to their relative positions on the range chart when stratigraphic succession data for sample sites are generated using probabilities of occurrence (P_1, P_2 and P_3). If exits or entries are considered on their own, this bias will not show up. However, if they are mixed, some exits will probably assume final positions, in any type of optimum sequence, below entries of other taxa which occur above them on the range chart. Although smaller tau-values are to be expected for sequences of mixed entries and exits, these differences were almost negligibly small in the results of Harper's experiments. Harper's experiments were limited to a single type of artificial dataset. It may be expected that different specific conclusions would result from other datasets. Nevertheless, the preceding discussions illustrated that valid generalizations can be derived from computer simulation experiments.

7.5 Precision of the scaled optimum sequence

On the basis of computer simulation experiments, it was concluded in Section 6.5 that, in general, it is possible to obtain unbiased estimates of the cumulative RASC distances of the scaled optimum sequence, provided that the order of events in the scaled optimum sequence is close to the true order of the events. On the other hand, it was not possible to obtain unbiased estimates of the standard deviations of the intervals between successive events along the relative time scale used for the scaling. It was pointed out (cf. Eq. 6.17) that the indirect distances used for estimating each interval are not stochastically independent. Consequently, it would not be a promising approach to add biased variances for the intervals in order to estimate precision of any cumulative RASC distance which is the sum of many intervals. It will be shown in this section that, in general, the jackknife method can be used to obtain approximately unbiased estimates of the standard deviations of the cumulative RASC distances if the order of events in the scaled optimum sequence is close to the true order of the events. The mathematical background of the jackknife method will be given in Chapter 10. Here the purpose of this procedure will be discussed in qualitative terms only, using two of the abbreviated computer simulation experiments for example.

Table 7.6 shows the complete matrix of Z-values which led to the scaled optimum sequence of Figure 6.10. It should be remembered that in this experiment, there are 25 sequences for 20 events which, in each sequence, occupy values that are 0.1 units apart. The standard deviation which controls the scatter of individual events about their means is 0.7071 for all events. Because total distance between the expected location of events 1 and 20 is only $(1.9 \times 0.7071 =)$ 1.34 standard deviations for the difference between two events, none of the 20 events is likely to occur before of after one or more of the other events in all sections. This explains why $q_c = 2.054$ does not occur as a Z-value in Table 7.6. The largest Z-value for this experiment is 1.751 corresponding to $P = 0.96$, representing the situation that event 1 occurs before event 19 in 24 of the 25 sequences. Consequently, it is not necessary to make adjustments for truncation effects when distances between events are estimated from the Z-matrix and the following slightly different procedure can be followed.

The bottom row of Table 7.6 shows the average Z-value for each column. Each column average is based on 19 separate Z-values because the diagonal elements were not used. These averages can be regarded as estimates of the expected locations $E(X)$ of the events along the RASC scale. The origin is between events 11 and 16. If this origin is moved to the first event of the scaled optimum sequence by adding 0.709, the RASC distances of the first column of Table 7.7 are obtained. These values are approximately equal to the unweighted linear scaling values (X_0) for this experiment which are listed in the second column of Table 7.7. The slight differences between the values in the first two columns of Table 7.7 are due to the fact that direct distance estimates are weighted twice as much as indirect distance estimates when the procedure of Table 7.6 is followed. It was already noted (cf. Section 6.4) that doubling the weights of direct distance estimates gives slightly better results. As a procedure it is also

TABLE 7.6

Matrix of Z-values of computer simulation experiment of Tables 4.15, Fig. 6.10 and Table 7.4. The 20 events in 25 sequences have expected values which are closely spaced (at 0.1 intervals) along the RASC scale. The column averages provide estimates of these mean positions variant of unweighted scaling method, see text for further explanation). Successive values within any column are stochastically independent because they deviate randomly from their mean values. The latter are for distances from the mean position of the event labelling the column. The standard deviation of the column average, therefore, can be estimated, e.g. by the jackknife method, without distortion by autocorrelation effects. This property is preserved when the jacknife method is applied to unweighted or weighted distance estimation as in the RASC computer program.

	4	3	1	5	6	2	7	8	9	11	16	14	10	12	15	13	17	18	19	20
4	x	-.151	0.253	0.151	0.253	0.468	0.253	0.468	0.468	1.175	0.842	0.706	0.706	0.842	0.842	0.994	1.405	1.405	1.405	0.994
3	0.151	x	0.151	-.253	0.151	0.151	0.358	0.253	0.253	0.253	0.842	0.842	0.706	0.842	0.994	0.583	1.405	0.842	1.175	1.405
1	-.253	-.151	x	-.253	-.050	0.151	0.151	0.468	0.358	0.706	0.583	0.468	0.706	0.583	0.842	0.994	0.994	0.842	1.751	1.405
5	-.151	0.253	0.253	x	0.050	0.358	0.253	0.468	0.583	0.468	0.253	0.358	0.468	0.583	0.583	0.583	0.842	0.842	1.175	1.175
6	-.253	-.151	0.050	-.050	x	-.151	0.151	0.151	0.358	0.253	0.994	0.583	0.583	0.706	0.706	0.706	1.405	1.405	0.994	1.175
2	-.468	-.151	-.151	-.358	0.151	x	-.050	0.253	0.151	0.358	0.468	0.706	0.583	0.358	0.583	0.842	0.706	0.994	0.994	1.405
7	-.253	-.358	-.151	-.253	-.151	0.050	x	-.050	0.358	0.050	0.358	0.583	0.706	0.468	0.253	0.583	0.842	0.706	0.994	1.175
8	-.468	-.253	-.468	-.468	-.151	-.253	0.050	x	0.358	0.253	0.253	0.253	0.358	0.468	0.358	0.468	0.706	0.706	0.842	1.405
9	-.468	-.253	-.358	-.583	-.358	-.151	-.358	-.358	x	-.358	0.050	0.151	0.358	0.253	0.253	0.583	0.358	0.583	0.468	0.842
11	-1.18	-.253	-.706	-.468	-.253	-.358	-.050	-.253	0.358	x	0.151	-.050	-.050	0.253	0.468	0.358	0.583	0.253	0.468	0.583
16	-.842	-.842	-.583	-.253	-.994	-.468	-.358	-.253	-.050	-.151	x	-.253	0.253	0.358	0.358	-.050	0.842	0.706	0.706	0.583
14	-.706	-.842	-.468	-.358	-.583	-.706	-.583	-.253	-.151	0.050	0.253	x	-.050	-.050	0.050	0.253	0.468	0.583	0.706	0.994
10	-.706	-.706	-.706	-.468	-.583	-.583	-.706	-.358	-.358	0.050	-.253	0.050	x	0.151	0.253	0.358	0.583	0.358	0.151	0.706
12	-.842	-.842	-.583	-.583	-.706	-.358	-.468	-.468	-.253	-.253	-.358	0.050	-.151	x	0.358	0.050	0.151	0.151	0.468	0.994
15	-.842	-.994	-.842	-.583	-.706	-.583	-.253	-.358	-.253	-.468	-.358	-.050	-.253	-.358	x	0.253	0.050	0.253	0.253	0.994
13	-.994	-.583	-.994	-.583	-.706	-.842	-.583	-.468	-.583	-.358	0.050	-.253	-.358	-.050	-.253	x	0.151	0.253	0.050	0.706
17	-1.41	-1.41	-.994	-.842	-1.41	-.706	-.842	-.706	-.358	-.583	-.842	-.468	-.583	-.151	-.050	-.151	x	0.151	0.358	0.706
18	-1.41	-.842	-.842	-.842	-1.41	-.994	-.706	-.706	-.583	-.253	-.706	-.583	-.358	-.151	-.253	-.253	-.151	x	0.050	0.706
19	-1.41	-1.18	-1.75	-1.18	-.994	-.994	-.994	-.842	-.468	-.468	-.706	-.706	-.151	-.468	-.253	-.050	-.358	-.050	x	0.468
20	-.994	-1.41	-1.41	-1.18	-1.18	-1.41	-1.18	-1.41	-.842	-.583	-.583	-.994	-.706	-.994	-.994	-.706	-.706	-.706	-.468	x
Ave	-.709	-.584	-.528	-.473	-.482	-.345	-.277	-.184	-.005	0.093	0.128	0.140	0.215	0.260	0.318	0.381	0.599	0.591	0.707	0.971

TABLE 7.7

Comparison of four scaling methods applied to example of Table 7.6. Ave represents column average of Table 7.6 after addition of 0.709 (= minus first column average). X_0 and X are RASC computer program unweighted and weighted scaling results. E (X) represents true mean value which is multiple of 0.0707. Q and $s(Q)$ are jackknife estimate and jackknife standard deviation using RASC weighted scaling method. t (X) is studentized deviation of X from true mean value. Penalty points (pp) for event numbers of column 1 are shown in last column.

	Ave	X_0	X	E(X)	Q	$s(Q)$	$t(X)$	pp
4	0.000	0.000	0.000	0.212	0.000	0.000	***	3
3	0.125	0.133	0.117	0.141	0.170	0.057	-.429	1
1	0.181	0.168	0.172	0.000	0.179	0.040	4.298**	2
5	0.236	0.228	0.185	0.283	0.200	0.064	-1.53	1
6	0.227	0.214	0.204	0.354	0.215	0.052	-2.88*	1
2	0.365	0.340	0.306	0.071	0.319	0.049	4.821**	4
7	0.433	0.420	0.375	0.424	0.417	0.054	-.920	0
8	0.525	0.501	0.453	0.495	0.488	0.054	-.781	0
9	0.705	0.680	0.634	0.566	0.677	0.067	1.019	0
11	0.803	0.741	0.663	0.707	0.636	0.067	-.651	1
16	0.838	0.793	0.726	1.061	0.727	0.059	-5.66**	5
14	0.849	0.812	0.736	0.919	0.774	0.036	-5.14**	2
10	0.924	0.887	0.803	0.636	0.837	0.048	3.499**	3
12	0.970	0.925	0.851	0.778	0.890	0.053	-1.39	2
15	1.027	0.983	0.923	0.990	0.972	0.059	-1.12	0
13	1.090	1.083	0.986	0.849	1.019	0.057	2.441*	3
17	1.308	1.234	1.154	1.131	1.170	0.057	0.394	0
18	1.300	1.226	1.170	1.202	1.188	0.056	0.578	0
19	1.417	1.343	1.265	1.273	1.281	0.065	0.124	0
20	1.680	1.628	1.598	1.344	1.644	0.063	4.072**	0

invoked in weighted distance estimation option of the RASC computer program.

The weighted scaling values (X) previously used for constructing the diagram of Figure 6.10 (also see Table 6.14) are shown in the third column of Table 7.7 in comparison with the theoretical mean positions E(X). Jackknife estimates (Q) for weighted scaling are presented in the next column. If the jackknife estimates (Q) are close to the weighted scaling values (X), their standard deviations can be used as standard deviations of X.

In general, the jackknife provides a non-parametric method of estimating the mean and its standard deviation for a sample of n

independent and identically distributed random variables. In the situation of ungrouped data, each of the n values is successively deleted from the sample and a pseudovalue is computed from each reduced data set with (n-1) values. The jackknife estimate is the mean of the n pseudovalues. In the situation of Table 7.6, each column average is based on n (=19) values for separate events. These values can be regarded as realizations of stochastically independent random variables for individual events. Every event corresponds to a set of 25 random normal numbers with its own mean value. Deletion of an event results in a reduced Z-matrix without the row and column of the deleted event. The Z-values for the remaining n-1 (=18) events are not changed by the process of deleting an event. The 19 pseudovalues are not necessarily stochastically independent but this hypothesis can be tested in the computer simulation experiment because all deviations from the true means are known. Studentized residuals $t(X)$ were obtained by dividing each difference X-E(X) by its corresponding standard deviation $s(Q)$ (see Table 7.7).

The 20 studentized residuals of Table 7.7 should have zero mean and deviate from zero according to the t-distribution with n-1 (=18) degrees of freedom. Consequently, it would be expected that, on average, only 1 out of 20 values, in absolute value, deviates by more than 2.101 from zero, and 1 in 100 values by more than 2.878. Most of the studentized residuals in Table 7.7 are within these confidence levels of Student's t-distribution for 18 degrees of freedom. However, a number of the studentized residuals are too large in absolute value indicating that locally the hypothesis of stochastical independence of the pseudovalues was not satisfied.

One problem here is that the origin of a RASC scale is set automatically at the first event of the scaled optimum sequence. All pseudovalues are forced to be zero at this point and this results in the artificial result $s(Q)$ = 0 for first events. This problem generally cannot be avoided in practical applications. Another problem indicated by the results of Table 6.15 is that anomalously large values occur at positions in the scaled optimum sequence for events that are out of position with respect to the true squence of expected values. The last column in Table 7.7 shows number of penalty points for each event. For example, event 16 ended up in position 11 of the scaled optimum sequence. For this reason, it was assigned (16-11=) 5 penalty points. Its studentized residual (= -5.66) is nearly twice as large as the significance limit (=2.878) for α=0.01. This suggests that $s(Q)$ (=0.059) for this event is too small by a factor of two or more. It is noted that the jackknife procedure applied to standard deviations obtained by means of Equation (6.13) does not remove bias from these estimates as illustrated in Table 7.8.

The preceding computations were repeated for the example of Table 4.13 and Figure 6.8 with expected interval equal to 0.5 instead of 0.1. The results are shown in Table 7.9. RASC distances (X) near the top and bottom of the scaled optimum sequence now are based on fewer data (N^*) than those in the middle. In general, it does not make sense to let the jackknife estimator of position of an event be affected by events that are clearly above or below this event. For this reason, a window equal to $X \pm 2$ was applied to each cumulative RASC distance ($= X$) and events outside this window were not used to compute Q and $s(Q)$. The reduced number of pseudovalues ($=N$) used is also shown in Table 7.9. The width of the window is such that N is approximately equal to N^*. Setting the width

TABLE 7.8

Comparison of differences between successive values for example of Table 7.7. D and $s(D)$ are intervals and their standard deviations estimated by weighted scaling in RASC computer program. D_1 and $s(D_1)$ are corresponding jackknife estimates.

	D	$s(D)$	D_1	$s(D_1)$
4-3	0.117	0.066	0.174	0.060
3-1	0.055	0.060	0.021	0.058
1-5	0.013	0.056	0.046	0.054
5-6	0.019	0.072	0.012	0.076
6-2	0.102	0.055	0.138	0.049
2-7	0.069	0.045	0.096	0.044
7-8	0.078	0.040	0.096	0.040
8-9	0.181	0.049	0.157	0.047
9-11	0.029	0.064	0.033	0.051
11-16	0.063	0.072	0.071	0.075
16-14	0.011	0.059	0.070	0.054
14-10	0.066	0.050	0.074	0.053
10-12	0.048	0 048	0.047	0.049
12-15	0.072	0.039	0.045	0.035
15-13	0.063	0.053	0.026	0.054
13-17	0.167	0.071	0.169	0.075
17-18	0.016	0.048	-0.000	0.050
18-19	0.095	0.055	0.095	0.054
19-20	0.334	0.069	0.348	0.064

TABLE 7.9

Jackknife method applied to computer simulation experiment of Table 4.13 and Fig. 6.8. The 20 events in 25 sequences have expected values *E(X)* spaced at intervals which are 5 times wider than those used in the previous example of Tables 7.6 to 7.8. *X, E(X), Q* and *s(Q)* as in Table 7.7. The weighted distance results *X* and *Q* were based on *N** and *N* differences between successive *Z*-values, respectively. *t(Y)* is studentized deviation of *Y*=*X*–*E (X)*+0.559.

	X	*N**	E(*X*)	*Q*	s(*Q*)	*N*	Y	*t*(*Y*)
1	0.000		0.000	0.000	0.000	7	0.559	***
3	0.492	7	0.707	0.540	0.063	8	0.343	5.439**
2	0.507	9	0.354	0.637	0.096	9	0.712	7.439**
4	0.708	9	1.061	0.797	0.107	9	0.206	1.925
5	1.190	10	1.414	1.247	0.095	11	0.334	3.502**
6	1.393	10	1.768	1.442	0.160	12	0.184	1.149
7	1.843	14	2.121	1.951	0.131	14	0.280	2.145*
8	2.069	13	2.475	2.146	0.164	14	0.153	0.908
9	2.505	14	2.828	2.476	0.168	15	0.236	1.399
10	2.871	13	3.182	2.953	0.148	13	0.247	1.665
11	2.977	13	3.536	3.053	0.139	13	0.000	0.000
12	3.287	13	3.889	3.342	0.158	14	-.044	-.277
13	3.696	11	4.243	3.706	0.134	14	0.012	0.090
14	3.753	13	4.596	3.805	0.130	14	-.284	-2.20*
15	4.234	13	4.950	4.407	0.096	12	-.157	-1.63
16	4.261	13	5.303	4.406	0.111	12	-.484	-4.63**
17	5.104	11	5.657	5.349	0.189	9	0.006	0.031
18	5.153	8	6.010	5.413	0.162	9	-.299	-1.84
19	5.567	8	6.364	5.804	0.140	10	-.148	-1.06
20	6.265	6	6.718	6.509	0.220	4	0.106	0.481

equal to 2 is equivalent to excluding events that occur above or below the deleted event with a probability greater than 95 percent. In micro-RASC (see Chapter 10), the user can change the width from its default value (=2) to any other value.

Both *X* and *Q* are relatively poor estimates of E(*X*) at positions near the top of the scaled optimum sequence. Because these poor estimates affect the other estimates lower down in the scaled optimum sequence and the choice of origin is arbitrary, it was decided to reset the origin to the position of event 11 near the midpoint of the scaled optimum sequence. Consequently, studentized residuals *t*(*Y*) were estimated for *Y* = *X*-E(*X*) + 0.559 (see Table 7.9). As in Table 7.6, the majority of the studentized residuals are within the 95 percent confidence limits. By means of

TABLE 7.10

Jackknife method applied to Hay example. X, Q and $s(Q)$ are weighted scaling results for cumulative RASC distance, its jackknife estimate and jackknife standard deviation, respectively.

	X	Q	$s(Q)$
9	0.000	0.000	0.000
10	0.317	0.435	0.049
8	0.493	-.064	0.302
6	1.263	1.064	0.642
4	1.529	1.929	0.657
7	1.686	1.930	0.638
5	1.843	2.170	0.677
1	2.038	2.347	0.684
2	2.156	2.470	0.693
3	2.162	2.469	0.668

asterisks it is shown that some values of $s(Q)$, especially those near the top of Table 7.9, are too small. Although this indicates that, locally, there are statistically significant discrepancies between X and $E(X)$, these differences are rather small in relative terms. In Table 7.7 the maximum difference between X and $E(X)$ is 0.254 or about 16 percent of the total range (=1.598) of the RASC scale. In Table 7.9, the maximum difference is 0.897 or 13 percent of total range (=6.718). It may be concluded that, on the whole, the jackknife method yields good estimates of the positions of the events in the scaled optimum sequence provided that the initial ranking was good.

Table 7.10 shows Q and $s(Q)$ in comparison with X for the Hay example. The six events in the lower part of the scaled optimum sequence are not only subject to strong clustering but also have relatively large standard deviations. Events 8, 9 and 10 clearly are above the other events with events 8 and 10 having relatively small standard deviations. Event 6 may be intermediate between the preceding two groups. Differences between X and Q for the Hay example are larger than those in Table 7.7 and 7.9. More research would be needed to determine which estimate (X or Q) is better than the other. It is known that jackknife estimators in parametric estimation frequently are superior because bias of order n^{-1} (i.e. inversly proportional to sample size) tends to be eliminated (see e.g.

Miller, 1974). On the other hand, this advantage may be offset by the introduction of bias related to lack of stochastical independence of the pseudovalues.

CHAPTER 8

NORMALITY TESTING AND THE MODIFIED RASC METHOD

8.1 Introduction

The normality test of the RASC computer program was briefly described in Section 6.6. In this chapter, it will be explained in more detail. The problem of estimating the autocorrelation of the second-order differences used in this test will be discussed first. A simple method will be introduced by which it is possible to determine statistically whether or not anomalous events belong to the normal distribution of the second-order differences. For comparison with results obtained by Guex and Davaud (1984) for a reworked bed using the Unitary Associations method, the normality test will be applied to Drobne's (1977) alveolinids from Yugoslavia. The RASC computer program with normality test also will be applied to Palmer's (1954) data for the fauna of the Riley Formation of the Llano Uplift in central Texas. Earlier, Shaw (1964) had constructed a composite standard from Palmer's database which involved the determination and elimination of what he considered to be anomalous events. It will be seen that the majority of the events deleted by Shaw are not anomalous when the normality test is applied and this difference in conclusions will be discussed.

The modified RASC method will be presented using the Gradstein-Thomas database for example. This procedure can be used to construct conservative range charts. Various types of range charts constructed by different methods will be compared with one another in the last two sections of this chapter. The modified RASC method can be very useful for defining marker events which have variances that are much smaller than the variances of other events. Modified RASC also provides new information on the shapes of the frequency distributions of stratigraphic events.

8.2 Autocorrelation of the second-order differences

The normality test was developed for two reasons: (a) to determine anomalous events which in a specific section occur much higher or lower than (at their average locations) in a regional standard developed on the basis of a number of sections in a region; and (b) to test the normality assumption used to transform cross-over frequencies into Z-values during scaling. The normality test contributes useful information with respect to both these objectives.

In the first few versions of the RASC computer program (Agterberg and Nel, 1982; Heller et al., 1983), the simplifying assumption was made that the second-order differences for stratigraphic events observed in specific sections would be approximately normally distributed with standard deviation equal to 2σ, if the original events are normally distributed along the RASC scale with standard deviation equal to σ. It was realized that this simple model yields results which were at best approximately true. In the original applications which were mainly to Cenozoic and Cretaceous foraminiferal databases for the northwestern Atlantic margin, the final histograms of the normality test showed observed frequencies that were, on the average, equal to the expected frequencies indicating that this simple model could be used. Three sets of frequencies for the original normality test are shown in Table 8.1.

Anomalous events would cause observed frequencies of the highest and lowest class (O_1 and O_{10}) to be greater than the expected frequency E_i ($i = 1, 2, ..., 10$) which is equal for all classes of i. During 1982 and 1983 when the RASC program was applied to other databases, several of which were listed in Appendix I of Gradstein et al. (1985), it turned out that the

TABLE 8.1

Normality test output from the original RASC program: Comparison of the observed frequencies (O_i) of second order difference-values in each of the ten classes $i = 1, 2, ..., 10$, with the expected frequencies (E_i) which are constant for each of the ten classes.

Source	E_i	O_1	O_2	O_3	O_4	O_5	O_6	O_7	O_8	O_9	O_{10}
Agterberg and Nel (1982, Table 6)	24.1	27	23	26	20	27	24	28	22	21	23
Heller et al. (1983, Table 6)	21.5	30	20	21	15	22	22	18	23	13	31
Gradstein (1984, Table 3)	39.8	50	36	32	41	43	31	39	42	38	46

TABLE 8.2

Normality test output for ten computer simulation experiments. Observed frequencies O_i are compared to the expected frequency (=90) for each of the ten classes $i = 1, 2, ..., 10$. E(D) represents the expected interval (or RASC distance) between event-positions along the RASC-scale in these experiments.

Original RASC	O_1	O_2	O_3	O_4	O_5	O_6	O_7	O_8	O_9	O_{10}
E(D) = 1.0, Set 1	156	55	32	69	127	145	54	39	48	175
E(D) = 1.0, Set 2	162	69	44	82	88	140	64	28	60	163
E(D) = 0.5, Set 1	119	98	77	52	78	117	55	80	104	120
E(D) = 0.5, Set 2	119	95	89	62	79	94	84	59	88	131
E(D) = 0.3, Set 1	89	111	75	84	87	88	85	77	107	97
E(D) = 0.3, Set 2	102	114	89	80	72	80	69	78	102	114
E(D) = 0.2, Set 1	84	101	83	76	80	98	106	100	88	84
E(D) = 0.2, Set 2	62	118	107	97	89	75	81	87	91	93
E(D) = 0.1, Set 1	18	77	91	135	115	123	153	111	62	15
E(D) = 0.1, Set 2	10	76	106	129	139	134	112	103	75	16

original normality test provided poor results in some situations because the frequencies of anomalous events were either much larger or much smaller than expected. For example, too many anomalous events were found in the database for Baumgartner's (1984) Jurassic Tethyan radiolarians, and too few in the Sullivan-Bramlette database for Paleogene Californian nannofossils (cf. Section 4.2). It became difficult or even impossible in these situations to define anomalous events on the basis of the normal distribution model originally assumed to hold approximately true for the second-order differences. It was decided to assess the problem systematically by means of the computer simulation experiments previously described in Chapter 6.

Table 8.2 shows observed frequencies obtained by a pre-1985 version of the RASC program for 10 classes of 900 second-order differences created in ten of the computer simulation experiments previously described in Chapter 6. The expected frequency is 90 for all 100 entries for observed frequencies in Table 8.2. Clearly, the observed frequencies in the tails of these distributions for the second-order differences are too large when E(D) is greater than 0.5 and they are too small when E(D) is less than 0.2. It is noted that the runs for E(D) = 1.0, have a single greater than expected frequency near the center of their distributions. This phenomenon is related to the use of pairs of Z-values arbitrarily set equal to q_c (= 2.326) and with zero difference between them (see Chapter 6). This constitutes a minor problem which is not related to the problem at hand and does not arise for smaller values of E(D) in the experiments.

The applications of Table 8.1 may be compared to the experiments on artificial data sets, with E(D) between 0.2 and 0.5, for which the observed frequencies, on the average, are equal to the expected frequency E_i (= 90) in Table 8.2.

The present, revised normality test in the RASC computer program consists of fitting a doubly-truncated normal distribution to the second-order differences belonging to the classes with observed frequencies O_3 to O_8. If present, anomalous events are most likely to occur in the tails of an observed frequency distribution. Values in the classes of frequencies O_1, O_2, O_9 and O_{10} therefore were not used for estimating a theoretical normal distribution.

Each second-order difference value in the normality test is computed as follows. First, the difference of two successive values is calculated. If an event precedes the next event for a section in the SEQ file, their difference is corrected by subtracting a small amount. This correction is made because a gradual increase in distance from the origin is to be expected for successive events in each section. The small amount was set equal to the difference between the highest and lowest cumulative RASC distance values in the observed sequence for a section divided by the total number of times an event precedes the next event for this section in the SEQ file without being coeval to it. No correction is made for pairs of coeval events. Next, the successive difference of two resulting values is determined. This procedure resembles the calculation of a second derivative with respect to location for every event except those in the first or last positions of an observed sequence.

The second-order difference calculated in the RASC normality test is minus the difference between twice the RASC distance of an event on the one hand and the sum of the distance of its two neighboring events, on the other. If successive differences could be regarded as realizations of independent normal random variables with variances equal to $2\sigma^2$, the variance of the second-order difference would amount to $6\sigma^2$. This can be seen as follows. Suppose that three successive distance estimates $X_{k\text{-}1}$, X_k and X_{k+1} were normally distributed with zero mean and variance σ^2, then the second-order difference $-(2X_k - X_{k\text{-}1} - X_{k+1})$ would be normal with variance of $6\sigma^2$ because $\sigma^2(2X_k) = 4\sigma^2$ and $\sigma^2(X_{k\text{-}1}) = \sigma^2(X_{k+1}) = \sigma^2$. However, the successive distance estimates have become autocorrelated because of the various manipulations to which the data were subjected during ranking and scaling. Suppose that the autocorrelation coefficient of successive distances X_k and X_{k+1} is written as ρ with $\rho = \mathrm{Cov}\,(X_k, X_{k+1})/\sigma^2$. The variance of the second-order difference satisfies

$$\sigma_2^2 = E[(X_{k-1}-X_k)-(X_k-X_{k+1})]^2 \tag{8.1}$$

if $\quad E[(X_{k-1}-X_k)-(X_k-X_{k+1})] = 0$

It follows that

$$\sigma_2^2 = 2\sigma^2(\rho^2-4\rho+3) \tag{8.2}$$

if $\quad \mathrm{Cov}\,(X_{k-1}, X_{k+1}) = \rho^2\sigma^2$

The procedure followed in the RASC program consists of ordering the second-order differences from all sections from the smallest to the largest value. The standard deviation of the central 60 percent of the ordered values is estimated and assumed to represent a truncated normal distribution. The relationship between standard deviations of truncated normal and normal distributions is given in statistical tables. Their ratio amounts to 0.463 if 20 percent is truncated from each tail. Division by 0.463 yields the estimate $\hat{\sigma}_2$. Not all second-order differences are used for this estimation because if anomalous values are present, these are more likely to occur in the tails of the distribution. From $\sigma^2 = \frac{1}{2}$, it follows that ρ can be estimated from $\hat{\sigma}_2$ by

$$\rho = 2-(1+\sigma_2^2)^{\frac{1}{2}} \tag{8.3}$$

In general (cf. Agterberg, 1974, p. 302), it can be assumed that n autocorrelated values are equivalent to n' stochastically independent values with

$$1/n' = 1/n+2\rho\left[n/(1-\rho)-1/(1-\rho)^2\right]/n^2 \tag{8.4}$$

This allows us to estimate n' which is part of the output of the RASC program. In the chi-squared test for goodness of fit, expected frequencies E_i of stochastically independent data in p classes are related to the corresponding observed frequencies O_i by

$$\sum_{i=1}^{p}(E_i-O_i)^2/E_i = \hat{\chi}^2(p-3) \tag{8.5}$$

if two parameters of the fitted distribution were estimated. For autocorrelated data, the sum on the left-hand side of this equation may be multiplied by n'/n in order to obtain an approximate estimate of chi-squared.

The 10 classes of the normality test in the RASC program (cf. Section 6.6) were constructed by dividing the expected ordered sequence of second-order differences into 10 equal parts in order to obtain 10 equal expected frequencies for comparison to the corresponding observed frequencies. The class limits are given by the Z-values of the relative frequencies 0.1, 0.2, ..., 0.9 multiplied by $\hat{\sigma}_2$. This procedure provides a convenient normality test. The individual second-order differences (top part of normality test output as shown in Table 6.16) were compared to the 95% and 99% confidence intervals $\pm 1.960\ \hat{\sigma}_2$ and $\pm 2.576\ \hat{\sigma}_2$, respectively.

The preceding method generally yields sets of observed frequencies O_i ($i = 2,3,...,9$) which are equal to one another (and to E_i) except for random fluctuations. The frequencies (O_1 and O_{10}) in the tails of the distribution may be too high when anomalous events occur in several of the sections.

Results of applying the revised normality test for nine databases are shown in Table 8.3 and for six computer simulation experiments in Table 8.4. Other statistics for most of these computer runs are given in Tables 8.5 and 8.6.

The normal distribution model provides a good fit for 13 of the 15 tests in Table 8.3 according to the approximate chi-squared test (see last column of Table 8.3). The 95 and 99 percent confidence limits of $\hat{\chi}^2(7)$ which should not be exceeded if the normality assumption holds true (with levels of

TABLE 8.3

Revised normality test output for the nine databases in Agterberg et al. (1985) using RASC program. Table 4.9 is slightly improved version of database 1; Tables 4.13, 4.14 and 4.15 are same as databases 9A, 9B and 9C, respectively.

Data Base	E_i	O_1	O_2	O_3	O_4	O_5	O_6	O_7	O_8	O_9	O_{10}	$\hat{\chi}^2(7)$
1. Gradstein-Thomas	50.3	70	42	38	49	55	52	49	45	46	57	5.93
2. Gradstein	21.1	20	21	30	13	23	17	29	18	18	22	7.55
3. Doeven	64.1	78	53	65	68	53	64	70	67	53	70	3.36
4. Baumgartner	149.6	127	175	142	143	158	155	149	140	176	131	15.80
5. Blank	172.2	235	139	145	139	210	173	179	147	118	235	53.72
6A. Rubel, brachiopods	62.3	61	59	65	73	52	59	66	69	63	56	2.35
6B. Rubel, ostracods	36.8	43	37	21	36	41	46	33	30	39	42	5.07
6C. Rubel, thelodonts	35.9	39	37	39	29	37	40	36	32	27	42	2.31
6D. Rubel, combined	57.6	50	75	45	62	62	51	54	69	52	51	12.52
7. Sullivan	47.4	55	40	40	49	66	37	44	46	42	55	2.45
8A. Corliss, tops	1.8	1	1	3	1	2	2	2	1	3	2	2.94
8B. Corliss, bottoms	5.0	6	2	6	10	5	3	2	7	2	7	9.24
9A. Agterberg-Lew, E(D)=0.5	45.0	44	41	56	35	34	35	42	54	48	41	8.76
9B. Agterberg-Lew, E(D)=0.3	45.0	43	45	45	44	38	57	36	56	50	36	6.20
9C. Agterberg-Lew, E(D)=0.1	45.0	62	29	34	46	43	46	53	47	50	40	3.51

TABLE 8.4

Normality test output for six computer simulation experiments. See text for further explanation.

A. Revised RASC (Set 1 only)	O_1	O_2	O_3	O_4	O_5	O_6	O_7	O_8	O_9	O_{10}	$\hat{\chi}^2(7)$
E(D) = 0.5	70	117	93	58	86	132	66	107	95	76	52.6
E(D) = 0.3	81	91	90	90	94	90	93	94	96	81	1.9
E(D) = 0.2	84	102	82	78	78	98	106	100	87	85	6.3
E(D) = 0.1	85	88	100	79	84	86	103	107	97	71	3.2
B. Revised RASC	O_1	O_2	O_3	O_4	O_5	O_6	O_7	O_8	O_9	O_{10}	$\hat{\chi}^2(7)$
E(D) = 0.0, Set 1	98	90	73	86	94	98	83	120	85	73	0.5
E(D) = 0.0, Set 2	86	81	98	90	86	94	76	108	106	75	0.2

TABLE 8.5

Some statistics for RASC results for 9 databases of Table 8.3. The equivalent number (n') of stochastically independent values was derived from number of second-order differences (n), standard deviation $\hat{\sigma}_2$ of Gaussian curve fitted to second-order differences (large values were not used, see text), and estimated autocorrelation coefficient ($\hat{\rho}$).

Data Base	k_c	No. of Events	No. of Sections	n	$\hat{\sigma}_2$	$\hat{\rho}$	n'
1. Gradstein-Thomas	7	44	24	503	1.223	0.420	206
2. Gradstein	5	31	20	211	1.471	0.222	135
3. Doeven	7	77	10	641	1.108	0.508	210
4. Baumgartner	13	86	43	1496	1.701	0.027	1419
5. Blank	15	80	81	1722	1.419	0.264	1003
6A. Rubel, brachiopods	8	54	20	632	1.234	0.412	260
6B. Rubel, ostracods	8	40	12	368	1.192	0.444	142
6C. Rubel, thelodonts	8	34	20	359	1.188	0.447	137
6D. Rubel, combined	13	43	35	576	1.659	0.063	507
7. Sullivan	9	52	10	474	0.791	0.725	76
8A. Corliss, tops	3	9	6	18	1.686	0.040	17
8B. Corliss, bottoms	4	15	6	50	1.516	0.184	35
9A. Agterberg-Lew, E(D)=0.5	25	20	25	450	1.512	0.187	309
9B. Agterberg-Lew, E(D)=0.3	25	20	25	450	1.388	0.289	248
9C. Agterberg-Lew, E(D)=0.1	25	20	25	450	0.881	0.668	90

TABLE 8.6

Autocorrelation statistics for RASC runs of five computer simulation experiments. If the original values along the RASC-scale were stochastically independent, the ratio $\hat{\sigma}_2 / \sigma$ would be equal to 1. Note extreme reduction from n to n' for E(D) = 0.0. The negative autocorrelation coefficients $\hat{\rho}_1$ apply to second-order differences (see text).

E(D)	n	$\hat{\sigma}_2$	$\hat{\sigma}_2/\sigma$	$\hat{\rho}$	n'	$\hat{\rho}_1$
0.5	900	1.698	0.98	0.030	848	-0.658
0.3	900	1.528	0.88	0.173	634	-0.621
0.2	900	1.408	0.87	0.273	514	-0.597
0.1	900	0.966	0.56	0.609	219	-0.532
0.0	900	0.327	0.19	0.948	25	-0.501

significance equal to 5 and 1 percent), amount to 14.1 and 18.5, respectively. Only $\hat{\chi}^2(7) = 53.7$ of database no. 5 clearly exceeds both confidence limits. According to Blank (1984, p. 65) a number of events in this database were determined to be anomalous because of four main reasons: (1) taxonomic problems with Mesozoic events, (2) short sections that were artificially truncated at coring gaps, (3) contamination due to reworking, and (4) provinciality because of the large latitudinal spread of control sites. The chi-squared value for database no. 4 exceeds the 95 percent confidence limit but is below the 99 percent confidence limit. There is the possibility that the tail frequencies O_1 (= 127) and O_{10} (= 131) are slightly too small (in comparison with $E_i = 149.6$).

The run for E(D) = 0.5 in Table 8.4 gave $\hat{\chi}^2(7) = 52.6$ indicating nonnormality. It is likely that the central frequency O_6 (= 132) is significantly greater than its expected value ($E_i = 90$) for the same reason that O_6 was too high in the computer simulation experiment with E(D) = 1.0 (see Table 8.2).

In part B of Table 8.4, the values of $\hat{\chi}^2(7)$ are equal to 0.5 and 0.2, respectively. The 1 and 5 percent confidence limits of $\hat{\chi}^2(7)$ amount to 0.6 and 1.6, respectively. This suggests a degree of fit which is too good to be true. The approximate chi-squared test is based on the assumption that n autocorrelated values are equivalent to n' independent values (see before). As shown in Table 8.6, this reduction becomes very large (from $n = 900$ to $n' = 25$) when E(D) = 0. There are no definite trends in the two sets of O_i-values in Table 8.6. It may therefore be assumed that the procedure used for estimating the observed and expected frequencies remains valid when E(D) approaches 0 but that the reduction from n to n' has become too large.

Finally, it is noted that the autocorrelation coefficient $\hat{\rho}$ estimated from $\hat{\sigma}_2/\sigma$ applies to the successive distances X_k and not to the second-order differences $(X_{k-1}-X_k)-(X_k-X_{k+1})$. Suppose that the autocorrelation coefficient of the second- order differences is called ρ_1. Then,

$$\rho_1 = E\left[\left\{(X_{k-1}-X_k)-(X_k-X_{k+1})\right\} \cdot \left\{X_k-X_{k+1})(X_{k+1}-X_{k+2})\right\}\right]/\sigma_2^2 \tag{8.6}$$

It follows that

$$\rho_1 = \frac{\rho^3-4\rho^2+7\rho-4}{2\rho^2-8\rho+6} \tag{8.7}$$

if $\quad \mathrm{Cov}(X_{k+1}, X_k) = \rho^i \sigma^2 \quad i = 1, 2, 3$

The latter condition would imply that the X_k satisfy a first-order Markov process (Agterberg, 1974). The autocorrelation coefficient ρ_1 of the second-order differences is negative and ranges from -0.6667 for $\rho = 0$ to -0.5 in the limit for $\rho \rightarrow 1$. Its values in five computer simulation experiments are shown in Table 8.6. It is noted that the estimation of the autocorrelation coefficients ρ and ρ_1 has no bearing on the calculation of the observed and expected frequencies of the normality test. The theory of autocorrelation only was used to provide an approximate chi-squared test for comparing the observed and expected frequencies with one another.

D'Iorio (1988) has performed experiments on the effect of increasing the threshold value q_c (=largest Z-value corresponding to P=1.00) on the RASC scaled optimum sequence for an integrated databank of Cenozoic foraminifers and dinoflagellates on the Labrador Shelf–Grand Banks. The total length for the scaled optimum sequence (=maximum cumulative RASC distance) increased from 7.781 to 12.351 when q_c was enlarged from its default value 1.645 (for P=0.95) to 2.576 (for P=0.995). When all RASC distances, after enlarging q_c, were reduced in length by the ratio (7.781/12.351=) 0.630, there was little change in the shape of the dendrogram. D'Iorio concluded that the scaled optimum sequence is not sensitive to changes in the choice of q_c. The large increase in q_c in the preceding experiment not only had an undesirable effect on the total length of the scaled optimum sequence, it also resulted in a slight but significant distortion of the shape of the normal distribution of the second-order differences. The estimated value of $\hat{\sigma}_2$ (cf. Eq. 8.2), which amounted to 1.454 (with $\hat{\rho}$=0.236) for D'Iorio's 860 second-order differences with q_c=1.645, increased to the unrealistically large value of $\hat{\sigma}_2$=2.413 for q_c=2.576.

The latter value is too large because there is no reason to expect that ρ in Equation (8.2) is much less than zero when n is too large. Consequently, the upper bound of σ_2 is approximately $\sqrt{3}$=1.732 which is less than $\hat{\sigma}_2$=2.413. By using q_c-values that are too large, both σ and σ_2 become too large and Equations (8.3) and (8.4) are no longer valid. As a result, the corrected sum used in the chi-squared test (cf. Eq. 8.5) was overestimated. On the other hand, the 95% and 99% confidence limits for second-order differences (used to indicate possibly anomalous events in the normality test for individual sections) are not sensitive to the choice of q_c.

8.3 Unitary Associations and RASC methods applied to Drobne's alveolinids

Guex (1981) has coded biostratigraphic information on alveolinids collected by Drobne (1977) and applied the Unitary Associations method to these data. Information on 15 species in 11 sections as used by Guex (1981) is shown in Figure 8.1 and Table 8.7. Figure 8.2 from Drobne (1977, Figs. 54 and 55, pp. 88-89) shows the original stratigraphic data for one of the sections (II, Dane near Divača), for example. Forbidden structures (see Chapter 3) have to be identified and eliminated before an interval graph with Unitary Associations can be constructed from the observed co-occurrences. The computer program of Guex and Davaud (1984) initially detected a strong component in the biostratigraphical graph for the Drobne data thus providing useful information on biostratigraphical inconsistencies. This strong component involved fossils 1, 3, 4, 11 and 13. The frequencies of arcs of the strong component belonging to cycles C_3 were tabulated by Guex and Davaud (1984) and the *s*-ratio (see Section 3.5) was determined. The arc from 4 to 3 which occurs only in Section I (Fatji hrib) has the highest *s*-ratio (= 3.00).

Other tabulations in the output from Guex and Davaud's (1984) computer program indicated that an abnormally large proportion of the inconsistencies is due to the occurrence of fossils 3, 4 and 8 in this same section. In the original plot for individual sections (Fig. 8.1) it can be seen that species 3 occurs higher in Section I than in the other sections where it was observed. Drobne (1977, p. 83) specifically stated that bed no. 5 in the Fatji hrib section which contains fossils 3 and 8 was reworked. For this reason, Guex and Davaud (1984) decided to delete fossil 3 from their level no. 4 in Section 1 and to repeat the analysis. Final results for the modified computer run (without species 3 in Section 1) are shown in Table 8.8. The method followed to obtain the unitary associations in the resulting "range chart" was as described in Section 3.5. The five U.A.'s of Table 8.8 which resulted from the union of some I.U.A.'s correspond closely to the original definition of Oppel zones (cf. Section 2.2).

In order to illustrate the normality test, I previously applied it to Drobne's alveolinids as follows (cf. Gradstein et al., 1985, pp. 253-262).

P	SAM	1	2	3	4	5	6	7	8	9	10	11	12	13	14	15	L
I	5			1					1								4
	7				1			1					1				3
	8						1										2
	11		1														1
II	23									1					1		4
	20	1					1	1		1		1	1				3
	16						1	1						1			2
	7			1		1		1									1
III	21									1	1				1		7
	19									1					1	1	6
	15											1				1	5
	14												1			1	4
	13						1	1						1			3
	12			1			1	1									2
	8		1														1
IV	5/2									1	1				1	1	4
	5/1				1					1	1				1		3
	6/1	1							1	1		1	1				2
	6/2	1						1		1		1					1
V	48				1												2
	26					1											1
VI	8/1										1				1		4
	7/4				1					1							3
	7/3	1						1									2
	7/1							1						1			1
VII	4							1						1		1	4
	8			1				1									3
	6					1											2
	3		1														1
VIII	53										1				1		6
	46								1						1		5
	36													1			4
	34						1										3
	33							1									2
	20		1			1											1
IX	12						1				1		1				2
	4										1					1	1
X	8								1								3
	7										1						2
	1·b	1					1					1	1				1
XI	p 92									1	1				1		1

(1) *A. moussoulensis*
(2) *A. aramaea*
(3) *A. solida*
(4) *A. globosa*
(5) *A. avellana*
(6) *A. pisiformis*
(7) *A. pasticillata*
(8) *A. leupoldi*
(9) *A. montanarii*
(10) *A. aragonensis*
(11) *A. dedolia*
(12) *A. subpyreneica*
(13) *A. laxa*
(14) *A. guidonis*
(15) *A. decipiens*

Fig. 8.1 Occurrence of 15 alveolinids (1 to 15) from Yugoslavia (data from Drobne, 1977) in 11 sections (I to XI). SAM: Sample numbers originally used by Drobne. Successive maximal horizons are numbered in the stratigraphically upward direction for each section (see last column). Section XI is an isolated occurrence described on page 92 of Drobne (1977). See Table 8.7 for names of sections.

TABLE 8.7

List of sections for Drobne's dataset (cf. Fig. 8.1).

I.	Fatji hrib	VII.	Kozina-Socerb
II.	Dane near Divača	VIII.	Golež
III.	Veliko Gradišče	IX.	Žbevnica
IV.	Ritomeče near Gradišče	X.	Dane-Istria
V.	Podgorje	XI.	Jelšane (isolated sample)
VI.	Podgrad-Hrušica		

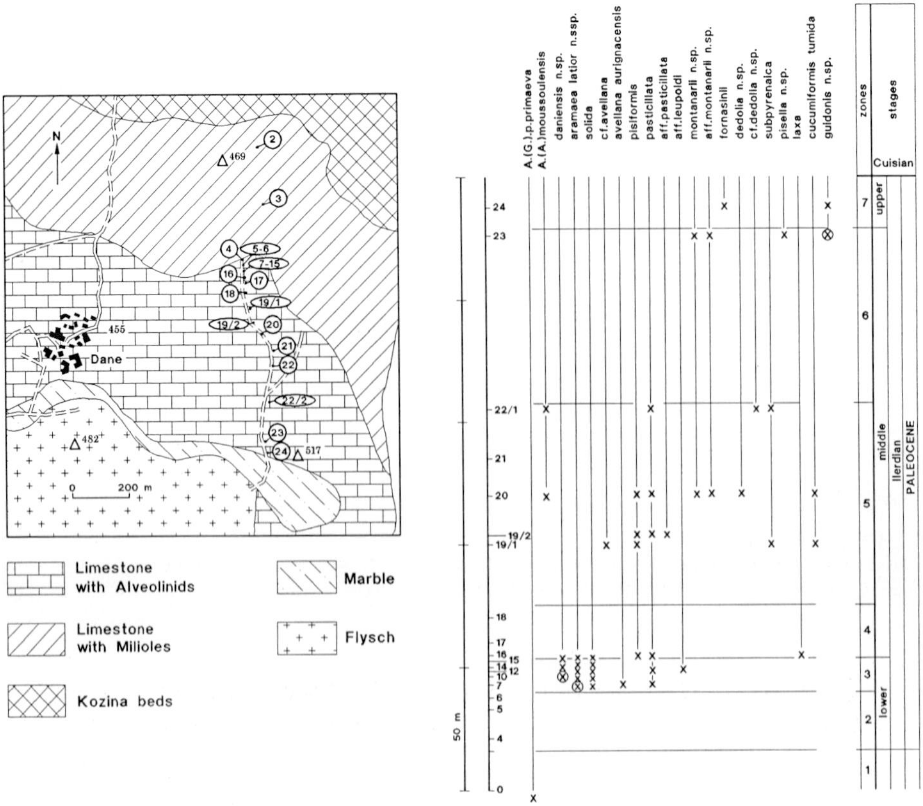

Fig. 8.2 Drobne's (1977) original stratigraphic data for Section II in Fig. 8.1 (Dane near Divača). Circled cross indicates *stratum typicum* of new species. Samples 7, 16, 20 and 23 are for maximal horizons (Guex levels).

The information of Table 8.1 was converted into RASC input by replacing each fossil number i (= 1, 2,...,15) by two numbers ($2i$-1) for highest occurrences and $2i$ for lowest occurrences, respectively. RASC was run on the resulting data set with $k_c = 4$, $m_{c1} = 1$ and $m_{c2} = 2$. Setting $k_c = 4$ ensured that no events were eliminated as in the U.A. computer program. However, it became immediately apparent that 7 of the 15 species were observed in one bed only in the sections containing them. Because the highest and lowest occurrences of these 7 species coincided everywhere, I decided to maintain a single number for each of these species indicating occurrence only. (The odd numbers for these taxa indicate coinciding highest and lowest occurrences.) Probabilistic ranking was applied and followed by the modified Hay method. Three cycles occurred and each of these involved the species 3 and 4. Based on $m_{c2} = 2$, 42 out of 253 pairs of

TABLE 8.8

Final Unitary Associations (U.A.) for Drobne's alveolinids as derived by Guex and Davaud (1984); upper part of table is range chart with ones for taxa belonging to a particular Unitary Association; lower part of table shows in which sections the final U.A.'s were identified.

U.A.	Taxa: (5)	(2)	(6)	(7)	(3)	(15)	(13)	(4)	(8)	(9)	(10)	(1)	(11)	(12)	(14)
1	0	0	0	0	0	1	0	1	1	1	1	0	0	0	1
2	0	0	1	1	0	1	0	1	1	1	1	1	1	1	0
3	0	0	1	1	0	1	1	0	0	0	0	0	0	0	0
4	1	0	1	1	1	0	0	0	0	0	0	0	0	0	0
5	1	1	0	0	0	0	0	0	0	0	0	0	0	0	0
U.A.	Sections: 1	2	3	4	5	6	7	8	9	10	11				
1	0	1	1	1	0	1	0	1	0	0	1				
2	1	1	1	1	0	1	0	0	1	1	0				
3	0	1	1	0	0	1	1	1	0	0	0				
4	0	1	1	0	0	0	1	0	0	0	0				
5	1	0	1	0	0	0	1	1	0	0	0				

Explanation of numbers used for taxa: (1) *A. moussoulensis*; (2) *A. aramaea*; (3) *A. solida*; (4) *A. globosa*; (5) *A. avellana*; (6) *A. pisiformis*; (7) *A. pasticillata*; (8) *A. leupoldi*; (9) *A. montanarii*; (10) *A. aragonensis*; (11) *A. dedolia*; (12) *A. subpyreneica*; (13) *A. laxa*; (14) *A. guidonis*; (15) *A. decipiens*.

matrix elements were zeroed for scaling. Weighted distance analysis was applied. From the results of the normality test (see Table 8.9), it may be concluded that species 3 (*A. solida*) occurs too high in Section I (because of reworking). In Table 8.9, *A. solida* has event number 5 for its lowest occurrence (LO) which coincides with its highest occurrence (see before).

TABLE 8.9

RASC normality test output for Drobne's Fatji hrib section with reworked bed at top (events 15 and 5 respresenting highest occurrences of fossils 8 and 3, respectively); the second-order differences were tested for statistical significance; events with two asterisks are out of place with a probability of 99%; those with one asterisk with a probability of 95%.

Event name	Event number	RASC distance	Second-order difference
LO *A. leupoldi*	15	0.626	
LO *A. solida*	−5	2.660	−4.390 **
LO *A. subpyreneica*	23	1.550	2.977 *
HI *A. pasticillata*	−14	2.172	0.023
LO *A. pasticillata*	−13	2.816	−2.589 *
LO *A. globosa*	−7	0.871	1.871
HI *A. pisiformis*	12	2.044	0.492
HI *A. pisiformis*	−11	2.962	0.239
LO *A. aramaea*	3	4.366	

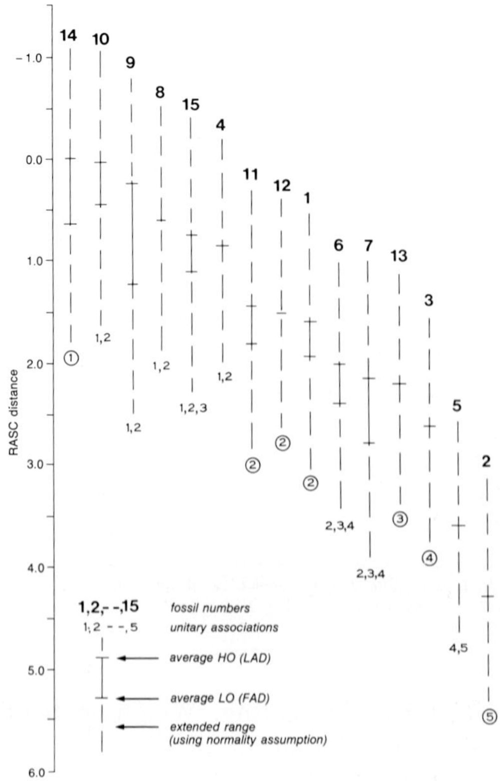

Fig. 8.3 Comparison of RASC results to Unitary Associations for Drobne's alveolinids. Fossils were ordered according to increasing RASC distance of their highest occurrence (HO or LAD).

Its RASC distance (=2.660) is larger than those of its neighbors in this section. This discrepancy was brought out by computation of the second-order difference (=-4.390**) in Table 8.9. The two asterisks indicate that the event is out of place with a probability of more than 99 percent.

Figure 8.3 shows a comparison of the 5 Unitary Associations of Table 8.8 with the scaled optimum sequence used for obtaining Table 8.9. The highest occurrences of the 15 fossils were ordered in Figure 8.3 according to their RASC distances. Because average highest and lowest occurrences are estimated by scaling, the distances between them on the RASC scale are less than their true stratigraphic ranges. According to the original scaling model, events in sections are normally distributed about their average position with standard deviations equal to $\sigma = 0.7071$. Consequently, the observed highest occurrence of a fossil in a section would occur with a probability of 95 percent below its RASC value decreased by $1.645 \times \sigma = 1.16$. This value provides a more reasonable estimate of the true highest occurrence or last appearance datum (LAD) than the original RASC value. Likewise 1.16 can be added to the RASC distance estimated for a lowest occurrence in order to obtain a more conservative estimate of this lowest occurrence or first appearance datum (FAD) along the RASC scale. The resulting enlargements of the RASC ranges are shown as dashed lines in Figure 8.3. According to the probabilistic range chart of Figure 8.3, fossil 14 probably co-occurred with 3 and probably not with 2. The dashed lines are based on the assumption that all events satisfy a normal distribution with the same standard deviation along the RASC scale. I pointed before (Gradstein et al., 1985, p. 255) that this assumption may not hold true in reality and care should be taken in interpreting the ranges of Figure 8.3. For example, Guex (personal information, 1984) had advised me that fossil 5 probably never coexisted with 11 although their ranges overlap in Figure 8.3.

The U.A. numbers of the fossils are also shown in Figure 8.3 and circled if a fossil belongs to a single U.A. only. The order of the overlapping U.A.'s is very similar to that of the sequence of RASC ranges for the fossils. The only discrepancy is that fossil 15 which belongs to U.A. 3 occurs in fifth position in Figure 8.3 while the other fossils of U.A. 3 (6, 7 and 13) occupy positions 10, 11 and 12, respectively.

The preceding comparison using Drobne's alveolinids is interesting in that similar results for ranking as well as stratigraphic "normality" were obtained by means of two methods (U.A. and RASC) which are built upon different premises. In the U.A. method, observed co-occurrences of fossils are augmented by virtual occurrences partly to resolve inconsistencies (forbidden structures) in order to obtain assemblage zones. In the RASC model, the observed highest and lowest occurrences of fossils in sections are considered to be realizations of random variables with fixed average positions along a linear scale. The two methods have in common that each provides a way of eliminating inconsistencies and filling in the gaps due to missing data. In the U.A. method, this is done by adopting rules based on graph theory whereas in the RASC method the observed data are considered to belong to small samples derived from (infinitely large) statistical populations of which the parameters (rankings, means and standard deviations) can be estimated.

The "zones" resulting from the U.A. method are primarily based on observed and inferred co-occurrences of fossil species while the "zones" resulting from the RASC method are primarily based on estimated proximity of stratigraphic events in time. Nevertheless, the two approaches can yield similar results for anomalous occurrences and groupings for correlation as shown in this section. It is noted that Guex's maximal horizon method (cf. Section 4.5) was used for coding the biostratigraphic information which implies loss of information from the sequence file. During the past three years, the Drobne data have been further discussed and re-analyzed by Guex (1987) and Brower (1989). Moreover, because of the development of the modified RASC method, it has become possible to construct range zones which are more representative of the observed superpositional relations than the 95 percent confidence interval ranges shown in Figure 8.3. For these

TABLE 8.10

Alphabetic DIC file for Palmer's database. Numbers are for highest occurrences. Subtraction of one gives code numbers for corresponding lowest occurrences. For example, 99 LO *Angulotreta triangularis* is lowest occurrence corresponding to first entry (= 100) listed.

100	HI	ANGULOTRETA TRIANGULARIS	44	HI	HOLCACEPHALUS TENERUS
98	HI	ANGULOTRETA TRIANGULARIS DIGITALIS	56	HI	KINGSTONIA PONTOTOCENSIS
102	HI	APHELASPIS CONSTRICTA	22	HI	KINSABIA VARIGATA
104	HI	APHELASPIS LONGIFRONS	8	HI	KORMAGNOSTUS SIMPLEX
88	HI	APHELASPIS SPINOSA	108	HI	LABIOSTRIA CONVEXIMARGINATA
82	HI	APHELASPIS WALCOTTI	116	HI	LABIOSTRIA PLATIFRONS
120	HI	APSOTRETA EXPANSUS	122	HI	LABIOSTRIA SIGMOIDALIS
20	HI	APSOTRETA ORIFERA	60	HI	LLANOASPIS MODESTA
50	HI	ARCUOLIMBUS CONVEXUS	78	HI	LLANOASPIS PECULIARIS
6	HI	BOLASPIDELLA BURNETENSIS	66	HI	LLANOASPIS UNDULATA
4	HI	BOLASPIDELLA WELLSVILLENSIS	74	HI	LLANOASPIS UNDULATA GRANULATA
10	HI	CEDARINA CORDILLERAE	58	HI	LLANOASPIS VIRGINICA
14	HI	CEDARINA EURYCHEILOS	68	HI	MARYVILLIA CF. M. ARISTON
84	HI	CHEILOCEPHALUS BREVILOBA	76	HI	METEORASPIS CF. M. LOISI
94	HI	CHEILOCEPHALUS MINUTUS	16	HI	METEORASPIS CF. M. ROBUSTA
34	HI	COOSELLA BELTENSIS	54	HI	METEORASPIS METRA
62	HI	COOSELLA CF. C. WIDNERENSIS	12	HI	MODOCIA CF. M. CENTRALIS
30	HI	COOSELLA GRANULOSA	2	HI	MODOCIA CF. M. OWENI
70	HI	COOSIA CF. C. ALBERTENSIS	26	HI	NORWOODIA QUADRANGULARIS
28	HI	COOSIA CONNATA	40	HI	OPISTHOTRETA DEPRESSA
64	HI	CREPICEPHALUS AUSTRALIS	72	HI	PEMPHIGASPIS INEXPECTANS
80	HI	CREPICEPHALUS CF. C. IOWENSIS	106	HI	PSEUDAGNOSTUS COMMUNIS
86	HI	CREPICEPHALUS? PERPLEXUS	110	HI	PSEUDAGNOSTUS JOSEPHUS
90	HI	DICTYONINA PERFORATA	48	HI	PSEUDAGNOSTUS? NORDICUS
52	HI	DIERACEPHALUS ASTER	92	HI	RAASCHELLA ORNATA
114	HI	DUNDERBERGIA VARIAGRANULA	38	HI	SPICULE A
124	HI	DYSORISTUS LOCHMANAE	24	HI	SPICULE B
118	HI	DYTREMACEPHALUS GRANULOSUS	46	HI	SPICULE C
112	HI	DYTREMACEPHALUS LAEVIS	18	HI	SYSPACHEILUS CF. S. CAMURUS
32	HI	GENEVIEVELLA CF. G. SPINOSA	42	HI	TRICREPICEPHALUS CORIA
96	HI	GERAGNOSTUS CF. G. TUMIDOSUS	36	HI	TRICREPICEPHALUS TEXANUS

reasons, the Drobne example will be recoded and subjected to modified RASC later in this chapter.

TABLE 8.11

SEQ file for 7 sections of Palmer's database. The event code numbers are explained in Table 8.10.

```
MORGAN CREEK
119 -120 -123 -124   84 -100 -108 -114   82 -105 -106  101 -102 -103 -104 -113   90  -99 -107   87
-88  -92   81   42  -68  -83  -85  -86  -89  -91   69  -70  -77  -78  -79  -80   24  -65  -66  -67
-73  -74   40    8  -54   60  -64   38  -56   59  -62   63   22   23  -61   34  -49  -50  -52  -55
 30  -39  -43  -44  -51  -53   19  -20  -25  -26  -27  -28  -29  -31  -32  -33  -35  -36  -37  -41
 13  -14  -15  -16  -17  -18  -21    7  -10    9    5   -6

WHITE CREEK
120  113 -114 -117 -118 -121 -122  119  100  107 -108   82 -115 -116   99   92  -98   89  -90  -91
-97   45  -46  -81   24  -40   42  -56  -65  -66  -67  -68   59  -60   54    8  -36   22   33  -34
-41  -47  -48  -53  -55  -57  -58   35   27  -28  -39   21  -23    7  -13  -14    4    2   -3    1

JAMES RIVER
117 -118  100   82 -108   90  -97  -98 -107   81  -89  -99   24  -47  -48  -56  -68  -70   40   42
-77  -78   55  -63  -64  -65  -66  -67  -69  -71  -72   60  -61  -62   23  -59    8  -22  -30  -34
-50   29  -33  -35  -36  -39  -41  -49    7  -15  -16  -17  -18  -19  -20  -21

LITTLE LLANO RIVER
 82  113 -114   99 -100   90   46  -89  -92  -93  -94  -95  -96   45  -83  -84  -91   42  -68  -69
-70  -77  -78  -81  -85  -86   40   24  -65  -66  -67  -73  -74   53  -54  -63  -64   56   34  -39
-41  -55    7   -8  -21  -22  -23  -33  -47  -48   10    9    5   -6

LION MOUNTAIN
 84 -114 -118 -119 -120   82 -100 -106 -108 -112 -117  102 -104   99 -101 -103 -105 -107 -111 -113
 81  -83  -87  -88  -91  -92   42  -68  -69  -70   67    7   -8  -31  -32  -34  -47  -48  -49  -50
-53  -54  -55  -56   29  -30  -33  -35  -36  -39  -40  -41  -43  -44  -45  -46

PONTOTOC
 82 -100   99  107 -108 -109 -110   45  -46  -91  -92  -97  -98   83  -84   87  -88   81   41  -42
-68   67  -70  -75  -76   64   39  -40  -63  -69   22   21  -33  -34    8  -10    7   11  -12    9
  6    5    3   -4

STREETER
 91   82   99 -100   92   81  -89  -90   40  -41  -42  -47  -48  -67  -68  -69  -70  -77  -78   24
-61  -62   33  -34  -53  -54   22  -23  -39   16  -18  -21   15  -17   14    7   -8  -13    9  -10
```

8.4 Application of RASC and normality test to Palmer's database for the Riley Formation in central Texas

Shaw's (1964) book contains detailed documentation including a 126-page appendix on construction of a composite standard for the fauna (mostly trilobites) of the Cambrian Riley Formation of Texas originally described by Palmer (1955). Various authors including Edwards and Beaver (1978), Hudson and Agterberg (1982), Edwards (1982) and Guex

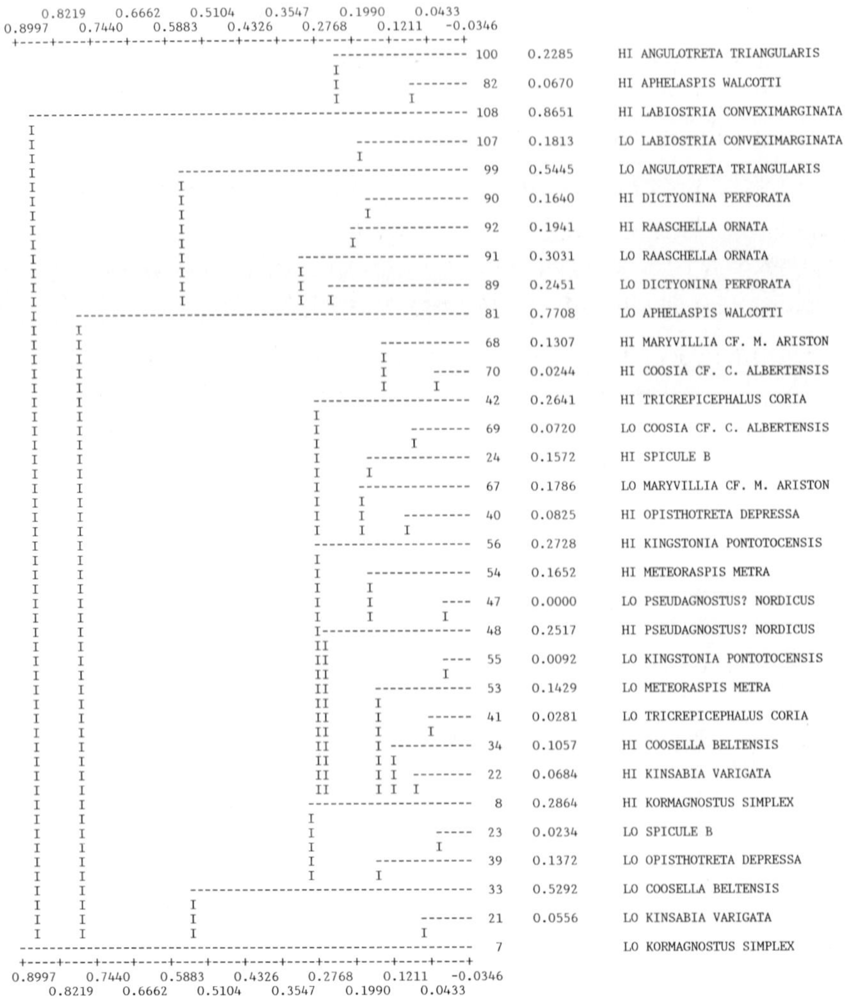

Fig. 8.4 Scaled optimum sequence (RASC 5/1/3 run) for Palmer's database for the Riley Formation in central Texas.

(1987) have used this database to compare results obtained by other methods with one another and to Shaw's composite standard.

Tables 8.10 and 8.11 contain DIC and SEQ files constructed from Shaw's Table A-1 (Shaw, 1964, pp. 230-232). Table 8.10 is an alphabetic listing of highest occurrences of all fossils. The corresponding dictionary numbers of the lowest occurrences are one unit less. Table 8.11 was obtained after pre-processing of a DAT file (not shown here) with input format as in Shaw's table, and retaining only those events that occur in five or more of the seven sections. Figure 8.4 shows the scaled optimum sequence obtained after final reordering in a RASC 5/1/3 run. Input to scaling was the optimum sequence resulting from probabilistic ranking. (Although the modified Hay method also was applied, this did not affect the probabilistic ranking results).

Table 8.12 gives the values of Kendall's tau for the 7 sections in comparison with the scaled optimum sequence. The seven tau-values range from 0.74 to 0.86 suggesting that all sections are correlated to the average ranking with nearly the same strength.

Table 8.13 shows results of the overall normality test applied to the 180 second-order differences for events occurring in 5, 6 or 7 sections. The sum of the values in the last column is 3.163. This chi-squared value is not statistically significant indicating that if there are anomalous events in the sections, these are rare. Table 8.14 shows RASC normality test output for the Morgan Creek, White Creek and Pontotoc sections.

TABLE 8.12

Kendall's rank correlation coefficients for sequences of 7 sections correlated with scaled optimum sequence of Fig. 8.4.

Section	Tau
Morgan Creek	0.86
White Creek	0.81
James River	0.79
Little Llano River	0.80
Lion Mountain	0.74
Pontotoc	0.82
Streeter	0.75

TABLE 8.13

Overall normality test applied to Palmer's database using taxa that occur in at least 5 of the 7 sections. No significant departures from normality are indicated.

Class No.	O	*E*	*O-E*	$(O-E)^2/E$
1	14	18	-4	0.415
2	19	18	1	0.026
3	26	18	8	1.659
4	18	18	0	0.000
5	16	18	-2	0.104
6	16	18	-2	0.104
7	17	18	-1	0.026
8	22	18	4	0.415
9	14	18	-4	0.415
10	18	18	0	0.000

TABLE 8.14

RASC normality test output for 3 sections in Palmer's database. Only the lowest occurrences of *Tricrepicephalus coria* and *Opisthotreta depressa* would be "too high" in the Pontotoc section. (Note that both fossils occur in single beds in this section). Within the context of the entire database, these events are not anomalous because, on the average, 4 single star events and 1 double star event are expected to occur in every set of 100 events.

MORGAN CREEK		CUM. DIST.	2ND ORDER DIFF.
HI ANGULOTRETA TRIANGULARIS	100	0.0000	
HI LABIOSTRIA CONVEXIMARGINATA	-108	0.2955	-0.7477
HI APHELASPIS WALCOTTI	82	0.2285	1.7249
HI DICTYONINA PERFORATA	90	1.8865	-1.8172
LO ANGULOTRETA TRIANGULARIS	-99	1.3420	0.3631
LO LABIOSTRIA CONVEXIMARGINATA	-107	1.1606	0.6859
HI RAASCHELLA ORNATA	92	2.0504	-0.1476
LO APHELASPIS WALCOTTI	81	2.7926	0.1837
HI TRICREPICEPHALUS CORIA	42	3.7185	-0.6957
HI MARYVILLIA CF. M. ARISTON	-68	3.5635	-0.8609
LO DICTYONINA PERFORATA	-89	2.5476	0.7128
LO RAASCHELLA ORNATA	-91	2.2445	1.6560
LO COOSIA CF. C. ALBERTENSIS	69	3.9826	-1.6414
HI COOSIA CF. C. ALBERTENSIS	-70	3.6942	0.2637
HI SPICULE B	24	4.0546	0.1820
LO MARYVILLIA CF. M. ARISTON	-67	4.2118	-0.3638
HI OPISTHOTRETA DEPRESSA	40	4.3905	0.9479
HI KORMAGNOSTUS SIMPLEX	8	5.5170	-1.5125
HI METEORASPIS METRA	-54	4.7457	0.1132
HI KINGSTONIA PONTOTOCENSIS	56	4.4730	1.2483
HI KINSABIA VARIGATA	22	5.4485	-0.6207
LO SPICULE B	23	5.8034	-0.8154
HI COOSELLA BELTENSIS	34	5.3429	0.6655
LO KINGSTONIA PONTOTOCENSIS	-55	5.1626	0.4592
LO OPISTHOTRETA DEPRESSA	39	5.8268	-0.9340
LO METEORASPIS METRA	-53	5.1719	1.0620
LO COOSELLA BELTENSIS	33	5.9641	-1.0563
LO TRICREPICEPHALUS CORIA	-41	5.3148	1.4425
LO KINSABIA VARIGATA	21	6.4933	-1.1228
LO KORMAGNOSTUS SIMPLEX	7	6.5489	

TABLE 8.14 (continued)

WHITE CREEK		CUM. DIST.	2ND ORDER DIFF.
HI ANGULOTRETA TRIANGULARIS	100	0.0000	
LO LABIOSTRIA CONVEXIMARGINATA	107	1.1606	-1.5892
HI LABIOSTRIA CONVEXIMARGINATA	-108	0.2955	0.3616
HI APHELASPIS WALCOTTI	82	0.2285	1.1804
LO ANGULOTRETA TRIANGULARIS	99	1.3420	-0.4050
HI RAASCHELLA ORNATA	92	2.0504	-0.2113
LO DICTYONINA PERFORATA	89	2.5476	-0.7217
HI DICTYONINA PERFORATA	-90	1.8865	1.0191
LO RAASCHELLA ORNATA	-91	2.2445	-0.2465
LO APHELASPIS WALCOTTI	81	2.7926	0.7138
HI SPICULE B	24	4.0546	-0.4895
HI OPISTHOTRETA DEPRESSA	-40	4.3905	-1.4444
HI TRICREPICEPHALUS CORIA	42	3.7185	1.8630
HI KINGSTONIA PONTOTOCENSIS	-56	4.4730	-1.0156
LO MARYVILLIA CF. M. ARISTON	-67	4.2118	-0.3872
HI MARYVILLIA CF. M. ARISTON	-68	3.5635	1.3940
HI METEORASPIS METRA	54	4.7457	-0.4111
HI KORMAGNOSTUS SIMPLEX	8	5.5170	-0.8396
HI KINSABIA VARIGATA	22	5.4485	0.5840
LO COOSELLA BELTENSIS	33	5.9641	-0.7001
HI COOSELLA BELTENSIS	-34	5.3429	0.5931
LO TRICREPICEPHALUS CORIA	-41	5.3148	-0.3758
LO PSEUDAGNOSTUS? NORDICUS	-47	4.9109	0.4038
HI PSEUDAGNOSTUS? NORDICUS	-48	4.9109	0.2609
LO METEORASPIS METRA	-53	5.1719	-0.2701
LO KINGSTONIA PONTOTOCENSIS	-55	5.1626	0.2368
LO OPISTHOTRETA DEPRESSA	39	5.8268	0.0022
LO KINSABIA VARIGATA	21	6.4933	-0.9197
LO SPICULE B	-23	5.8034	0.9988
LO KORMAGNOSTUS SIMPLEX	7	6.5489	

PONTOTOC		CUM. DIST.	2ND ORDER DIFF.
HI APHELASPIS WALCOTTI	82	0.2285	
HI ANGULOTRETA TRIANGULARIS	-100	0.0000	0.9959
LO ANGULOTRETA TRIANGULARIS	99	1.3420	-1.5233
LO LABIOSTRIA CONVEXIMARGINATA	107	1.1606	-0.1092
HI LABIOSTRIA CONVEXIMARGINATA	-108	0.2955	2.2396
LO RAASCHELLA ORNATA	91	2.2445	-1.5685
HI RAASCHELLA ORNATA	-92	2.0504	0.3617
LO APHELASPIS WALCOTTI	81	2.7926	1.7799
LO TRICREPICEPHALUS CORIA	41	5.3148	-3.5439 **
HI TRICREPICEPHALUS CORIA	-42	3.7185	1.4412
HI MARYVILLIA CF. M. ARISTON	-68	3.5635	0.2288
LO MARYVILLIA CF. M. ARISTON	67	4.2118	-0.5914
HI COOSIA CF. C. ALBERTENSIS	-70	3.6942	2.0758
LO OPISTHOTRETA DEPRESSA	39	5.8268	-2.9945 *
HI OPISTHOTRETA DEPRESSA	-40	4.3905	1.0286
LO COOSIA CF. C. ALBERTENSIS	-69	3.9826	1.2991
HI KINSABIA VARIGATA	22	5.4485	-0.4212
LO KINSABIA VARIGATA	21	6.4933	-0.9993
LO COOSELLA BELTENSIS	-33	5.9641	-0.0920
HI COOSELLA BELTENSIS	-34	5.3429	0.2207
HI KORMAGNOSTUS SIMPLEX	8	5.5170	0.8579
LO KORMAGNOSTUS SIMPLEX	7	6.5489	

To those who have read Shaw's (1964) book, the preceding evaluation of Palmer's database may seem surprising in that during his construction of the composite standard, Shaw frequently did not use events which were deviating more than other events from the straight lines fitted by the method of least squares to events initially in two sections plotted against one another, and later in other sections plotted against the composite of two or more sections. However, most of these unused events appear not to be anomalous in a statistical sense. It may be concluded that Shaw was trimming the data in order to improve least-squares estimation of the lines of correlation. Trimming is a statistical procedure in which estimates are restricted to measurements which are relatively close to the quantity to be estimated. Such methods now are widely used in exploratory data analysis (Tukey, 1977). It is noted that, in order to obtain the normal distribution of the second-order differences, only 60 percent of the observations were used (see Section 8.2). This can be regarded as another example of trimming.

It will be shown in Section 8.9 that Shaw's composite standard method, because of trimming, yields a range chart with ranges that, for some taxa in length are intermediate between those in the scaled optimum sequence of Figure 8.4 and extended ranges resulting from the modified RASC method with use of all observations. On the whole, however, the ranges obtained by modified RASC are very similar to those obtained by other "conservative" range chart construction methods including the composite standard method.

8.5 Modified RASC Method

Although robustness is increased by combining events with one another (application of central limit theorem, see Chapter 6), ordinary scaling is based on the assumption that all events have normal distributions with equal variance along the interval scale. It is noted that the assumption of equality of variance for different events frequently has been made in quantitative stratigraphy in an implicit manner. For example, Shaw's (1964) lines of correlation were fitted assuming that this condition is satisfied.

By comparing individual sequences with the scaled optimum sequence and collecting deviations from smoothing splines fitted for different sections, it is possible to estimate the frequency distribution of each event separately. The RASC scaling algorithm can be modified to allow for different variances of the events. An iterative procedure has been developed (cf. Agterberg and D'Iorio, in press; D'Iorio, 1988; D'Iorio and Agterberg, 1989) in which the methods of (1) weighted spline fitting, and (2) modified scaling are applied alternately until a stable solution is reached upon convergence. In these two methods, the variances of the events are not assumed to be equal to one another. Application of this method to highest occurrences of Cenozoic foraminifers along the northwestern Atlantic Margin (Gradstein-Thomas database) showed (1) unequality of variances for different events; and (2) minor departures from normality of the frequency distributions for separate events. Changes in the scaled optimum sequence resulting from the iterative procedure were negligibly small. The new approach allows identification of small-variance events which disappeared approximately simultaneously from different sections in the same study region.

The RASC method for ranking and scaling consists of (1) forming a single, optimum sequence from mutually inconsistent sequences of observed events for different stratigraphic sections, and (2) positioning these events along a relative time interval scale. In modified RASC, the scaling part of the RASC method is generalized to account for possible differences in uncertainty associated with the positioning of different events along the RASC interval scale. The original scaling model was illustrated in Figure 6.4. Each of a group of biostratigraphic events (A, B, ..., G) was assumed to be a random variable (X_A, X_B, ..., X_G) with Gaussian probability distribution along the RASC scale. These Gaussian curves have different means (EX_A, EX_B, ..., EX_G) but their variances (σ^2) are assumed to be equal to one another. By means of this model it became possible to estimate the intervals between the successive mean values denoted as EX_A, EX_B, ..., EX_G. The model of Figure 6.4 can be generalized by allowing the variances of the events to be different. Such an extension of the method only is possible if the variances σ_A, σ_B, ..., σ_G of the frequency distributions $f(x_A)$, $f(x_B)$, ..., $f(x_G)$ of the events can be estimated. A possible estimation procedure is described here.

The original RASC method provides estimates x_i of EX_i where i denotes events. In each stratigraphic section x_i can be plotted against u_i, representing relative position of event i in the so-called event level scale of the section. New estimates $\hat{x}_i$ of EX_i in the section can be obtained by fitting a cubic spline curve with u as the independent variable. The differences ($\hat{x}_i$-x_i) can be collected from all sections in which event i occurs and plotted as a histogram that provides an approximation of $f(x_i\text{-}EX_i)$. The shape of the latter distribution is the same as that of $f(x_i)$. The standard deviation s_i of the differences provides an estimate of σ_i.

In the application to Cenozoic Foraminifera from 24 wells on the Labrador Shelf and Grand Banks to be discussed in the next two sections, distinct differences were found in the widths of the probability distributions $f(x_i)$ for different events. The number of differences per event (sample size, n) varies from 7 to 22 in this application. Most observed frequency distributions are unimodal and slightly skewed to the right or to the left. A few distributions may be bimodal. The sample sizes are too small to demonstrate statistical significance of the possible departures from the Gaussian model. However, each event can be assumed to have its own variance because the widths of the $f(x_i)$ are clearly different. This led to the modified RASC model to be explained in this section. Application of modified RASC with different variances for different events, results in a new set of estimates of the positions of EX_A, EX_B, ..., EX_G. Spline-curves can again be fitted to data for individual sections. Repetition of these steps results in an iterative procedure which converges toward a final solution. The histograms of the differences $(\hat{x}_i\text{-}x_i)$ after convergence provide better approximations of $f(x_i)$ than the histograms at the beginning of the iterative process.

Suppose that the x-axis for relative time interval scale points in the stratigraphically upward direction. For example, the events A, B, ..., G in reversed order, may represent highest occurrences encountered successively in a well drilled downward in a basin where age increases with depth. The location of each stratigraphic event is represented as a random variable (X_A, X_B, ..., X_G) that in each well may assume a specific value along the x-axis with probabilities controlled by its Gaussian curve. Suppose that two events (e.g. A and B) both occur in R wells. In R_A wells A is observed above B and in R_B wells B above A. When A and B are observed to be coeval in a well, 0.5 is added to R_A as well as to R_B. Setting $R_A + R_B = R$, the ratio $P_{AB} = R_A/R$ can be set equal to the probability that A is observed before B in a randomly selected well and used to estimate the interval $\Delta_{AB} = EX_B\text{-}EX_A$. The difference Δ_{AB} is the mean of a random variable $D_{AB} = X_B\text{-}X_A$ for difference between the random variables X_B and X_A. If Δ_{AB} is positive, D_{AB} would turn out to be positive in most sections. However, the model also allows B to be observed before A in some sections with negative D_{AB}. If the Gaussian curves of two events were to coincide, the probability that one of these two events is observed before the other, is exactly 0.5. If the variances of the Gaussian curves in Figure 6.4 are all equal to σ^2, P_{AB} estimates

$$P(D_{AB}>0) = \Phi(\Delta_{AB}/\sigma\sqrt{2}) \tag{8.8}$$

In this equation, which is equivalent to Equation (6.1), the mean interval Δ_{AB} is divided by $\sigma\sqrt{2}$ representing the standard deviation of the random variable D_{AB}. If the RASC model, it is not possible to estimate both Δ_{AB} and σ. For this reason, σ was set equal to an arbitrary constant ($\sigma = 0.7071$). A different choice of σ would be equivalent to rescaling the axis for the distance estimates (x-axis). From Equation (8.8) it follows that $\Delta_{AB} = \Phi^{-1}\ \{P(D_{AB}>0)\}$. Consequently, $Z_{AB} = \Phi^{-1}(P_{AB})$ where P_{AB} is converted into Z_{AB} representing a fractile of the normal distribution in standard form. Suppose now that events A and B have different variances σ^2_A and σ^2_B. Then the variance of D_{AB} becomes $\sigma^2_{AB} = \sigma^2_A + \sigma^2_B$. The corresponding standard derivation σ_{AB} reduces to $\sigma\sqrt{2} = 1$ only if $\sigma^2_A = \sigma^2_B = \sigma^2$. In the modified RASC model, Equation (8.8) is replaced by

$$P(D_{AB}>0) = \Phi(\Delta_{AB}/\sigma_{AB}) \tag{8.9}$$

and Z_{AB} is replaced by $G_{AB} = Z_{AB}{\cdot}s_{AB}$. Thus, the Z_{AB}-value of a relative frequency P_{AB} must be multiplied by s_{AB} representing an estimate of σ_{AB} before it can be interpreted as an estimate of the interval EX_B-EX_A. As pointed out before, the precision of a Z-value depends on relative frequency P as well as sample size R. More weight w can be given to G-values with larger R by using the equation

$$w = \frac{1}{s^2(G)} = \frac{R\exp(-G^2)}{2\pi P(1-P)} \tag{8.10}$$

where $s^2(G)$ denotes estimated variance of G. These weights may be used when sets of G-values are combined with one another in order to improve the estimate of the interval between two events. For example, because $(EX_C\text{-}EX_A) - (EX_C\text{-}EX_B)$ reduces to $EX_B\text{-}EX_A$, $G_{AB{\cdot}C} = G_{AC} - G_{BC}$ provides an indirect estimate of $EX_B\text{-}EX_A$ with weight $w_{AB{\cdot}C} = (w_{AC}\times w_{BC})/(w_{AC} + w_{BC})$. The direct estimate G_{AB} can be combined with $G_{AB{\cdot}C}$ and other differences between G-values according to the equations (e.g. Eq. 6.2) previously used for the Z-values.

8.6 Application of modified RASC to the Gradstein - Thomas database

The database used in this example is for highest occurrences of Cenozoic Foraminifera in 24 exploration wells on the Labrador Shelf and Grand Banks previously introduced in Section 4.6 (see Tables 4.7 and 4.9). Table 8.15 shows estimated RASC distances for 44 events each occurring in at least 7 wells. This RASC distance is plotted against event level in Figure 8.5A for one of the wells (Adolphus D-50). The horizontal scale for relative event levels increases with depth. The Adolphus D-50 well was sampled by taking cuttings at a regular interval of 30 ft (approximately 10 m). Only 23 distinct levels to a depth of about 9000 ft showed one or more highest occurrences for the 44 species considered. These levels were numbered from 1 to 23 in Figure 8.5. In total, only 30 of the 44 species were encountered in Adolphus D-50.

A cubic spline curve was fitted to the data shown in Figure 8.5A with smoothing factor set equal to $\sigma = 0.7071$ representing the standard deviation of events along the distance scale in the ordinary RASC model (see before). In general, the smoothing factor (SF) is the square root of the mean squared deviation for the deviations between points and spline curve (measured along the RASC distance scale). SF is selected in advance and the best-fitting spline curve will have SF as standard deviation (biased estimate) of its residuals. This standard deviation is "biased" because the sum of squares of the deviations was divided by n instead of its number of degrees of freedom. For example, the number of degrees of freedom for a best-fitting straight line is n-2. Division of the sum of squared deviations by n-2 then results in an "unbiased" estimate. The best-fitting straight line is the smoothest possible spline-curve. This solution always is obtained if SF exceeds the standard deviation of the residuals from the best-fitting straight line. If the spline-curve is not a straight line, the number of degrees of freedom is not readily determined. An unbiased estimate of SF could be obtained by cross-validation (see Section 9.5) but this method is not used here.

In the original RASC model, it is assumed that all events have the same standard deviation (σ). In modified RASC, each event i has its own standard deviation σ_i estimated from the n deviations of the event in the wells where it occurs. The sum of squared deviations for each event was divided by (n-1) to obtain the estimated variance s_i^2 (see Table 8.15, 3rd column). This is an "unbiased" estimate because, in general, the

TABLE 8.15

RASC distances and variances s_i^2 estimated for 44 species (event numbers as in Gradstein et al., 1985) before (First run) and after (Fifth and Sixth runs with refinement) convergence.

FIRST RUN			FIFTH RUN			SIXTH RUN		
Event number	RASC dist.	Unbiased variance (0 mean)	Event number	RASC dist.	Unbiased variance (0 mean)	Event number	RASC dist.	Unbiased variance (0 mean)
10	0.000	0.978	10	0.000	1.067	10	0.000	1.057
17	0.288	0.688	17	0.441	0.699	17	0.439	0.702
16	1.016	0.344	16	1.137	0.266	16	1.138	0.261
67	1.237	0.511	67	1.216	0.557	67	1.215	0.524
18	1.616	0.202	18	1.669	0.095	18	1.665	0.093
21	1.858	0.085	21	1.722	0.016	21	1.715	0.009
71	1.865	0.427	20	1.837	0.073	20	1.830	0.070
20	1.946	0.164	71	1.855	0.370	71	1.848	0.372
26	2.087	0.366	26	1.983	0.409	26	1.976	0.411
70	2.337	0.145	70	2.171	0.121	70	2.167	0.135
15	2.370	0.446	15	2.206	0.412	15	2.199	0.419
24	2.754	0.199	24	2.573	0.173	24	2.567	0.180
27	2.768	0.649	27	2.724	0.725	27	2.720	0.735
69	2.988	0.649	69	2.869	0.636	69	2.862	0.632
25	3.084	0.319	25	2.894	0.235	25	2.890	0.238
81	3.168	0.552	81	3.007	0.615	81	3.000	0.624
202	3.289	0.285	202	3.144	0.110	202	3.141	0.093
259	3.400	0.151	259	3.236	0.092	259	3.233	0.092
34	3.834	0.449	147	3.668	0.173	147	3.667	0.166
147	3.898	0.413	34	3.718	0.537	34	3.717	0.554
33	4.018	1.030	33	3.833	1.111	33	3.861	1.142
260	4.114	0.190	260	4.007	0.149	260	4.007	0.151
261	4.155	0.134	261	4.133	0.068	261	4.134	0.070
263	4.297	0.347	263	4.187	0.339	263	4.188	0.350
29	4.520	0.232	29	4.322	0.136	29	4.321	0.135
32	4.603	0.289	32	4.419	0.218	32	4.420	0.220
40	4.662	0.554	40	4.440	0.426	40	4.437	0.433
264	4.869	0.461	42	4.682	0.824	42	4.680	0.843
42	4.882	0.779	264	4.691	0.355	264	4.691	0.359
30	4.921	0.468	41	4.735	0.352	41	4.735	0.361
41	4.947	0.496	30	4.798	0.399	30	4.799	0.416
90	5.235	0.368	90	5.041	0.384	90	5.043	0.413
86	5.249	0.175	86	5.053	0.042	36	5.052	0.377
36	5.315	0.332	36	5.056	0.356	86	5.053	0.033
57	5.352	0.500	57	5.095	0.544	57	5.095	0.557
45	5.906	0.849	45	5.655	0.916	45	5.653	0.925
50	6.110	0.204	50	5.886	0.008	50	5.885	0.002
46	6.227	0.597	46	5.925	0.397	46	5.923	0.393
230	6.325	0.432	230	6.053	0.397	230	6.051	0.395
52	6.426	0.550	54	6.067	0.217	54	6.067	0.222
54	6.473	0.267	52	6.130	0.174	52	6.128	0.168
56	6.925	0.352	56	6.386	0.195	56	6.385	0.189
55	7.405	0.274	55	6.937	0.261	55	6.940	0.276
59	7.780	0.576	59	7.164	0.517	59	7.162	0.515

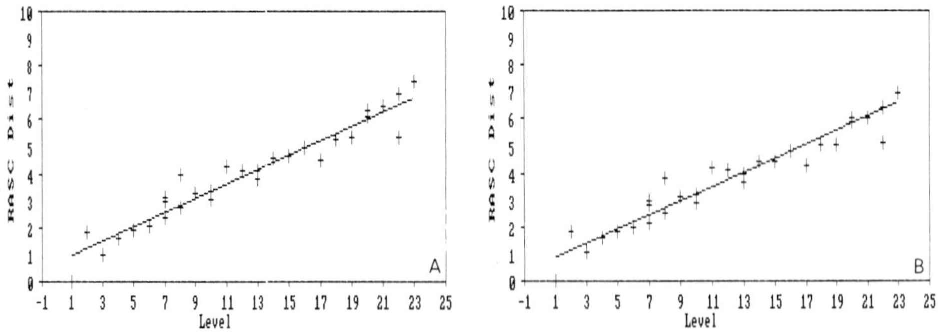

Fig. 8.5 Results of fitting a spline-curve to data for Adolphus D-50 well before (A) and after (B) iteration. For Fig. 8.5A, the smoothing factor (SF) was set equal to SF=0.7071 and standard deviations for individual data (s_i) were kept equal to 1.000. This procedure provides results identical to setting SF=1.000 and s_i=0.7071 for all i). For Fig. 8.5B, the smoothing factor was set equal to SF=1.000 and use was made of s_i-values obtained after convergence. In both diagrams, SF exceeded the standard deviation of the residuals so that the spline-curve became a best-fitting stratight line.

number of degrees of freedom for n deviations from a mean is equal to n-1. The values of s_i^2 could be used to run the modified RASC program. This would give a different set of RASC distances which, in turn, might be used to estimate new variances from new spline-curves. However, the values of s_i^2 also can be used to repeat the spline-curve fitting stage without first going through modified RASC.

In weighted spline-curve fitting, the observations are weighted according to the inverse of their variance. Application to Adolphus D-50 using the values of s_i^2 in Table 8.15 (3rd column) yielded an improved best-fitting straight line. Deviations from this line and spline-curves for the 23 other wells gave improved estimates s_i^2 which were used as input for modified RASC. This extra step is only taken at the beginning of the iterative process. During later steps, weighted spline-curve fitting is used only. It was found that the iterative process converged to the same final solution with and without the extra step at its beginning. With this refinement, the final solution was reached faster. Modified RASC distances and the variances used to obtain them are shown in Table 8.15 for steps 5 and 6 of the iterative process with refinement. These estimates are preceded by their fossil event numbers because of minor reordering with regard to the original sequence order (Table 8.15, column 1). The weighted spline-curve fitted after step 5 of the iterative process with refinement for Adolphus D-50 is shown in Figure 8.5B.

At the beginning of the iterative process, the average variance for the 44 species is equal to 0.500. At the end of the process the overall variance has become 0.351. This implies that the standard deviation $\sigma = 0.70$ was reduced to 0.59. The total range for the species along the RASC scale was reduced from 7.78 (original RASC output) to 7.16 after steps 5 and 6 (cf. Table 8.15). This shrinking is related to the reduction in the standard deviation.

The mean deviation of the species in individual wells from their spline-curves was computed at each step of the iterative process. In Figure 8.6, this mean deviation is plotted against RASC distance at the beginning (RASC output) and end of the iterative process (modified RASC output). Clearly, there is a systematic departure from zero near the top and bottom of the stratigraphic sequence. The average deviation of the first 3 species amounts to -0.65 and that of the last 9 species is 0.28 in Figure 8.6B. The discrepancies for these 12 events were not significantly reduced during the iterative process. It indicates that, on the average, the fitted spline-curves slightly underestimated RASC distances near the tops of the sections and overestimated them near the bottoms. This effect would be reduced if more weight were given to the 12 events, e.g. by centering their variances with respect to the average deviations. However, this also would result in a further decrease of the overall variance with increased shrinking of the total range for the species along the RASC scale.

8.7 Frequency distributions of stratigraphic events

As mentioned in the previous section, most frequency distributions for individual species are unimodal and slightly skewed to the right or to the left. A few distributions seem to be bimodal. All distributions change shape during the iterative process. We will restrict our presentation mainly to the final result obtained after convergence.

Figure 8.7 shows histograms for taxon 42 (*Cibicidoides alleni*) and taxon 50 (*Subbotina patagonica*) before and after convergence. *S. patagonica* which is an abundant planktonic species was already a relatively good marker at the beginning of the iterative process because its variance (= 0.204) was less than 0.5. After convergence, its variance has become very small. The corresponding histogram is a narrow peak indicating that the final spline-curves for the nine wells with *S.*

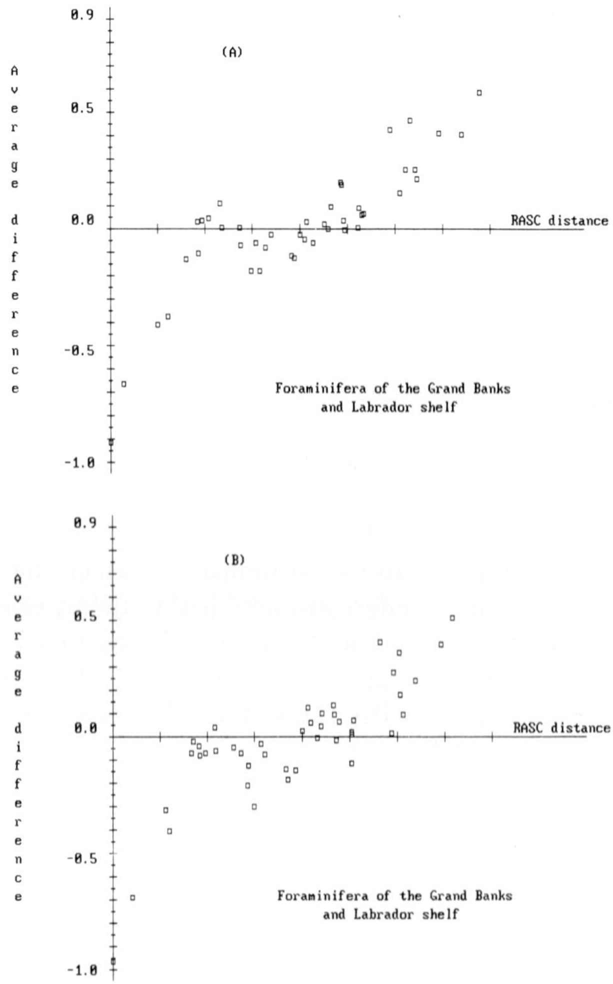

Fig. 8.6 Mean deviation from spline-curves per species plotted against RASC distance before (A) and after (B) convergence. For further explanation see text.

patagonica passed almost exactly through the points for this taxon. It may be concluded that *S. patagonica* is an excellent marker, whose position in individual sections is everywhere close to its position in the scaled optimum sequence. This property is enhanced when modified RASC is used. On the other hand, *Cibicidoides alleni* which is a rare benthonic species has a variance above 0.5, both before and after iteration. Its histogram also has not changed significantly (see Fig. 8.7). This taxon seems to have a bimodal frequency distribution. According to F.M. Gradstein (personal communication, 1987), *C. alleni* is not well defined taxonomically and may actually represent two different forms.

An unsolved problem of considerable interest regards the shapes of unimodal frequency distributions of biostratigraphic events. It is unlikely that such frequency distributions are exactly symmetrical. Two models with asymmetry for highest occurrences were suggested in Section 2.6:

Model A – The species disappeared in most places at approximately the same time but, perhaps due to lack of preservation, had already disappeared earlier in some places. This is the most likely model for exits as explained in Section 2.6. A "mass extinction" or a hiatus would create frequency distributions of this type. *Model A* predicts negative skewness (cf. Fig. 2.10D).

Model B – The species disappeared in most places (from most sections) at approximately the same time but remained in existence longer in a few places due to favorable conditions or was subjected to localized reworking.

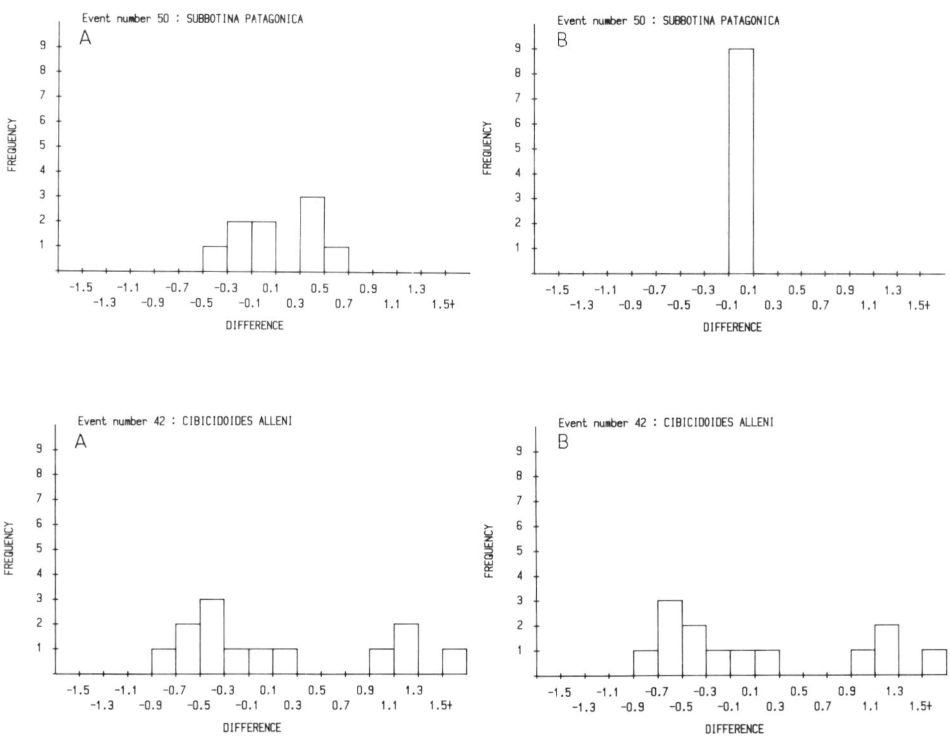

Fig. 8.7 Histograms of *Cibicidoides alleni* and *Subbotina patagonica* before (A) and after (B) iteration. After iteration, the bimodal histogram of *C. alleni* has remained approximately the same, whereas the histogram of *S. patagonica* has become very narrow.

The tail of the frequency distribution then extends in the stratigraphically upward direction with predicted positive skewness of the frequency distribution (cf. Fig. 2.10D).

The skewness of the histograms for 44 Cenozoic foraminifers along the northwestern Atlantic Margin has been determined by computing their (unbiased) sample skewness statistics (see Table 8.16). (The "unbiased" skewness was obtained by multiplying the sum of cubes of standardized deviations from the mean by n/n-1)(n-2)). In column 3 of Table 8.16 the skewness was estimated for deviations from the best-fitting spline-curves. Although individual estimates of skewness are not significantly different from zero (= symmetry), because sample sizes are small (from 7 to 22 only), column 3 shows a pattern in that the events in the upper half of the table display almost exclusively negative values for skewness, whereas those in the lower half are almost all positive. This pattern partly can be explained by the fact that RASC distances near the tops of the sections were underestimated whereas those near the bottoms were overestimated (cf. Fig. 8.6). Bias introduced by use of estimated means which are too low or too high can be eliminated by substituting the mean deviations plotted in Figure 8.6B for the sample mean in the equation used for estimating skewness. The resulting revised estimates are shown in column 4 of Table 8.16. Clearly skewness was increased near the top of this table and decreased near its bottom. However, the pattern remains that in the upper half of the table, most skewnesses are negative, whereas those in the lower half are mostly positive. It is noted, that 6 of 8 species at the bottom of the table have negative skewness in column 4 of Table 8.16.

Comparison of the RASC distance scale to the geological time scale shows that the positive skewness values are largely restricted to the Eocene which extends approximately from event 56 to event 259 (cf. Gradstein et al., p. 339) corresponding to a time interval of about 21 Ma (from 58 to 37 Ma). The total range of RASC distances in Tables 8.15 and

TABLE 8.16

Selected statistics for the 44 species after convergence. Degrees of freedom $f_i = n_i$-1 where n_i represents sample size for event i. Skewness 1 and 2 are sample statistics per species using zero mean and sample mean for deviations from spline-curves, respectively. The pooled variance s^2 is equal to 0.351. Variance ratio s_i^2/s^2 has asterisk if its value is below 0.005 fractile or above 0.995 fractile of corresponding χ^2/f distribution. Last column shows individual terms added to give Bartlett's $\hat{\chi}^2 = 180.734$ (see text). Constant $C = 1.034$ was computed by formula in Hald (1975, p. 291).

Event	f_i	Skewness 1	Skewness 2	s^2_i/s^2	f_i*ln (s^2_i/s^2) / C
10	9	-1.367	-0.059	3.900*	-9.589
17	11	-1.678	-1.276	1.999	-7.367
16	21	-1.392	0.205	0.745	5.983
67	7	-2.375	-1.297	1.492	-2.710
18	21	-1.140	-0.451	0.264	27.034
21	9	-1.074	-0.507	0.025*	32.066
20	19	-1.542	-1.108	0.198*	29.681
71	12	-1.040	-0.617	1.061	-0.683
26	12	-0.016	0.368	1.172	-1.838
70	6	-0.479	-0.965	0.384	5.556
15	21	-1.548	-1.284	1.192	-3.570
24	16	-0.792	-0.469	0.512	10.370
27	12	-1.313	-1.045	2.094	-8.575
69	10	-1.139	-0.253	1.799	-5.680
25	18	-0.586	0.233	0.677	6.778
81	11	-1.652	-0.563	1.776	-6.109
202	6	-1.499	-1.153	0.266	7.689
259	13	-0.357	0.495	0.263*	16.782
147	6	-0.812	0.601	0.472	4.359
34	14	-0.727	0.103	1.578	-6.172
33	6	-0.404	0.148	3.251*	-6.841
260	14	1.681	1.442	0.431	11.399
261	14	1.920	0.809	0.199*	21.836
263	12	0.791	0.425	0.998	0.038
29	18	-0.034	-0.027	0.385	16.633
32	17	-0.481	-0.836	0.627	7.672
40	9	1.207	0.651	1.232	-1.816
42	12	1.356	0.859	2.399*	-10.157
264	6	2.403	1.808	1.023	-0.131
41	11	0.358	0.429	1.029	-0.307
30	11	0.600	0.229	1.185	-1.816
90	6	1.084	1.894	1.175	-0.936
36	10	0.511	0.424	1.072	-0.676
86	6	0.830	0.271	0.093*	13.789
57	18	0.469	0.150	1.586	-8.030
45	9	1.511	0.185	2.634*	-8.429
50	8	0.118	-1.394	0.006*	39.602
46	13	1.361	0.038	1.119	-1.414
230	6	1.466	-0.675	1.124	-0.677
54	12	1.659	0.573	0.632	5.334
52	6	-0.333	-1.424	0.478	4.285
56	13	1.486	-0.046	0.539	7.764
55	8	1.388	-1.278	0.790	1.821
59	7	1.321	-1.597	1.465	-2.587

8.16 corresponds to about 63 Ma. The species with positive skewness, therefore, tend to occur during the epoch (Eocene) that is represented by relatively many species in our application. It seems that *Model A* predominated during this time interval, whereas *Model B* predominated after and possibly before the Eocene. This result is corroborated by the observation that tests usually are reworked in the younger Neogene section of the Labrador Shelf (cf. Section 4.7).

It was assumed in the previous section that variances s_i^2 obtained for the species are significantly different from one another. This assumption has been tested statistically with the results shown in the last two columns of Table 8.16. Column 5 shows species variances s_i^2 divided by $s^2 = 0.351$ representing the pooled variance for all 44 species (see before). If the variances are equal, this ratio is approximately distributed as $\chi^2/f = s^2/\sigma^2$ where the chi-squared (χ^2) has f degrees of freedom. The fractiles of this distribution have been tabulated for different values of f by Hald (1960, p. 44). In Table 8.16, an asterisk was given to values below the 0.005 or above the 0.995 fractile. Such values would occur with probability $\alpha = 0.01$. This test indicates that six variances are probably too small and four are too large in Table 8.16. Bartlett's χ^2-test for equality of variances (see e.g. Hald, 1957, p. 291) has also been applied. According to this test, the quantities in the last column of Table 8.16 would add up to χ^2 with $(k-1) = 43$ degrees of freedom. The total chi-squared value is equal to 180.734 which far exceeds the corresponding 99% confidence limit (= 67.5). Bartlett's chi-squared test, therefore, also indicates that the variances s_i^2 are not equal to one another.

Another statistical experiment conducted for this example is as follows. From the preceding results, it may be concluded that the variances of the 44 species are not equal to one another. For this reason, the values used for the histograms of individual species were standardized by dividing them by s_i. Consequently, 44 sets of values were obtained with means equal to zero and standard deviations equal to one. These 44 sets of values were combined with one another to give a single new set of 550 standardized values of which the histogram is shown in Figure 8.8. This composite frequency distribution would be positively or negatively skew if the frequency distributions for individual species would all tend to be asymmetric, e.g. according to *Model A* or *B* (see before). Instead of this, the composite distribution (Fig. 8.8) seems to be approximately symmetric. When the last two classes in upper and lower tail are combined with each other, 13 observed frequencies are retained for the histogram of Figure 8.8

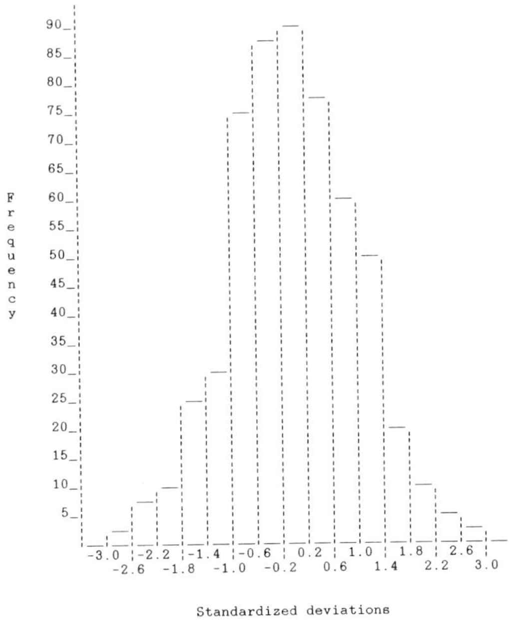

Fig. 8.8 Histogram of 550 standardized differences from all spline-curves for all species after convergence. Standardization was achieved by dividing each difference by the standard deviation s_i for its species.

which can be compared to 13 theoretical frequencies obtained from the normal distribution in standard form. Application of the chi-squared test for goodness of fit gave $\hat{\chi}^2(10) = 12.03$ for the difference between observed and theoretical normal distribution. For 10 degrees of freedom, the corresponding 95% and 99% fractiles of the χ^2-distribution are 18.3 and 23.2, respectively. Because the $\hat{\chi}^2$-value estimated for Figure 8.8 is less than these values, it may be concluded that the composite distribution of Figure 8.8 is approximately normal (Gaussian).

Earlier in this section, positive and negative skewness of individual frequency distributions was discussed. Although sample sizes are too small to establish that the individual skewness values of Table 8.16 are significantly different from zero, the sign of skewness changed through time according to a regular (nonrandom) pattern. Obviously, this pattern is too weak to show up as a systematic departure from normality in the composite frequency distribution of Figure 8.8.

The modified RASC method consists of alternately obtaining two different estimates (x_i and $\hat{x}_i$) of the mean position EX_i of each event i along the relative time interval scale. This iterative process converges to a final solution which does not differ greatly from the ordinary RASC scaled optimum sequence. The differences ($\hat{x}_i$-x_i) provide an estimate of the frequency distribution for event i. It has been demonstrated that the highest occurrences of Cenozoic Foraminifera along the northwestern Atlantic margin have different variances. The histogram of standardized values for all species was shown to be approximately normal. The possibility to identify good markers with small variance (e.g. *Subbotina patagonica*) is a new feature of modified RASC not previously provided by ordinary RASC. Likewise, it has become possible to identify relatively poor markers with relatively large variance and perhaps bimodal distribution (e.g. *Cibicidoides alleni*). Although x_i and $\hat{x}_i$ both provide good approximations of EX_i, some bias was introduced during the iterative process consisting of reduction of average variance as well as non-zero mean values of ($\hat{x}_i$-x_i) for events near top and bottom of the stratigraphic sequence. The method also provides a way to construct conservative range charts in which the ranges of the fossils are extended to the highest occurrences in individual sections.

For example, in Figure 8.7B, the largest (positive) deviations on the right side of the frequency curves are plotted at 0.1 and 1.7, respectively. These values can be added to the RASC distances (sixth run, Table 8.14) in order to obtain conservative ranges. (The maximum positive deviation exceeded 1.5 for only two of the 550 values used in the histograms for separate events. In these two situations, the range extension was set equal to 1.7). Figure 8.7 shows highest occurrences based on cumulative (modified) RASC distances (A) as well as highest occurrences for individual sections (C) obtained by subtracting the largest positive deviations. For comparison, the mean deviations (B) of Figure 8.6B also are shown in Figure 8.9 in the form of positive or negative deviations from the RASC distance (A).

If all variances were equal to 0.5, 95 percent of the positive deviations would be less than 1.163. This was the value previously used for the range extensions in the Drobne example of Figure 8.4. It was shown by analysis of variance that the variances of the taxa in the Gradstein-Thomas database are not equal to one another. Thus the shorter range extensions in Figure 8.9 are for taxa with variances which are significantly less than the average variance. On the other hand, it should be kept in mind that

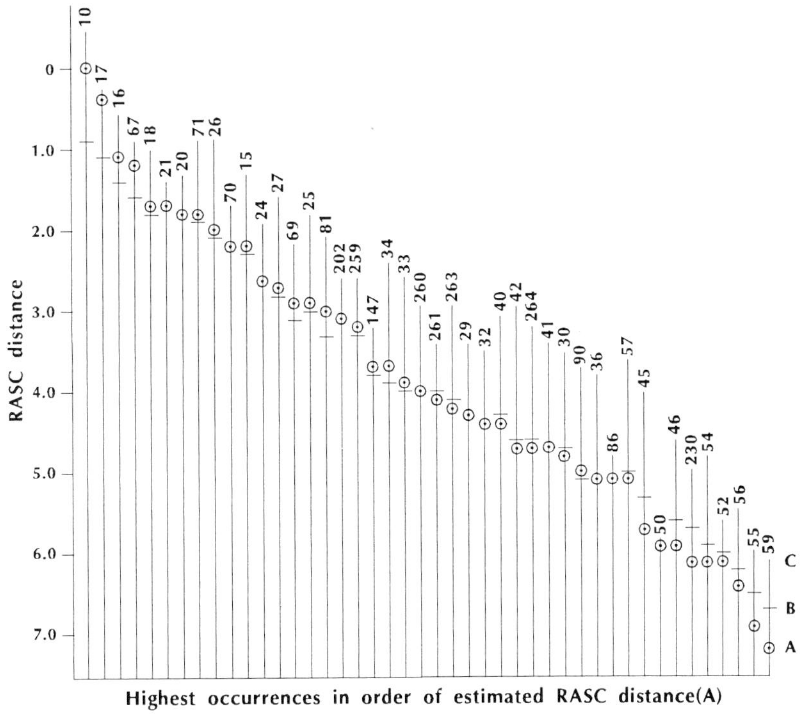

Fig. 8.9 Extended RASC ranges for Cenozoic Foraminifera in Gradstein–Thomas database. Letters for taxon 59 on the right represent (A) estimated RASC distance, (B) mean deviation from spline-curve, and (C) highest occurrence of species (i.e. maximum deviation from spline-curve). B is shown only if it differs from A. Good markers such as taxon 50 (*Subbotina patagonica*) have approximately coinciding positions for A, B and C. Note that as a first approximation it could be assumed that the highest occurrences (C) have RASC distances which are about 1.16 units less than the average position (cf. Section 8.3). This systematic difference in distance is equivalent to approximately 10 m.y. (cf. Fig. 9.2, see later).

the range extensions have their own variances and are subject to more uncertainty than the RASC distances themselves. The subject of conservative range charts also will be discussed in the next two sections with applications to smaller datasets.

8.8 Application of modified RASC to Drobne's alveolinids

The Drobne example (cf. Section 8.3) was subjected to modified RASC instead of RASC with results shown in Tables 8.17 and 8.18. Sections V, IX and XI have only one or two event levels (see Fig. 8.1) and could not be used in modified RASC because at least 3 event levels are needed for curve-fitting. The scaled optimum sequence previously obtained by RASC

TABLE 8.17

Modified RASC method applied to original Drobne example of Section 8.3. After 4 iterations, the RASC distances ($\bar{x}_4$) are close to the original RASC distances ($\bar{x}_1$). The event variances ($s^2{}_4$) are for zero mean deviations and differ from one another. Degrees of freedom (d.f.) in last column are equal to 3 or 4 for nearly all events. For 3 degrees of freedom the 95% confidence interval of the sample variance ranges from $0.32\sigma^2$ to $3.12\sigma^2$. Here σ^2 is the expected value of the variance which is approximately equal to 0.5 in this application. According to this single variance test, the variance of event 15 would be too large and those of events 20, 27, 22, 2, 23, 1 and 3 would be smaller than average. However, modified RASC gives results that are approximate if samples sizes are very small. It will be seen later (see Table 8.21) that only the variances of events 27, 2 and 1 are again much smaller than average after enlarging the dataset and re-running modified RASC.

Event	$\bar{x}_1$	$\bar{x}_4$	$s^2{}_4$	d.f.
28	0.00	0.00	0.31	4
20	0.02	0.11	0.05	4
19	0.30	0.32	0.14	4
18	0.45	0.45	0.45	3
27	0.88	0.76	0.06	4
15	1.16	1.16	3.04	3
17	2.00	2.02	0.76	3
22	2.02	2.07	0.07	3
2	2.16	2.20	0.03	3
23	2.16	2.18	0.05	4
21	2.32	2.33	0.26	3
1	2.47	2.45	0.13	3
14	2.69	2.69	0.30	6
12	2.70	2.70	0.26	4
25	2.89	2.89	0.33	4
11	3.33	3.33	0.44	4
5	3.33	3.32	0.96	3
13	3.52	3.53	0.43	6
3	4.60	4.60	0.00	3

is shown as $\bar{x}_1$ in Table 8.17. It was the starting point for modified RASC which, after four iterations, produced nearly the same scaled optimum sequence ($\bar{x}_4$ in Table 8.17).

It is noted that on the basis of the results by modified RASC described in the previous section (also see D'Iorio, 1988) indicating that the order of events does not change significantly when this method is applied, it was

TABLE 8.18

Deviations of observed relative positions of events from spline-curves after 4 iterations. Numbers along top indicate the eight sections used. Event numbers are given in first column. Events 15, 23, 25, 5 and 3 have asterisk for coinciding highest and lowest occurrences in all sections. The variances of Table 8.17 were based on these numbers. Largest deviations for even code numbers (=highest occurrences) and lowest deviations for odd code numbers (=lowest occurrences) were used for range chart of Fig. 8.10. These numbers are shown in bold print. Rows with asterisks have two bold numbers.

	1	2	3	4	6	7	8	10
28	x	-0.97	-0.23	**-0.04**	-0.47	x	-0.07	x
20	x	x	-0.12	**0.07**	-0.37	x	0.04	-0.22
19	x	x	0.08	**-0.68**	-0.16	x	x	x
18	x	-0.52	0.21	**0.40**	-0.93	x	-0.21	x
27	x	-0.20	-0.19	**-0.23**	0.29	x	-0.21	x
15*	**-0.98**	x	x	-0.78	x	x	0.18	**2.74**
17	x	-0.09	1.06	**-0.86**	0.64	x	x	x
22	x	-0.03	**0.39**	0.13	x	x	x	-0.17
2	x	0.10	x	**0.26**	-0.07	x	x	-0.05
23*	**-0.27**	0.08	-0.23	**0.24**	x	x	x	-0.06
21	x	0.23	0.64	**-0.55**	x	x	x	0.09
1	x	0.34	x	**-0.44**	0.17	x	x	0.20
14	0.24	**0.59**	-0.45	-0.19	0.42	-0.04	-1.00	x
12	-0.54	**0.60**	-0.44	x	x	x	-0.08	0.46
25*	x	**-0.34**	-0.25	x	-0.28	0.16	**1.01**	x
11	0.08	0.09	**-0.54**	x	x	x	0.54	1.08
5*	**1.19**	**-1.04**	-0.54	x	x	-0.34	x	x
13	1.08	**-0.83**	-0.34	0.65	0.36	-0.13	-0.16	x
3*	0.00	x	**0.01**	x	x	**0.00**	0.00	x

decided to change the procedure slightly as follows. Instead of taking the scaled optimum sequence without final reordering as the starting point, it is now possible to take the scaled optimum sequence after final reordering as the starting point. On the other hand, the order of events is not allowed to change during successive iterations in modified RASC. The order of events in $\bar{x}_4$ in Table 8.17 is identical to that in $\bar{x}_1$ except for events 11 and 5 which are nearly coeval on the average.

The variances of the events (s^2_4) had not completely converged after 4 iterations. Because the number of degrees of freedom for s^2_4 is small for all events ranging from 3 to 6, these results are subject to considerable uncertainty. According to Table 8.17, events 2 and 3, corresponding to the highest occurrence of species 1 (*A. moussoulensis*) and the lowest occurrence of species 2 (*A. aramaea*) have variances closest to zero and could be good marker horizons. However, these two events each occur in 4 sections only. The fact that their positions are on the fitted spline-curves may not be significant because there are so few data. It should be kept in mind that small variance events receive relatively more weight than other events in spline-curve fitting. In fact, zero-variance events have the property (cf. Section 3.11) that the best-fitting spline-curve is forced to pass exactly through their points on the scattergram. The possibility, therefore, exists that an event which happens to have a small variance because it occurs in so few sections, obtains zero-variance during the convergence process which involves repeated spline-curve fitting for all sections.

The final deviations of the 19 events from the 8 fitted spline-curves are shown in Table 8.18. If all variances are assumed to be equal, numbers with absolute value greater than 1.16 denote events out of position with probability greater than 95%. The two events with this property are event 15 (species 8) and event 5 (species 3). The latter event occurs in a reworked bed as discussed in Section 8.3. According to the preceding equal variance test applied to Table 8.18, species 8 would occur too high in Section X. However, this result would need confirmation by additional evidence or other experiments because there are too few event levels per section in this dataset for a fully convincing application of modified RASC.

Brower (1990) has carried out a method comparison study on the Drobne dataset. Figure 8.10 shows ranges for 12 species obtained by 5 methods. The ranges resulting from the Unitary Associations (U.A.) method, seriation (SER) and RASC were calculated by Brower and plotted along a relative time-scale with 10 units. The RASC distances $\bar{x}_4$ of Table 8.17 were enlarged by the factor (10/4.16=) 2.40 so that their largest value (for lowest occurrence of event 2) became 10 instead of 4.16 in Table 8.17. These RASC distances are shown as tick marks on the left of the ranges for each species in Figure 8.10. Species with coinciding highest and lowest occurrence in all sections have a single tick mark only.

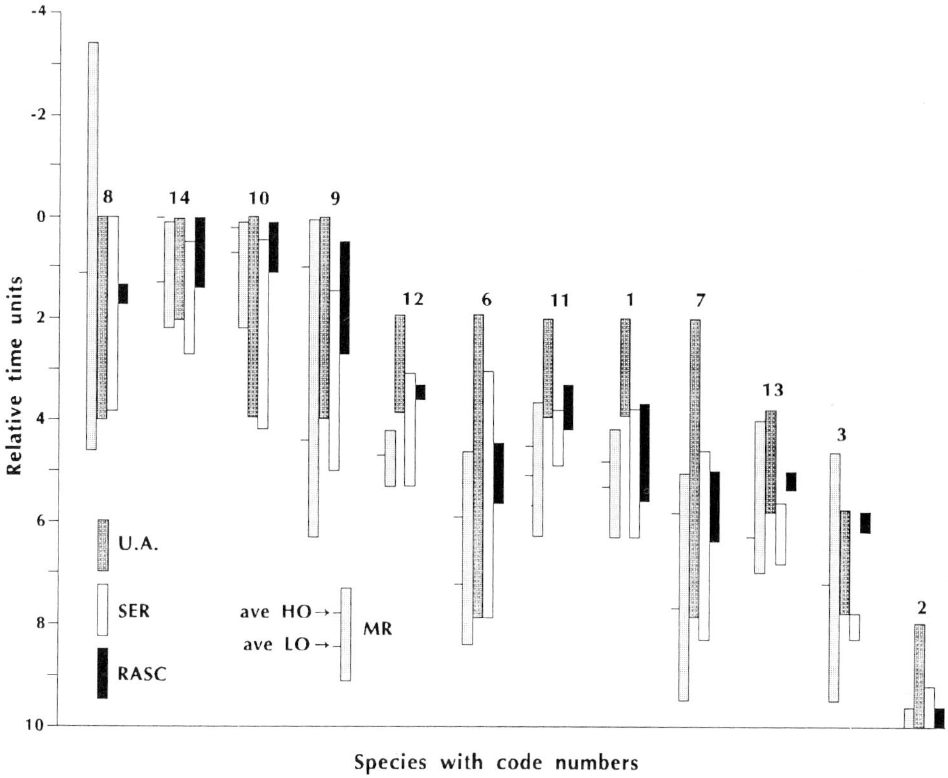

Fig. 8.10 Comparison of five types of ranges for Drobne's alveonilids along relative time scale of Brower (1990) who pointed out that RASC ranges are significantly shorter than Unitary Associations (U.A.) and Seriation (SER) ranges. These results are compared to the modified RASC (MR) ranges and the average highest occurrences (ave HO) and average lowest occurrences (ave LO) on which these MR ranges are based. The relative time scales used for U.A., SER, RASC and MR, respectively, have different units and are not completely comparable (cf. Brower, 1990). However, on the whole, the MR ranges are about as wide as the U.A. and SER ranges.

The ranges between tick marks were extended by adding deviations from Table 8.18 as follows. For highest occurrences (even numbers in Table 8.18), the largest deviation was subtracted from the RASC distance; for lowest occurrences, the absolute value of the smallest deviation was added to the RASC distance; and for species with coinciding highest and lowest occurrence, both the largest and the smallest deviations were used. The resulting extended ranges are shown in Figure 8.10.

Brower (1990) used his own computer algorithms for U.A. and RASC which differ somewhat from those used by Davaud and Guex (1984) and in Gradstein et al. (1985). Also, because different methods have different time-scales, plotting all ranges along a single time-scale may distort some results. However, Brower (1990) correctly concluded that the average ranges obtained by RASC were significantly shorter than the ranges obtained by U.A. and seriation. The distances between ave HO and ave LO are very close to the Brower's RASC ranges, and the extended modified RASC (MR) ranges are approximately as wide as the U.A. and SER ranges. For species 8, 9 and 3, the MR ranges are wider than the other ranges. These wider extensions are in part due to the "anomalous" values (greater than 1.16) for species 8 and 3.

The number of event levels per section can be enlarged by not using the maximal horizons method for data reduction. Table 8.19 is based on use of all stratigraphic information on relative positions of highest and lowest occurrences. For example, Section II (2) for Figure 8.2 has 9 event levels in Table 8.19 versus 4 maximal horizons in Figure 8.1. The reworked bed (level 4 in Section I of Fig. 8.1) was not included in the SEQ file of Table 8.19. The new scaled optimum sequence obtained after final reordering is shown as $\bar{x}_1$ in Table 8.21.

Table 8.20 shows normality test results for the 3 sections with events that are anomalous with a probability of 99% (2 asterisks for second-order

TABLE 8.19

SEQ file for recoded Drobne dataset. Most sections have more event levels than in Fig. 8.1. Section 2 (Dane near Divača, see Fig. 8.2) has 9 event levels which were reduced to 4 maximal horizons in Fig. 8.1. The number – 999 denotes end of section in SEQ file.

```
SECTION 1
  15 -16   7  -8 -13 -14 -23 -24  11 -12   3  -4-999   0   0   0   0   0   0   0
SECTION 2
  28  18 -27   2 -14 -24   1 -12 -17 -21 -22  23  11 -25 -26   4  -6  15 -16   3
  -5  -9 -10 -13-999   0   0   0   0   0   0   0   0   0   0   0   0   0   0   0
SECTION 3
  18 -20  28  19  30  27  17  21 -22  23 -24 -29  14 -26  12 -25   6 -11   5 -13
   3  -4-999   0   0   0   0   0   0   0   0   0   0   0   0   0   0   0   0   0
SECTION 4
  20 -28  18  29 -30   7  -8 -19 -27   2 -15 -16 -22 -23 -24   1 -13 -14 -17 -21
-999   0   0   0   0   0   0   0   0   0   0   0   0   0   0   0   0   0   0   0
SECTION 5
   7  -8   9 -10-999
SECTION 6
  19 -20 -27 -28   7  -8 -15 -16   1  -2 -14  13 -25 -26-999   0   0   0   0   0
SECTION 7
  14 -25 -26 -29 -30   5  -6 -13   9 -10   3  -4-999   0   0   0   0   0   0   0
SECTION 8
  20 -28  19  15 -16 -27  11 -12 -25 -26  13 -14   4 -10   3   9-999   0   0   0
SECTION 10
  23  15 -16  19 -20  24   1  -2 -11 -12 -21 -22-999   0   0   0   0   0   0   0
SECTION 11
  19 -20 -27 -28   7  -8 -17 -18   1  -2 -14  13 -25 -26-999   0   0   0   0   0
```

TABLE 8.20

RASC normality test output for the 3 sections in the recoded Drobne dataset with one or more events with double asterisks.

SECTION 1			CUM. DIST.	2ND ORDER DIFF.
LO	A. LEUPOLDI	15	1.9144	
HI	A. LEUPOLDI	-16	1.9144	-1.3814
LO	A. GLOBOSA	7	1.3920	1.3814
HI	A. GLOBOSA	-8	1.3920	2.3005 *
LO	A. PASTICILLATA	-13	3.6925	-3.4994 **
HI	A. PASTICILLATA	-14	2.4935	0.6432
LO	A. SUBPYRENEICA	-23	1.9377	0.4902
HI	A. SUBPYRENEICA	-24	1.8722	0.5950
LO	A. PISIFORMIS	11	3.2605	-0.9526
HI	A. PISIFORMIS	-12	2.8374	1.7864
LO	A. ARAMAEA	3	5.0596	-1.9317
HI	A. ARAMAEA	-4	4.4912	

SECTION 2			CUM. DIST.	2ND ORDER DIFF.
HI	A. GUIDONIS	28	0.0000	
HI	A. MONTANARII	18	0.5247	0.1032
LO	A. GUIDONIS	-27	0.6910	0.6962
HI	A. MOUSSOULENSIS	2	2.0151	-0.3842
HI	A. PASTICILLATA	-14	2.4935	-1.0997
HI	A. SUBPYRENEICA	-24	1.8722	0.8393
LO	A. MOUSSOULENSIS	1	2.5517	0.0678
HI	A. PISIFORMIS	-12	2.8374	-1.1310
LO	A. MONTANARII	-17	1.9921	1.2539
LO	A. DEDOLIA	-21	2.4006	-0.7461
HI	A. DEDOLIA	-22	2.0631	-0.2494
LO	A. SUBPYRENEICA	23	1.9377	1.4482
LO	A. PISIFORMIS	11	3.2605	-0.9277
LO	A. LAXA	-25	3.1941	-0.1121
HI	A. LAXA	-26	3.0156	1.1926
HI	A. ARAMAEA	4	4.4912	-4.0524 **
LO	A. LEUPOLDI	15	1.9144	3.0383 **
HI	A. LEUPOLDI	-16	1.9144	2.6836 *
LO	A. ARAMAEA	3	5.0596	-2.8790 *
LO	A. AVELLANA	-9	4.8642	-0.2161
HI	A. AVELLANA	-10	4.4527	-0.3487
LO	A. PASTICILLATA	-13	3.6925	

SECTION 8			CUM. DIST.	2ND ORDER DIFF.
HI	A. ARAGONENSIS	20	0.1267	
HI	A. GUIDONIS	-28	0.0000	0.1094
LO	A. ARAGONENSIS	19	0.6595	0.5955
LO	A. LEUPOLDI	15	1.9144	-0.5782
HI	A. LEUPOLDI	-16	1.9144	-1.2234
LO	A. GUIDONIS	-27	0.6910	3.1161 **
LO	A. PISIFORMIS	11	3.2605	-2.3158 *
HI	A. PISIFORMIS	-12	2.8374	0.7798
LO	A. LAXA	-25	3.1941	-0.5352
HI	A. LAXA	-26	3.0156	0.1786
LO	A. PASTICILLATA	13	3.6925	-1.1991
HI	A. PASTICILLATA	-14	2.4935	2.5199 *
HI	A. ARAMAEA	4	4.4912	-1.3594
HI	A. AVELLANA	-10	4.4527	-0.0314
LO	A. ARAMAEA	3	5.0596	-0.8023
LO	A. AVELLANA	9	4.8642	

TABLE 8.21

Modified RASC method applied to recoded Drobne dataset. n is number of sections in which event was observed. $\bar{x}_1$, $\bar{x}_3$ and $\bar{x}_4$ are RASC distances at beginning and after 3 and 4 iterations, respectively. Variances after 3 and 4 iterations are for zero mean deviation and are only approximately equal to one another. SK1 and SK2 are skewness statistics with and without zero means, respectively.

Event	n	$\bar{x}_1$	$\bar{x}_3$	$\bar{x}_4$	s^2_3	s^2_4	SK1	SK2
28	6	0.00	0.00	0.00	0.09	0.12	-1.61	0.30
20	6	0.13	0.12	0.13	0.41	0.42	-2.70	-1.55
18	4	0.52	0.51	0.52	0.43	0.45	-0.86	1.18
19	6	0.66	0.64	0.64	0.29	0.29	-1.83	-0.04
27	6	0.69	0.67	0.68	0.46	0.49	-2.13	-0.52
8	5	1.39	1.37	1.38	0.03	0.01	-1.57	-1.84
7	5	1.39	1.37	1.38	0.03	0.01	-1.53	-1.85
24	5	1.87	1.84	1.83	0.05	0.07	-1.77	-1.12
16	6	1.91	1.86	1.86	0.99	0.92	-2.28	-1.97
15	6	1.91	1.86	1.86	0.99	0.92	-2.28	-1.97
23	5	1.94	1.88	1.88	0.25	0.27	-0.91	-0.30
17	4	1.99	1.94	1.99	0.35	0.34	0.56	0.01
2	5	2.02	2.02	2.01	0.04	0.04	-0.12	0.60
22	4	2.06	2.00	2.00	0.06	0.05	-1.72	-0.08
21	4	2.40	2.27	2.27	0.07	0.06	-2.08	0.92
14	8	2.49	2.41	2.41	0.37	0.40	-0.53	-0.51
1	5	2.55	2.42	2.42	0.05	0.05	1.76	-1.39
12	5	2.84	2.72	2.72	0.15	0.14	1.15	0.07
26	6	3.01	2.91	2.91	0.03	0.02	1.15	0.94
25	6	3.20	3.02	3.02	0.02	0.02	2.84	1.76
11	5	3.26	3.09	3.08	0.28	0.28	1.84	-0.03
13	8	3.69	3.54	3.54	0.83	0.89	1.95	0.77
10	4	4.45	4.30	4.29	0.00	0.00	-2.17	-1.13
4	5	4.49	4.35	4.35	0.34	0.35	1.37	1.02
9	4	4.87	4.71	4.71	0.64	0.59	1.66	-2.00
3	5	5.06	4.92	4.92	0.24	0.23	2.05	-0.11

differences). The highest and lowest occurrence of species 7 (*A. pasticillata*) coincide in Section I (1). The lowest occurrence occurs too high in this section in comparison with its neighbors. On the contrary, species 8

(*A. leupoldi*), of which the highest and lowest occurrence coincide in all sections containing it, occurs too low in Section II (2). This is not immediately obvious from the pattern of asterisks for this section but follows when it is considered that its highest and lowest occurrence have the same cumulative RASC distance. Finally, the lowest occurrence of *A. guidonis* may be situated too low in Section VIII (8). The suggestion on the basis of Table 8.18 that *A. leupoldi* occurred too high in Section X (10) is not confirmed by the new normality test results.

Table 8.21 shows modified RASC results. The RASC distances ($\bar{x}_4$) after four iterations are nearly equal to those ($\bar{x}_3$) after three iterations. Comparison to the original scaled optimum sequence ($\bar{x}_1$) for this experiment shows that the modified RASC method left the scaled optimum sequence nearly unchanged. The variances (unbiased estimates of deviations from the spline-curves) have not yet fully stabilized after 4 iterations because the values of $s^2{}_4$ differ from those of $s^2{}_3$ in Table 8.21. Slightly more events have negative skewness after 4 iterations but there is no systematic pattern of change in the skewness. There is no evidence supporting prevalence of either *Model A* or *Model B* during a particular time interval (cf. Section 8.7).

Extended modified RASC ranges for the newly coded Drobne dataset are shown in Figure 8.11. The three events identified as possibly anomalous by the normality test (Table 8.20) had deviations from the spline-curves exceeding 1.16 representing the 95% confidence limit if all event variances are equal. This indicates good agreement between normality test and modified RASC results.

Elimination of anomalous events would shorten the extended ranges for species 14 and 8 by the amounts shown in Figure 8.11. The assumption that the lowest occurrence of species 7 occurs too high in Section I (cf. Table 8.20) does not change the length of the extended range for this species. Because the modified RASC range chart of Figure 8.11 is based on more information than the corresponding range chart of Figure 8.10 it is probably better. It is not possible to determine by how much the procedure of recoding the Drobne data followed by modified RASC has improved upon the original RASC extended range chart shown in Figure 8.3. The new result is closer to Drobne's subjective zonation. It also is interesting to recall Guex's remark (see Section 8.3) that species 5 probably never

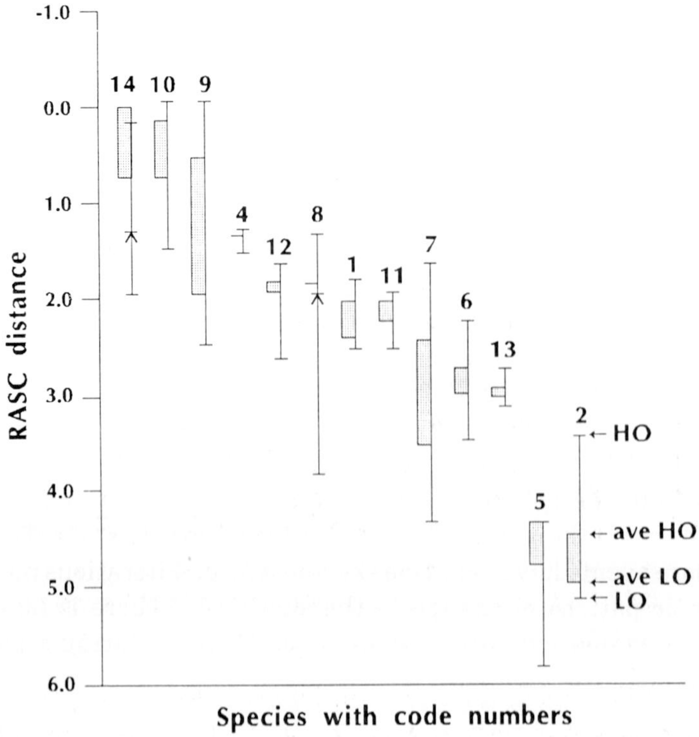

Fig. 8.11 Extended modified RASC ranges for Drobne's dataset. As in Fig. 8.3, the species were ordered on the basis of the RASC distances of their average highest occurrences. The sample sizes were small and this is the main reason for the random fluctuations in the positions of the highest (HO) and lowest (LO) occurrences. Deletion of events with double asterisks in normality test (see Table 8.20) would result in shorter ranges for species 8 and 14 as shown by arrows in Fig. 8.11.

coexisted with 11 although their ranges overlapped in Figure 8.3. In Figure 8.11, the extended ranges for these two sepcies are clearly separate.

It is good to keep in mind that the highest and lowest positions of fossils in individual sections are subject to more uncertainty than the positions of the average highest and lowest occurrences (cf. Chapter 2). In general, it is better to base the construction of isochrons in automated stratigraphic correlation (see Chapter 9) on average highest and lowest occurrences because these can be known with more precision than the (conservative) truly highest and lowest occurrences. On the other hand, if assemblages of fossils are used for subjective correlation, the extended range chart may provide a better tool than a range chart which is based on average stratigraphic events.

8.9 Comparison of range charts for Palmer's database

Application of the normality test to Palmer's (1954) database for the Riley Formation in central Texas was discussed in Section 8.4. This section contains results of the modified RASC method for this example. The scaled optimum sequence shown in the third column of Table 8.22 was taken as the starting point for a run with seven iterations in total. Approximate convergence was obtained as shown in Table 8.22. Figure 8.12 shows the final spline-curve in comparison with observations for the Morgan Creek section. This is output from the modified RASC module of micro-RASC (cf. Chapter 10)

Table 8.23 contains deviations used for the extended range chart with lowest and highest occurrences for each taxon. The deviations between points and curve graphically shown in Figure 8.12 for the Morgan Creek section correspond to the deviations listed in the column numbered 1 in Table 8.23. The mean deviation per event (in Table 8.22) was subtracted from the original deviations between points and curves before entering them into Table 8.23.

The extended ranges are shown in Figure 8.13 for comparison with 4 other range charts. The central 3 sets of ranges were taken from Edwards (1982) who used the same abbreviated dataset of 16 taxa representing those taxa occurring in 5, 6 or 7 sections. Edwards' comparison was for (1) Shaw's (1964) final composite standard, (2) Edwards' (1978) conservative method, and (3) Hay's (1972) method as applied by Edwards. The 32 lowest and highest occurrences were ranked from 1 to 32 for each method. The ranges obtained by the first two methods were considerably wider than those obtained by the Hay method which is comparable to RASC.

The RASC ranges plotted in Figure 8.13 are the final modified RASC distances ($\bar{x}_7$) of Table 8.22. Deviations for highest and lowest occurrences were taken from Table 8.23. For most taxa in Figure 8.13, the three ranges on the left side (modified RASC, Shaw and Edwards ranges) are approximately equally wide. The same holds true for the two ranges on the right (Hay and RASC ranges) which are considerably shorter than the three ranges on the left. On the whole, modified RASC has the widest ranges, partly because its ranges are clearly wider than the Shaw ranges for taxa 4, and 36 (with highest occurrences 8 and 70, respectively). The deviations in Table 8.22 corresponding to these two taxa are 1.27 (for event 8) in the Morgan Creek section and -1.07 (for event 69) in the Pontotoc

TABLE 8.22

Modified RASC method applied to Palmer's database. Approximate convergence was reached after 7 iterations. See Table 8.21 for explanations of column headings. The average deviation (Ave) is significantly less than zero for the first 6 events listed (see text for further discussion).

Event	n	$\bar{x}_1$	$\bar{x}_6$	$\bar{x}_7$	s^2_5	s^2_6	s^2_7	Ave	SK1	SK2
100	7	0.00	0.00	0.00	1.73	1.76	1.74	-1.12	-1.53	0.66
82	7	0.31	0.23	0.23	1.43	1.45	1.43	-0.97	-1.61	0.53
108	5	0.46	0.42	0.42	1.39	1.42	1.38	-0.99	-1.68	0.77
107	5	1.32	1.24	1.24	0.66	0.67	0.65	-0.59	-1.94	0.44
99	7	1.46	1.38	1.38	0.64	0.64	0.64	-0.49	-2.05	-0.54
90	5	2.02	1.95	1.95	0.55	0.55	0.55	-0.54	-2.13	-0.59
92	6	2.17	2.40	2.40	0.02	0.02	0.02	-0.08	-2.06	-0.32
89	5	2.82	2.83	2.82	0.07	0.07	0.07	-0.06	-0.44	0.80
91	6	2.85	2.89	2.88	0.06	0.07	0.07	0.04	0.90	0.28
81	7	3.08	3.11	3.10	0.07	0.07	0.07	0.08	1.28	0.16
68	7	3.86	3.63	3.62	0.08	0.08	0.08	0.05	0.72	0.02
70	6	3.99	3.80	3.80	0.14	0.15	0.15	0.13	-0.11	-1.84
42	7	4.02	3.81	3.81	0.18	0.18	0.19	0.10	0.36	-0.60
69	6	4.28	4.14	4.14	0.57	0.57	0.59	0.12	-0.19	-0.88
24	5	4.35	4.21	4.21	0.30	0.30	0.30	0.28	1.77	-0.43
67	7	4.51	4.40	4.42	0.23	0.19	0.20	0.21	1.29	-0.52
40	7	4.69	4.55	4.55	0.93	0.91	0.91	0.19	-0.64	-1.55
56	5	4.77	4.64	4.64	0.66	0.63	0.63	0.10	1.16	0.53
54	5	5.05	4.91	4.91	0.20	0.19	0.19	0.11	0.96	-0.20
48	5	5.21	5.07	5.07	1.44	1.40	1.39	0.23	1.27	0.36
47	5	5.21	5.07	5.07	1.44	1.40	139	0.23	1.27	0.36
55	5	5.46	5.37	5.36	0.12	0.12	0.12	-0.05	0.66	1.39
53	5	5.47	5.33	5.33	0.20	0.20	0.20	-0.02	-0.25	0.03
41	7	5.62	5.45	5.44	1.27	1.24	1.23	0.13	1.48	1.02
34	7	5.64	5.49	5.46	0.08	0.07	0.07	-0.10	-1.14	0.33
22	6	5.75	5.61	5.61	0.17	0.17	0.17	0.02	0.53	0.26
8	7	5.82	5.63	5.63	0.98	0.97	0.96	-0.06	-0.70	-0.44
23	5	6.10	5.91	5.90	0.25	0.25	0.25	0.10	1.04	0.04
39	7	6.13	5.91	5.90	0.16	0.15	0.14	0.07	2.36	1.70
33	7	6.26	6.04	6.03	0.17	0.16	0.16	0.13	2.35	1.33
21	6	6.79	6.53	6.52	0.11	0.11	0.11	0.18	2.28	0.80
7	7	6.85	6.68	6.69	0.47	0.46	0.47	0.21	1.96	0.94

section, respectively. In absolute value, these two numbers are close to 1.16, representing the 95% confidence limit for anomalous values if all event variances are equal. Shaw (1964) did not use these events for

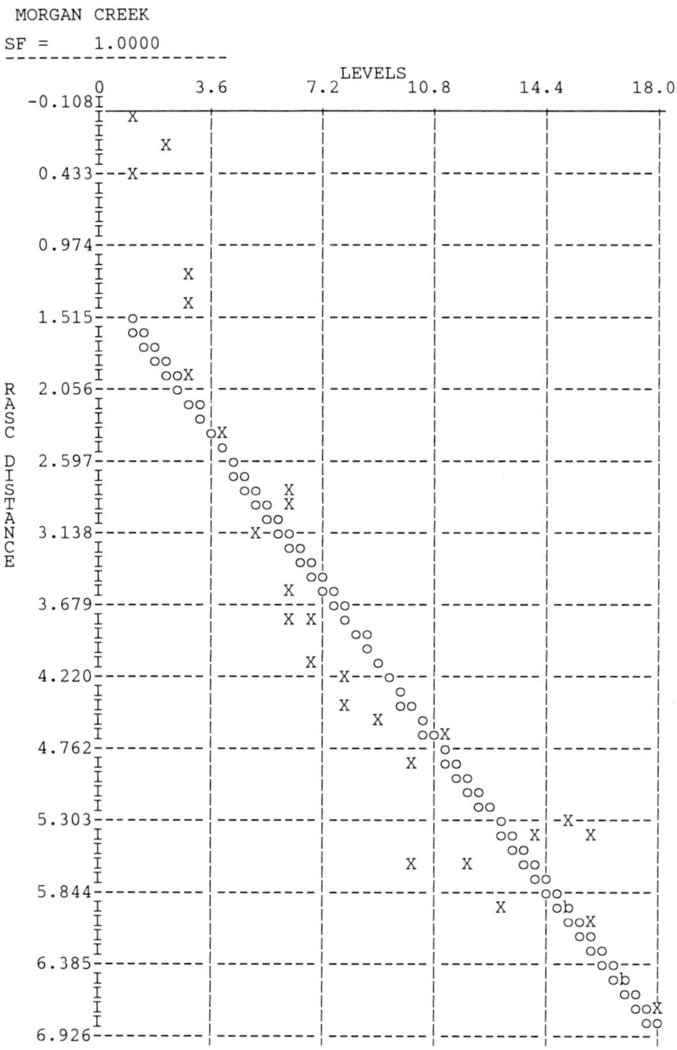

Fig. 8.12 Comparison of observed highest and lowest occurrences (shown as x-es) with best-fitting spline-curve (=straight line) after iteration. The line shows a relatively poor fit at the first two event levels. The RASC distances plotted in the vertical direction are close to those of the scaled optimum sequence used in Fig. 9.23 (see later). The spline-curve in Fig. 9.23 was obtained by cross-validation and provides a better fit than the straight-line fit of Fig. 8.12.

constructing his composite range chart. However, neither event was flagged as possibly anomalous in the normality test (Table 8.13).

The largest deviation (=1.60) in Table 8.23 is for event 41 in the Pontotoc section. The corresponding second-order difference has two asterisks in Table 8.13 suggesting a possible anomaly. However, because

TABLE 8.23

Deviations of observed relative positions of events after 7 iterations. Values were corrected for average deviation from spline-curve (Ave in Table 8.22). Numbers 1 to 7 for columns correspond to the 7 sections. Largest deviations in bold print were used to construct modified RASC range chart of Fig. 8.13.

Event	*n*	1	2	3	4	5	6	7
100	7	-0.41	0.34	-0.05	-0.49	0.26	**0.89**	-0.53
82	7	-0.66	-0.38	-0.57	0.04	0.34	**0.97**	0.26
108	5	-0.12	0.23	-0.36	x	**0.55**	-0.31	x
107	5	-0.35	0.66	**-0.52**	x	0.09	0.12	x
99	7	-0.32	-0.11	**-1.07**	0.26	0.13	0.89	0.22
90	5	0.31	-0.29	0.13	**0.43**	x	x	-0.58
92	6	-0.03	0.10	x	-0.03	-0.15	-0.00	**0.12**
89	5	**-0.27**	0.11	-0.04	0.38	x	x	-0.18
91	6	**-0.31**	0.06	x	-0.11	0.22	0.37	-0.23
81	7	0.19	-0.16	0.09	**-0.38**	0.40	-0.10	-0.04
68	7	**0.42**	-0.41	0.09	0.17	0.06	-0.14	-0.20
70	6	0.19	x	0.19	**0.26**	0.15	-0.68	-0.11
42	7	**0.56**	-0.27	-0.72	0.31	0.20	-0.00	-0.07
69	6	0.55	x	-0.85	0.62	0.51	**-1.07**	0.24
24	5	0.13	0.34	**0.45**	-0.37	x	x	-0.56
67	7	0.42	0.23	**-0.66**	-0.09	-0.19	-0.14	0.43
40	7	0.26	**0.78**	0.39	0.51	-1.81	-0.72	0.59
56	5	-0.19	0.55	**1.06**	-0.66	-0.75	x	x
54	5	0.39	**0.42**	x	0.06	-0.48	x	-0.28
48	5	x	-0.74	**1.37**	-1.25	-0.44	x	1.07
47	5	x	-0.74	1.37	**-1.25**	-0.44	x	1.07
55	5	-0.23	-0.18	0.54	**-0.24**	0.12	x	x
53	5	**-0.59**	-0.24	x	0.61	0.07	x	0.16
41	7	**-0.93**	-0.27	-0.75	-0.33	-0.86	1.60	1.54
34	7	-0.08	-0.02	-0.11	-0.08	0.28	-0.37	**0.37**
22	6	**0.55**	0.39	-0.09	-0.51	x	-0.02	-0.32
8	7	**1.27**	0.90	0.01	-0.41	0.40	-0.54	-1.63
23	5	0.46	**-0.59**	0.52	-0.29	x	x	-0.10
39	7	-0.11	-0.16	-0.24	0.18	**-0.34**	0.74	-0.07
33	7	**-0.35**	0.31	-0.17	-0.20	-0.27	-0.03	0.71
21	6	-0.21	-0.05	-0.13	0.25	x	0.41	**-0.27**
7	7	-0.37	-0.31	0.01	0.39	1.20	-0.08	**-0.84**

this regards a lowest occurrence which would be situated too high, the extended range chart is not affected by it. The modified RASC results had

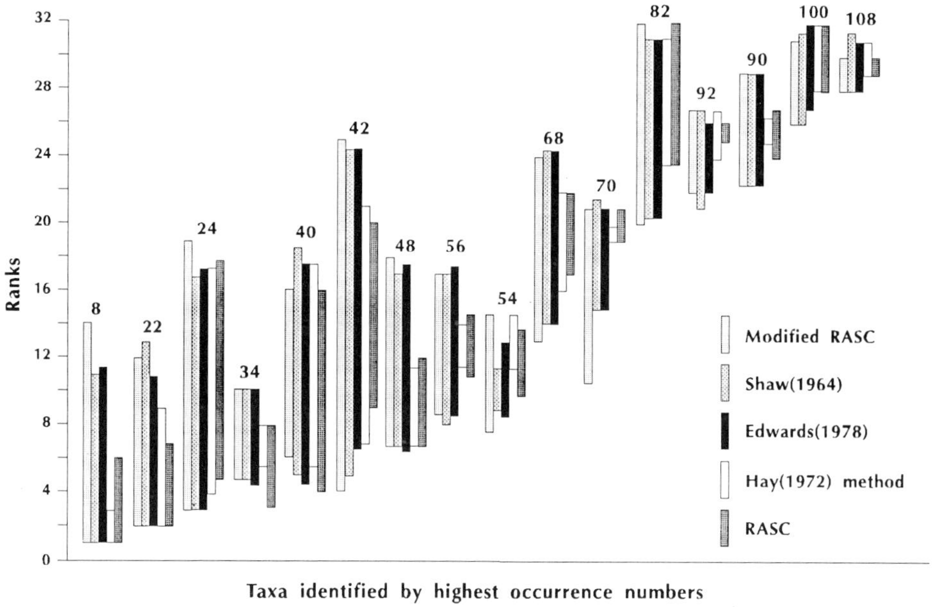

Fig. 8.13 Comparison of range charts obtained by five different methods for Palmer's database. Modified RASC and RASC results were added to ranges previously plotted by Edwards (1982). Lowest and highest occurrences were ranked for each method and these ranks were used to display the ranges. The modified RASC, Shaw (1964) and Edwards (1978) results are similar. The Hay (1972) and RASC ranges were based on average highest and lowest occurrences. These generally are shorter than the other (conservative) ranges.

converged almost completely after 7 iterations as can be seen in Table 8.22 by comparing $\bar{x}_6$ to $\bar{x}_7$. The three sets of event variances ($s^2{}_5$, $s^2{}_6$ and $s^2{}_7$) for deviations from the spline-curves after five, six and seven iterations are reasonably close. It is noted that the process of convergence shows oscillations for some events (i.e. those for which the value of $s^2{}_6$ is not between those of $s^2{}_5$ and $s^2{}_7$ in Table 8.22).

The average deviation is clearly negative for the first six events in Table 8.22. The same phenomenon was previously encountered for the average deviation of the Gradstein-Thomas database (see Fig. 8.6) where it was accompanied by a positive average deviation at the bottom of the stratigraphic sequence. The reason for this calibration problem can be understood by comparing spline-curve and data in Figure 8.12 for the Morgan Creek section. A decrease in the smoothing factor, which is equivalent to increasing all event standard deviations by the same factor, would result in a curve instead of a straight line for the Morgan Creek section. This curve would be closer to the first five events in this section (cf. Fig. 9.23 and later discussion in Section 9.10). It may be assumed that the calibration problem of lack of fit near the tops and bottoms of some sections is related to a slight overestimation of the smoothing factors which, in turn, is equivalent to a slight overestimation of the event variances in these sections.

The sample sizes (*n*) of the events in the Palmer's database are too small to decide which events have variances that are significantly smaller or larger than average. Neither can it be decided from the skewness statistics in the last column of Table 8.22 which events have an asymmetrical frequency distribution. It is interesting that the five largest (positive) skewness values (events 39, 33, 55, 41 and 7) are for lowest occurrences whereas the two smallest (negative) skewness values (events 70 and 40) are for highest occurrences. All remaining events have skewness values which are less than 0.90 in absolute value.

The preceding observation would support the hypothesis that Palmer's trilobites satisfy the model advocated by Edwards (see Fig. 2.13) and Baumgartner (see Fig. 2.14). For the latter model, a lowest occurrence has its longest tail pointing in the stratigraphically upward direction (positive skewness) whereas a highest occurrence has its longest tail in the stratigraphically downward direction (negative skewness).

CHAPTER 9

EVENT-DEPTH CURVES AND MULTI-WELL COMPARISON

9.1 Introduction

This chapter describes the theory and application in geological basin analysis of CASC for Correlation and SCaling in time. The CASC method of quantitative correlation is based on the RASC method and on the philosophical reviews and statistical methodology of several geologists and mathematicians including: Shaw (1964), Hay (1972), Drooger (1974), Blank (1979), Reinsch (1967), De Boor (1978) and Eubank (1988). The method provides a precise, automated and semi-objective means of correlation of rock sections for which an optimum sequence or a scaled optimum sequence of biostratigraphic events has been determined using the zonation method and computer program RASC. The next two sections on principles of correlation and scaling in time and generalized description of CASC method consist of material only slightly modified from Gradstein et al. (1985), Agterberg et al. (1985), and Agterberg and Gradstein (1985). This introductory material is followed by explanations of the cross-validation and jackknife methods. Next this chapter contains a number of regional applications of CASC. These include a comparison of CASC with other methods of biostratigraphic correlation using Palmer's database for example. Comparisons between automated stratigraphic and more subjective (manual) correlation are given for the Mesozoic and Cenozoic microfossil record of the NW Atlantic margin. Other topics to be discussed are integration of biostratigraphic and lithostratigraphic information for the Central and Viking Grabens, North Sea, and integrated CASC correlation of foraminiferal and dinoflagellate datasets for the Labrador Shelf and Grand Banks.

The use of RASC and CASC provides the stratigrapher with an integrated biostratigraphic method, particularly suitable for exploiting the considerable amount of micropaleontological data that accumulates during sedimentary basin analysis. The method starts with a data file of the original observations on the distribution in time and in space in wells or outcrop sections of all taxa identified. Next, this data file is reduced to biozonations that best explain the regional and temporal trends. Finally, geologically reasonable correlations of the sections will be calculated. Segmentation and correlation of the original sections can be achieved by means of fossil events and RASC biozones. Interpolation of the scaled optimum sequence in linear time makes it feasible to correlate by means of isochrons. Each correlation line carries an uncertainty limit, which is a combined estimate of various original uncertainties in the data.

9.2 Principles of correlation and scaling in time and comparison with composite standard method

As previously discussed in Chapters 1 and 2, geological correlation traditionally is expressed in terms of: (1) rock type units such as formations or well log intervals (lithostratigraphic correlation), (2) fossil units such as biozones (biostratigraphic correlation), (3) relative age units or stages (chronostratigraphic correlation) and (4) linear time units or ages (geochronologic correlation). Instead of using units with a certain thickness or a duration in time, correlation frequently is based on events. Events or datum planes refer to fossilized, physical or organic occurrences of supposedly irreducible resolution along the geological time scale.

An important contention of geological correlation is that once the events or various types of rock, or relative age units have been properly determined and defined, these units can indeed be used for correlation. As pointed out by F.M. Gradstein in the Foreword, existing stratigraphic codes show how to define stratigraphic units but they do not define how to correlate them. The actual correlation generally takes place in the subjective domain of experts. Procedures for correlation or stratigraphic equivalence depend on subjective evaluation of the unique relationship of each individual record to the derived and accepted standard. It follows that correlation as practised in geology cannot be readily verified without a detailed review of all the underlying facts.

Traditionally there is no method of formulating the uncertainty in fixation of individual records to the standard. As Riedel (1979) stated: "Biostratigraphy will be continued to be regarded as an art rather than a science, until it is possible to attach confidence limits to suggested correlations". An improvement in definition of the zonation through increased numbers of observations and taxa may increase the number of correlation tiepoints, but still leaves the question of uncertainty unanswered. Such an uncertainty generally is couched in qualitative terms only. In many geological investigations such a subjective procedure yields satisfactory results, correlation being only a part of the scientific objectives. Situations do arise, however, where the quality of correlations determines the outcome of the study. This is particularly true in the field of operational biostratigraphy, where large and complex data sets may have to be reduced before they can be of assistance in deciphering basin history.

The problem of using subjective judgement only is not so much that it leads to right or wrong stratigraphy, but that a single solution is proposed. It should be attempted to establish reasonable criteria for successful correlation by providing insight into the actual uncertainty in correlation, either in millions of years or in depth in meters. In regional correlations there frequently is limited or no understanding of how much (in depth or in relative time units) the solution differs from alternate solutions, using the same data. In all likelihood it is difficult to propose or compare two alternative correlations, without major review or analysis of all underlying facts.

Biostratigraphic correlation depends on the probability that: (a) in each rock section the events defining a biostratigraphic increment have been detected and properly taxonomically determined; and (b) the true (or natural) sequence of events is known. This principle was succinctly stated by Hay (1972), who then went on to propose the principle of matrix permutation for construction of the most likely sequence of (nannofossil) events in time (see Chapter 5). In the ranked sequence each event position is an average of all the relative positions but no direct insight is available into the uncertainty of rank.

As early as 1964, Shaw not only proposed a simple ranking method for biostratigraphic events, but also a method for correlating the sections in which the events occur. The original method is as follows. From a number of individual geological sections (A, B, C, D, etc.), one (for example A) is selected that shows a relatively complete and reasonable "normal" order and spacing of events. This particular sequence of entries and exits of taxa is plotted along one axis and that of a comparable sequence B along the other axis of a conventional two-axis graph. Scale units are in feet or meters, as found in each section but, in a simplified procedure, order only can be used. The best fit of the resulting scattergram is called the line of correlation.

Shaw (1964) advocated regression analysis as a linear trend-fitting technique although A and B probably are both subject to uncertainty. (By subjectively deleting the larger deviations, Shaw avoided systematic discrepancies due to neglecting uncertainty in A as previously discussed in Section 8.4). The order and spacing of first and last occurrence events along the A-axis is now updated through projection of the homologous B-axis events, through the best-fit line, onto the A-axis. If the first occurrence of an event in B occurs relatively lower than in A, the range of this event in A is extended downward. If a last occurrence of an event in B occurs relatively higher than in A, the range of the event is extended upward. It is attempted to maximize the stratigraphic ranges. Next the updated A-axis (composite section) is compared to section C, in the same manner as A was compared to B and the process is continued by including an increasingly larger number of sections. In the final composite section, the scale of the successive events has become a composite of all spacing values between successive events. Because the final result depends on the order in which individual sections were added to the composite section, there may be a second or even a third round during which A, B, C, ... again are plotted against the composite section.

Actual correlation of events is achieved by making new bivariate plots for each individual section as a function of the final composite standard. For each bivariate scattergram a new best fit line or best fit channel is calculated which serves to project the composite events onto the individual section scale. In a mathematical sense, each value in the composite standard can be expressed as a function of its correlative (depth) value in the individual sections. Miller (1977) provides a good description of use of the composite standard method.

The CASC method of quantitative correlation combines average sequence methodology with bivariate correlation technique. Input for CASC is the RASC input file that shows the original sequence of events in each of the sections. In addition, the program requires a depth file, that shows the observed depth in feet or in meters for all the events in the original sequence file. The correlation and scaling in time (CASC) program first computes the RASC optimum sequence and RASC scaled optimum sequence of events. Using the three normality testing techniques in the RASC method (bivariate graphs, stepmodel and normality testing), outliers in the individual sections may be eliminated. Based on the filtered data file, a new optimum sequence may be calculated, after which each individual sequence of events is compared to the scaled

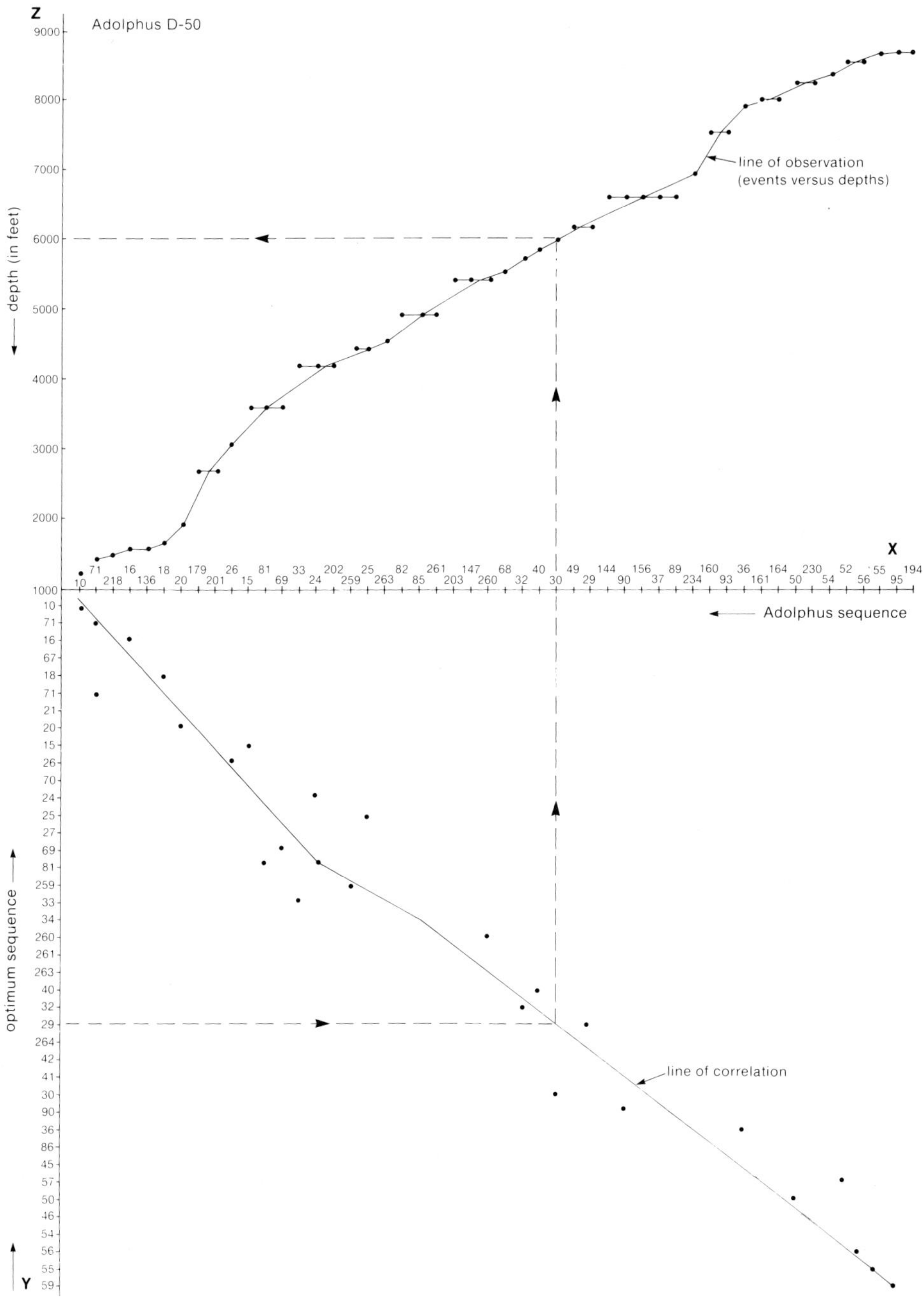

Adolphus D-50
Z
depth (in feet)
line of observation (events versus depths)
X
Adolphus sequence
optimum sequence
Y
line of correlation

optimum sequence, and best fit curves (smoothing splines, see later) are calculated. A spline fit yields a function such that, for each optimum sequence position, the most likely stratigraphic equivalent position can be found in the individual sequences. These normalized tiepoints then are correlated.

Figure 9.1 graphically depicts the principal steps, executed for the correlation of event 29 (top of *Cyclammina amplectens*) in the Adolphus D-50 well on the Grand Banks, which is part of the Gradstein-Thomas database. The y-axis is the optimum sequence in 21 of the wells ($k_c = 7$, $m_{c1} = 2$; probabilistic ranking followed by modified Hay method). Instead of the optimum sequence, the scaled optimum sequence can be used (see later). The x-axis is the observed sequence of events, whereas the z-axis is the common depth scale of the well. The lower scattergram expresses mismatch of the individual sequence and the optimum sequence. The best fit line for the graph (here visually estimated) is the line of correlation.

Working with event scales initially has the advantage that complications due to different rates of sedimentation in different places which may be hundreds of km apart are avoided. Moreover, equal spacing of values for the independent (x-axis) variable in spline-curve fitting has the considerable advantage that the possibility of unrealistic oscillations of the fitted curve between irregularly spaced control points is avoided. However, the number of levels in the event scale differs from section to section in a "random" manner. For correlation between wells it is necessary to replace the levels of the event scales by depths (in km). This replacement is shown in the upper part of the scattergram of Figure 9.1.

The individual sequence x is a function of the depth z at which the events were observed. This function is shown as the "line of observation". The most likely position of event 29 in the Adolphus well is found by projecting its optimum position via the line of correlation to the individual sequence and from there via the line of observation to the depth scale. Thus all optimum sequence events are scaled in (well) depth. In a multi-well comparison, the most likely depth value (z-axis) in each well is calculated for selected y-axis values (event positions) in the optimum sequence. In the example, event 29 in Adolphus should occur at 6050ft (observed 6200ft). In another well (Flying Foam, not shown here) event 29 was projected to occur at 4850ft (observed 5330ft). These depths then are the most likely correlation tiepoints and a line can be drawn to connect them.

The standard deviation (SD) of the events relative to the line of correlation (and the line of observation) in the y-direction (and parallel z-direction, which is the depth-axis in the well) provides an estimate of the mismatch of each event. Later on this will be called the local error. When it is geologically unreasonable to expect a continuous sedimentation rate in the vicinity of a certain depth, the local SD can be modified to account for changes in the sedimentation rate.

The same procedures as shown in Figure 9.1, using the RASC optimum sequence, can be applied when the scaled optimum sequence is chosen instead. The interfossil distances in the scaled optimum sequence reflect the average relative distance of the events in relative time. If it is possible to estimate the numerical geological ages of some of the events in the scaled optimum sequence, the relative distance estimates can be used to stretch the scaled sequence in linear time. This way the scaled optimum sequence becomes a (local) biochronology and hence isochrons can be traced through the wells or outcrop sections. For paleontologists this is a valuable method for finding the numerical age of the most likely position of principal zone boundaries in each well. Such boundaries, as argued in Gradstein et al. (1985, Chapter II.4), can delineate sedimentary cycles. The original standard deviations for the interfossil distances in the scaled optimum sequence now reflect the uncertainty in linear time between the events. This uncertainty can be expressed by means of the global error bar (see later).

As a first test of the validity and use of this numerical time interpolation for geological analysis, Figures 9.2 and 9.3 were constructed by Gradstein et al. (1985). Along the vertical axis of Figure 9.2 the interfossil distances are plotted for the Cenozoic foraminiferal events in a scaled optimum sequence for 21 wells. For some events, listed later in Section 9.8 on regional applications , which are the (regionally averaged) last occurrences of key planktonic and a few benthic Foraminifera, numerical ages can be estimated. This involves comparison of the regional to standard Atlantic zonations, details of which follow later. The horizontal scale is linear time in Ma, for the Cenozoic period. In principle, the numerical ages are for highest occurrences and not for average highest occurrences (see Fig. 8.9). On average, there is about 10 m.y. difference between these two types of stratigraphic events in the Atlantic zonations considered. A systematic discrepancy of this type, however, is eliminated automatically provided that is approximately the same for the different taxa used.

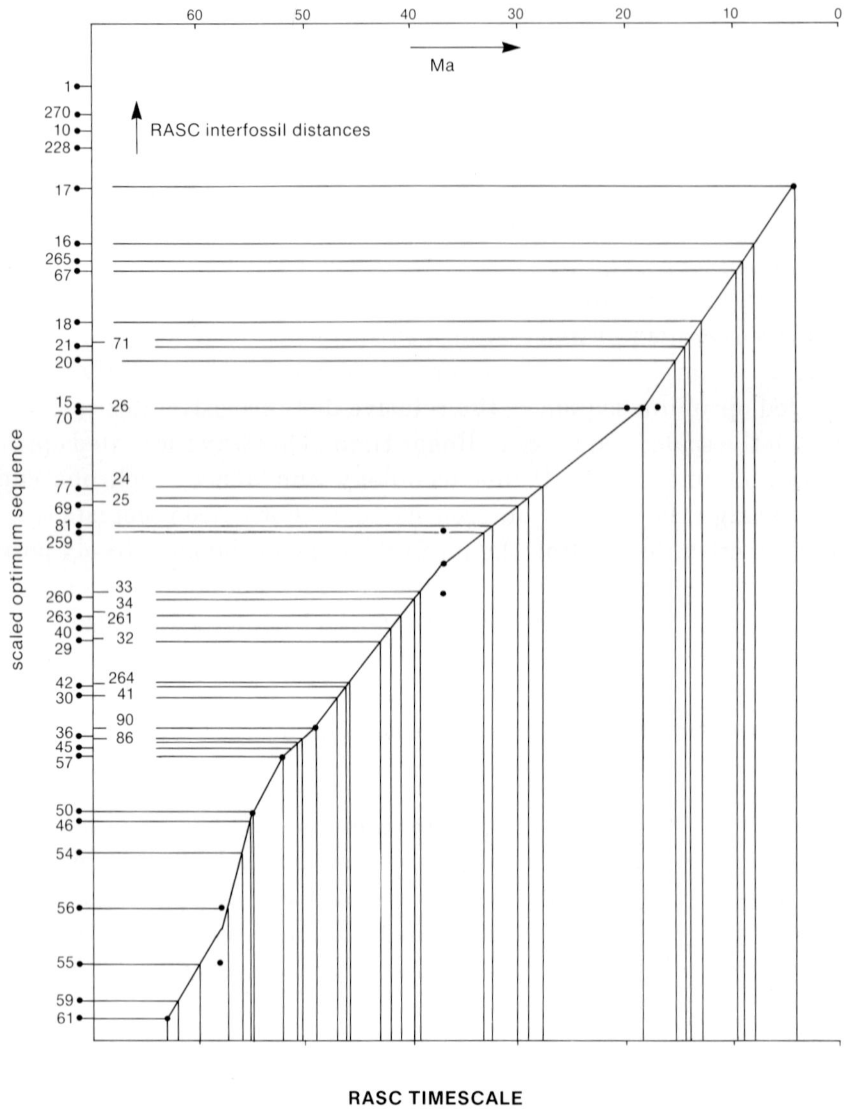

Fig. 9.2 Plot of the Cenozoic scaled optimum sequence (21 wells; 7/2/4 run) versus linear time in Ma. The inter-event distances are plotted cumulatively. For some selected events in the scaled optimum sequence the numerical age is known (dots), and this allows to scale the whole fossil sequence in linear time.

The calibrated events are used to form a nomogram for correlation, so that all events in the scaled optimum sequence can be dated. In Figure 9.3

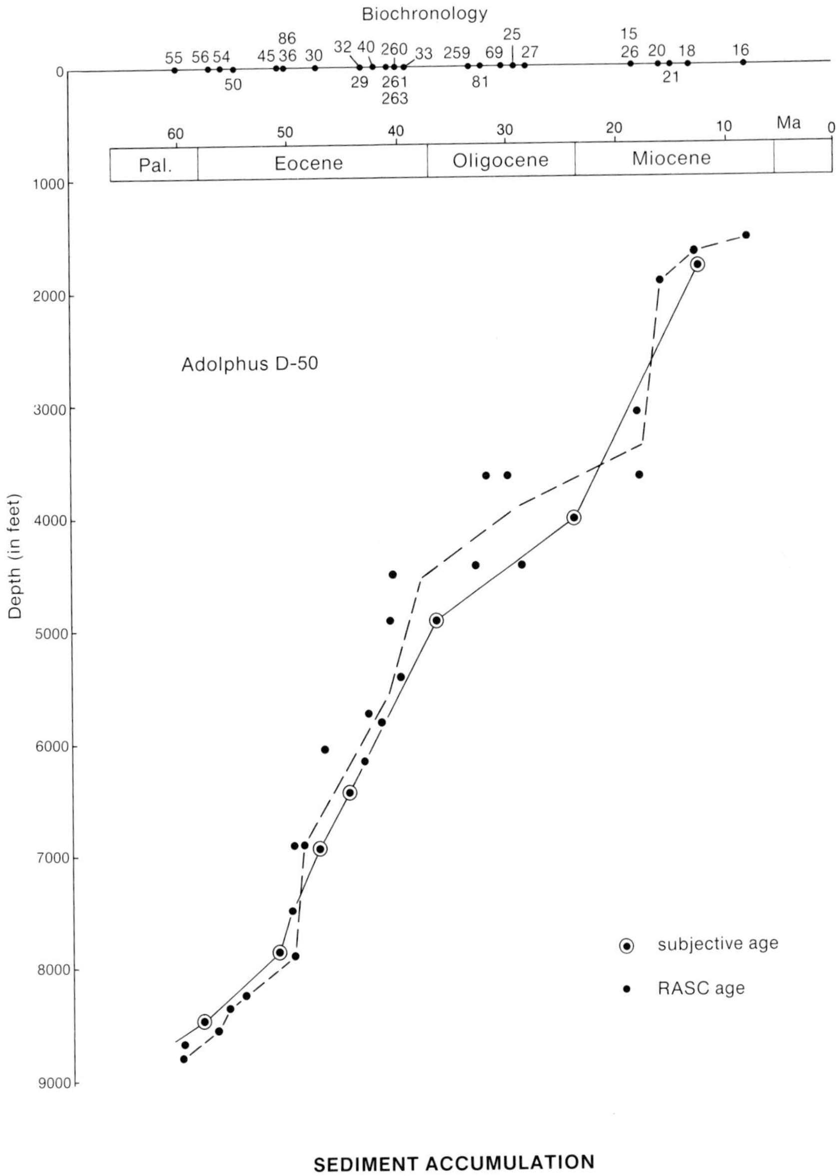

Fig. 9.3 The RASC biochronology of Fig. 9.2 is used to estimate rate of sedimentation (dashed line) in the Adolphus well. The solid line (subjective) shows approximately the same trend, using independent well history data (from Gradstein and Agterberg, 1985).

this new RASC biochronology (horizontal axis) is used to estimate the rate of sediment accumulation (dashed line) in the Adolphus D-50 well. Several years earlier, prior to development of RASC and CASC an approximate chronostratigraphy of this well section had been given, in system units from the Paleocene upward. As shown in Figure 9.3, there is a close approximation of the two, independently arrived at, sediment accumulations. The earlier interpretation obscured a possible late Oligocene-early Miocene hiatus. Scaling in time of the scaled optimum sequence is a practical way of erecting a regional time scale.

In summary, the CASC method of correlation is based on three conditions: (1) each individual stratigraphic sequence of events is a sample of the optimum sequence; (2) the observed depths of the events in a stratigraphic section are estimates of the true depths; and (3) the calculated relative interfossil distances of events in the scaled optimum sequence can be used to stretch this sequence along the numerical geological scale with known ages for index fossils in this sequence providing the necessary tiepoints. Input for automated correlation of fossil events (or zones) by means of CASC with confidence limits in depth or time units are: (a) depth in feet or in meters for all fossil events in all wells or outcrop sections. These events are the same as those used in the RASC method; (b) ages of index fossils to stretch the scaled optimum sequence in linear time; (c) events, clusters of events (zones) or ages to be correlated; and (d) wells or outcrop sections to be correlated.

9.3 Generalized description of the CASC method

Originally, the CASC program (Agterberg et al., 1985) was developed on a CDC Cyber 730 mainframe computer with a Tektronix 4014 terminal with code in FORTRAN Extended Version 4. Two computer libraries were required to use CASC: IMSL Library and Tektronix Advanced Graphing Library. Also, mass storage facilities were used. It was assumed that in order to obtain the geologically most satisfactory bivariate fits, mainframe CASC was best used interactively.

One of two different routes could be selected at the beginning of an interactive CASC session. The first route uses as a starting point the RASC optimum sequence, plotted against the so-called event scale. The latter has entries for the original sequence data in each stratigraphic section. Instead of the optimum sequence, the RASC scaled optimum sequence may be used. The latter combines average order and relative distance. The optimum sequence option is simpler than the so-called distance option based on the scaled optimum sequence, but not principally different. As an additional illustration the distance method will be applied to RASC results for the distribution of Cenozoic foraminifers in offshore wells on the Grand Banks and Labrador Shelf.

If RASC distances were used as input in mainframe CASC, it was required to replace them by ages (in Ma). This replacement is not required in the CASC modules of micro-RASC (see Chapter 10). The procedure in mainframe CASC is schematically shown in Figure 9.4. Assuming approximate ages are known for a subgroup of events in the scaled optimum sequence shown in Figure 9.4, the objective is to fit a curve to these data in order to be able to replace any RASC distance by its age (see Fig. 9.4d). First, RASC distances with the same age are averaged (see Fig. 9.4a). Then a cubic spline curve is fitted to the age-distance pairs

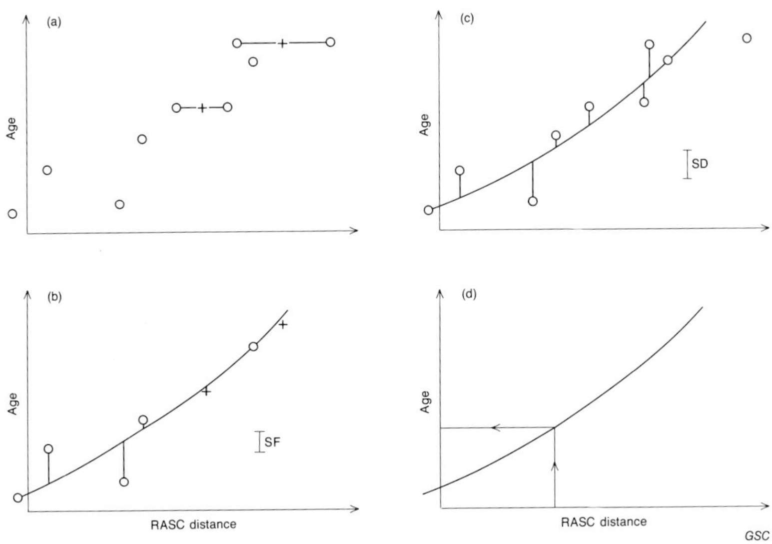

Fig. 9.4 Schematic illustration of method followed in CASC mainframe computer program to establish relation between RASC distance and age. (a) Two (or more) RASC distances for the same age are averaged. (b) Cubic spline-curve is fitted using age as the dependent variable; smoothing factor (SF) representing standard deviation of differences between event ages and curve is chosen in advance, before curve-fitting. (c) Standard deviation (SD) for differences between original values and curve is computed after curve fitting. (d) Fitted curve is used to convert any RASC distance into corresponding age.

minimizing the sum of squares of deviations between points and curve in the vertical (age-) direction of Figure 9.4b. The smoothing factor SF can be chosen beforehand by the user of the interactive CASC computer program. It is equal to the square root of the mean square deviation between points and curve. Because this standard error generally is not known beforehand, the user can determine it by trial and error while experimenting with different plots on the screen of the monitor. In Figure 9.4b a curve was fitted to 5 original values (o) and 2 averages of two values (+). The standard deviation of the original data in relation to this curve is also shown on the screen (SD in Fig. 9.4c). The fitted curve does not extrapolate outside the range of the RASC distances used for the curve fitting. Consequently, the circle with the highest RASC value is not considered for estimating SD in this example. It is noted that a curve also could be fitted directly through the 8 circles in Figures 9.4a and 9.4c. Then SD would be equal to SF.

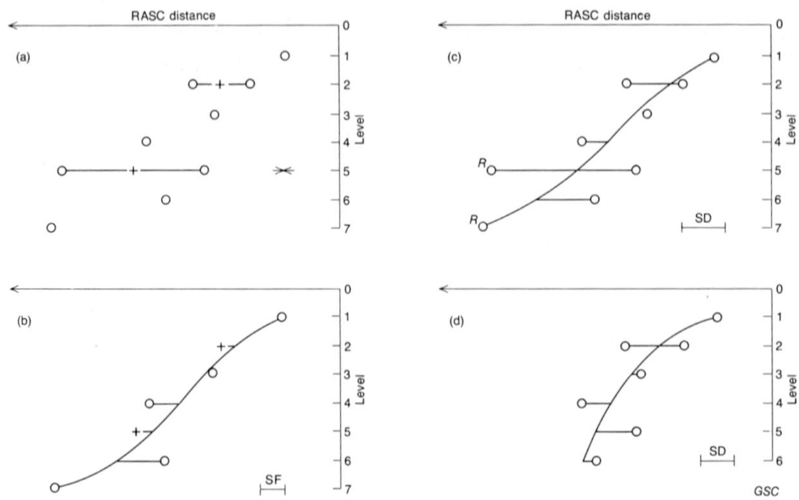

Fig. 9.5 Schematic illustration of preliminary computing and optional editing procedure at beginning of CASC mainframe computer program. (a) Events found to be anomalous with a probability of over 99 per cent (asterisk) may be omitted from spline-curve fitting and later plots; RASC distances of two (or more) coeval events are averaged. (b) Cubic spline-curve is fitted using RASC distance as the dependent variable; smoothing factor (SF) representing standard deviation of differences between RASC distances assigned to levels and curve is chosen in advance. (c) Standard deviation (SD) is computed from differences between original values and curve after curve-fitting; original values (e.g. those labelled R) can be deleted. (d) New curve with new standard deviation (SD) is obtained without use of deleted values.

Next, the CASC user can display and edit the RASC distances for any well from the set of the wells used. Editing options are schematically shown in Figure 9.5, which displays preliminary data analysis. The scale in the vertical direction is relative. It shows successive levels for the stratigraphic events in the well considered. RASC distances of 2 or more coeval events are averaged (see Fig. 9.5a) before cubic spline fitting (Fig. 9.5b). The user has the option of omitting events for which the second-order differences were anomalously high (i.e. shown by two asterisks in the RASC normality test). Such anomalous events are then displayed by use of a special symbol (single asterisk in Fig. 9.5a) and are not employed for curve fitting. The deviations are measured in the horizontal direction. SF and SD serve the same purpose as in Figure 9.4. The user may wish to remove other events considered to be anomalous, for example, those labelled R in Figure 9.5c. Then a new cubic spline-curve will be fitted for the reduced data set (Fig. 9.5d). If extreme values are deleted, SD will probably be decreased in value. The original RASC model is based on the assumption that positions of events in a well are distributed around their expected value, according to a normal probability distribution, with standard deviation set equal to $1/\sqrt{2} = 0.7071$. One, therefore, would expect SD to be approximately equal to 0.7 if the number of events in the stratigraphic section is sufficiently large.

For further analysis in preparation of automated correlation, RASC distance is replaced by age (see Fig. 9.6a) using the earlier derived relationship between RASC distance and age (see Fig. 9.4d). In the following discussion, the variables for event level, age and depth are denoted as x, y and z, respectively. A spline-curve can be fitted to express y as a function of x, as was done for distance in Figure 9.5d. It also is possible to replace the levels by their depths and fit a spline curve to express y as a function of z using depth as the independent variable. This leads directly to a plot similar to Figure 9.6f.

However, the rate of sedimentation may have changed significantly during geologic time at a well site and this can result in irregular distribution of the points along the z-axis. This, in turn, may make it difficult to obtain a spline-curve that extrapolates in a satisfactory manner across data gaps along the z-axis corresponding to short periods with increased sedimentation rates (also see Section 3.6). For this reason, the indirect method given in Figure 9.6 can be employed instead. Assume that the spline-curve of Figure 9.6a is written as $y = f(\underline{x}) + e_y$ where e_y represents a random deviation in the y-direction. The bar under x

indicates that y is regressed on x using data points which are regularly spaced along the x-axis. Depth (z) is plotted against x in Figure 9.5b, and a separate spline-curve with $z = g(\underline{x}) + e_z$ is obtained, using the same set of regularly spaced data points along the x-axis. The deviation e_z points in

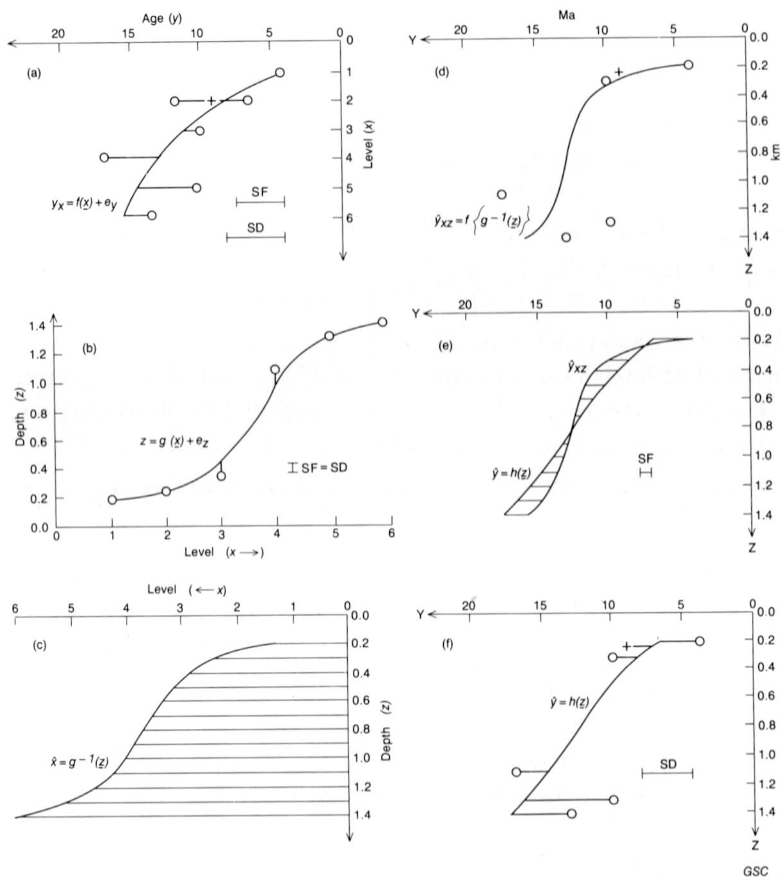

Fig. 9.6 Schematic illustration of calculation of an event-depth curve from RASC output for a well. (a) RASC distances have been replaced by ages using relation illustrated in Fig. 9.4d; new spline-curve $f(\underline{x})$ is fitted; bar in $\underline{x}$ denotes use of regular sampling interval for x; smoothing factor (SF), which was selected before curve-fitting using one age per level, is smaller than standard deviation (SD) for all original values. (b) Spline curve $g(\underline{x})$ is fitted to express depth as a function of level x; bar in $\underline{x}$ denotes use of regular sampling interval for x; SF = SD is equal to some small value. (c) $\hat{x}$ represents spline-curve $g(\underline{x})$ in Fig. 9.6b now coded as set of values for x at regular interval of z. (d) $\hat{y}_{xz}$ denotes curve passing through set of values of y at regular interval of z obtained by combining spline-curve of Fig. 9.6a with that of Fig. 9.6c. (e) $\hat{y}$ is spline-curve fitted to values y_{zx} of Fig. 9.6d using new smoothing factor SF. (f) Standard deviation SD is computed after curve-fitting, using one age per level.

the z-direction. Obviously, the curve $g(\underline{x})$ cannot decrease in the x-direction.

The curve for z in Figure 9.6b again is shown in Figure 9.6c. It has been rewritten in the form $\hat{x} = g^{-1}(\underline{z})$, to indicate that estimates $\hat{x}$ were obtained at points which are regularly spaced along the z-axis. Assume that $\hat{y}$ is obtained for the irregularly spaced values of x in Figure 9.6c using $f(x)$ shown in Figure 9.6a. This results in a set of values of $\hat{y}_{xz} = f(g^{-1}(\underline{z}))$ for regularly spaced points along the z-axis (see Fig. 9.6d). The function $f(g^{-1}(\underline{z}))$ is not a simple mathematical expression. For example, its first derivative is not readily available. A cubic spline $\hat{y} = h(\underline{z})$ can be fitted to the values $\hat{y}_{xz}$ (see Fig. 9.6e). In Figure 9.6e, $\hat{y}$ is considerably smoother than $\hat{y}_{xz}$. By using a smaller smoothing factor (SF), the difference between $\hat{y}$ and $\hat{y}_{xz}$ may be kept negligibly small (see curve to be used for example in Fig. 9.7a). The standard deviation SD for points used for fitting in Figure 9.6a with respect to the curve $\hat{y}$ is provided in Figure 9.6f. The deviations from $\hat{y}$ are measured in the y-direction. A similar age-depth diagram is shown in Figure 9.7a where less smoothing was applied. The spline-curve $\hat{y} = h(z)$ can be used to assign a probable age to any point along the well.

Figure 1.2 in Chapter 1 showed a so-called multi-well comparison for five wells. It is based on a 7/2/4 RASC run on 21 wells. Points with estimated ages of 10, 30, and 50 Ma along the five wells are connected by lines of correlation in Figure 1.2. The uncertainty in the position of these isochron contours is indicated by error bars, constructed according to one of three methods further explained in Figures 9.7 and 9.8. The displays of Figures 1.2a and 1.2c were redrafted from displays on the Tektronix terminal obtained during an interactive CASC session; the error bars in Figure 1.2b were obtained from event-depth plots according to the method explained in Figure 9.7d.

The local error bar in Figure 1.2a was obtained by multiplying $s(y)$ (=SD) along the y-axis by rate of sedimentation to obtain a modified error $s(z)$ along the z-axis, as shown in Figure 9.7b. The rate of sedimentation (Fig. 9.7c) is the first derivative dz/dy for z in $\hat{y} = h(z)$. In general, a cubic spline curve y, fitted to n data points, consists of $(n$-1) successive cubic polynomials

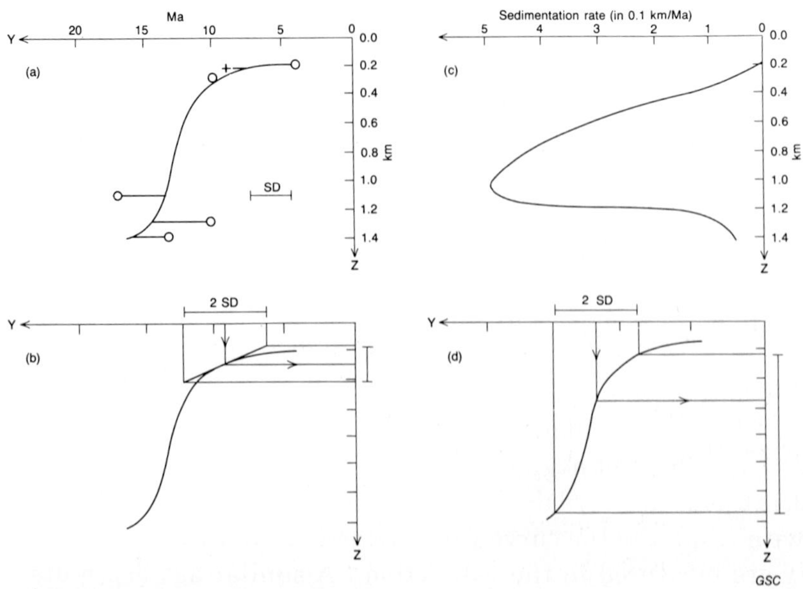

Fig. 9.7 Schematic illustration of estimation of local error bar and modified local error bar. (a) Standard deviation SD was computed after curve-fitting, using one age per level. (b) Error bar of age value plus or minus SD along Y-axis is transformed into error bar along Z-axis using first derivative (dz/dy) of age-depth curve. (c) Rate of sedimentation ($=dz/dy$) can be displayed on screen during CASC interactive session. (d) Modified local error bar is asymmetrical with respect to depth value for a given age.

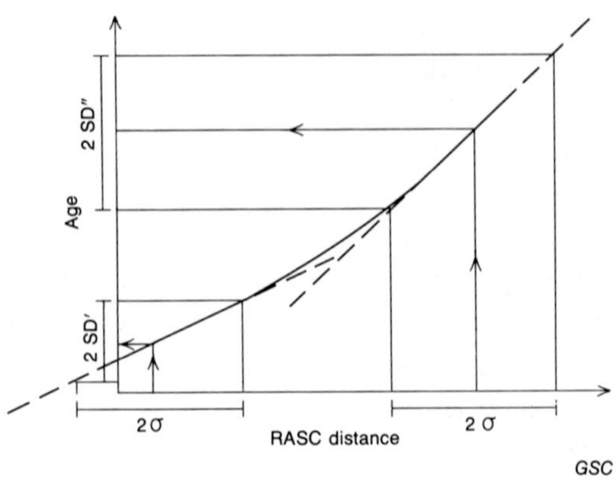

Fig. 9.8 Schematic illustration of estimation of global error bar. Theoretical standard deviation $\sigma (=0.7071)$ along RASC distance scale is assumed to remain constant. It is transformed into variable SD along age scale (e.g. SD' and SD").

$$y = y_i + c_{1i}d + c_{2i}d^2 + c_{3i}d^3 \tag{9.1}$$

with $d = z - z_i$, $z_i \leqq z < z_{i+1}$, where z_i and z_{i+1} ($i = 1, 2, ..., n$-1) represent the n depths used to convert $\hat{y}_x = f(x)$ into $\hat{y} = h(z)$. The coefficients c_{1i}, c_{2i} and c_{3i} can be used to calculate

$$dy/dz = c_{1i} + 2c_{2i}d + 3c_{3i}d^2 \tag{9.2}$$

at any point. Inversion of this expression gives dz/dy. The new standard error $s(z) = (dz/dy)\ s(y)$ can be displayed for any z as the local error bar $z \pm s(z)$ (see Fig. 1.2a). This propagation of error is based on the local rate of sedimentation, which is assumed to remain approximately constant over the interval $y \pm s(y)$. The latter condition frequently is not satisfied, especially when $\hat{y}$ has many inflection points (between local maxima and minima in sedimentation rate). Curvature of $\hat{y}$ is considered in the construction of a modified local error bar as illustrated in Figure 1.2d. For any point $z = h^{-1}(y)$, this bar extends from the point $h^{-1}\{y - s(y)\}$ to $h^{-1}\{y + s(y)\}$. It is asymmetrical with respect to z and is significantly shorter at places where the rate of sedimentation is high.

Finally, a global error bar (Fig. 1.2c) can be constructed as illustrated in Figure 9.8. The standard deviation $\sigma = 1/\sqrt{2}$ of events along the RASC linear scale for distance is changed into a variable standard error $s(y)$ along the age scale. This new, variable standard error is changed into $s(z)$ according to the method used for SD in Figure 9.7b. In global error bar estimation, it is assumed that a single RASC distance error σ can be applied to all wells. On the other hand, in local error bar estimation, use is made of a constant SD value along the age scale which was estimated from the deviations between the points used for spline-fitting and the spline curve itself (cf. Fig. 9.6e). Because of possible elimination of anomalous events and averaging of ages for events at the same levels, the local error bar is likely to be narrower than the global error bar. It is possible that the quality of the biostratigraphic information is not the same in all sections considered. Such differences would be considered in local error bars but not in global error bars.

The purpose of the error bar is to quantify the uncertainty of the observed depths of events with respect to their estimated depths in the wells. Each local or global error bar extends from the estimated depth minus one standard deviation to estimated depth plus one standard deviation. If the observed depth is normally distributed about the estimated depth, the error bar can be interpreted as a 68 percent confidence interval for single events. Then there is a 68 percent probability that the observed event falls within the range outlined by the error bar. Likewise, there is 95 percent probability that the observed event falls within an extended error bar, which is 1.96 times as wide as the error bar shown.

It will be shown in Section 9.7 that the actual precision of the estimated depth of an event in a well can be greater than that indicated by the error bar for single events. If a value on the spline-curve would be interpreted as the arithmetic average of all (n) values used for its estimation in a well, the standard deviation of this mean would be equal to the standard deviation of the single events, divided by $\sqrt{n}$. For example, in a well with 16 event levels, the standard deviation of the difference between estimated depth and "true" depth would be one-fourth the standard deviation of the 16 single events used to construct the error bar. The degree of validity of the assumption that a value on the spline-curve can be interpreted as an arithmetic mean of all values used for estimation is not precisely known, except when the smoothing factor is large. In the limit, which is reached when SF exceeds an upper threshold value, the spline-curve reduces to the best fitting straight line of least squares. Then the preceding assumption holds approximately true.

Output from a 7/2/4 RASC run on 21 wells and a 5/2/3 run on 24 wells were used as input for examples of actual CASC runs in the remainder of this section. Table 9.1 shows the optimum sequence, modified optimum sequence (after final reordering) and RASC distances for the 7/2/4 run on 21 wells (also see Fig. 6.2). Several events, occurring in fewer than seven wells, were later inserted as unique events. Table 9.2 shows estimated ages of 22 events, including these unique events. Average RASC distances for events with the same age are shown in the last column of Table 9.2. Figure 9.9a shows the ages plotted against the RASC distances. The displays in Figure 9.9 (and Figs. 9.10-12) were redrafted from hardcopy of displays on a Tektronix terminal. A cubic spline function with smoothing factor SF = 2.0 was fitted to the 15 ages, using the average distances shown in the last column of Table 9.2. The smoothing factor SF is the standard deviation of differences between the 15 ages and corresponding estimated ages on the spline-curve for the same RASC

TABLE 9.1

RASC output for 7/2/4 run on 21 wells (Grands Banks - Labrador Shelf) used as CASC input. Event levels (sequence position numbers 1-40) (A), optimum sequence of events identified by their dictionary numbers (B), modified optimum sequence after final reordering (C), and cumulative RASC distances for events in last column.

A	B	C	RASC Distance	A	B	C	RASC Distance
1	10	10	0.000	21	261	261	4.364
2	17	17	0.391	22	263	263	4.518
3	16	16	0.912	23	40	40	4.645
4	67	67	1.204	24	29	32	4.802
5	21	18	1.647	25	32	29	4.809
6	18	71	1.799	26	41	264	5.175
7	71	21	1.919	27	264	42	5.189
8	20	20	2.108	28	42	41	5.237
9	26	15	2.442	29	30	30	5.251
10	15	26	2.486	30	86	90	5.531
11	70	70	2.513	31	36	36	5.620
12	27	29	3.198	32	90	86	5.667
13	69	25	3.358	33	45	45	5.786
14	24	27	3.418	34	57	57	5.799
15	25	69	3.499	35	50	50	6.302
16	81	81	3.722	36	46	46	6.429
17	259	259	3.738	37	54	54	6.738
18	33	33	4.295	38	56	56	7.178
19	34	34	4.311	39	55	55	7.689
20	260	260	4.342	40	59	59	8.033

TABLE 9.2

Estimated ages for 22 events and calculation of average RASC distances for two or three events with same estimated age.

Event No.	Age (Ma)	RASC Distance	Average Distance	Event No.	Age (Ma)	RASC Distance	Average Distance
4	3.5	-0.585	-0.385	85	38	4.456	4.456
5	3.5	-0.476		29	40	4.809	4.809
269	3.5	-0.096		90	49	5.531	5.531
17	11	0.391	0.391	57	52	5.799	5.247
179	15	1.984	1.984	93	52	5.895	
15	17	2.442	2.442	50	55	6.302	6.302
26	20	2.486	2.427	194	57	7.073	7.073
137	20	2.368		55	58	7.689	7.434
24	28	3.198	3.198	56	58	7.178	
33	37	4.295	4.017	61	63	8.228	8.039
259	37	3.738		253	63	7.849	

distance values. The standard deviation of the 22 original ages before averaging of some RASC distances is also shown in Figure 9.9a.

The original sequence of events occurring in 7 or more wells is shown in Table 9.3 for Indian Harbour M-52 (Well No. 5), which will be used for

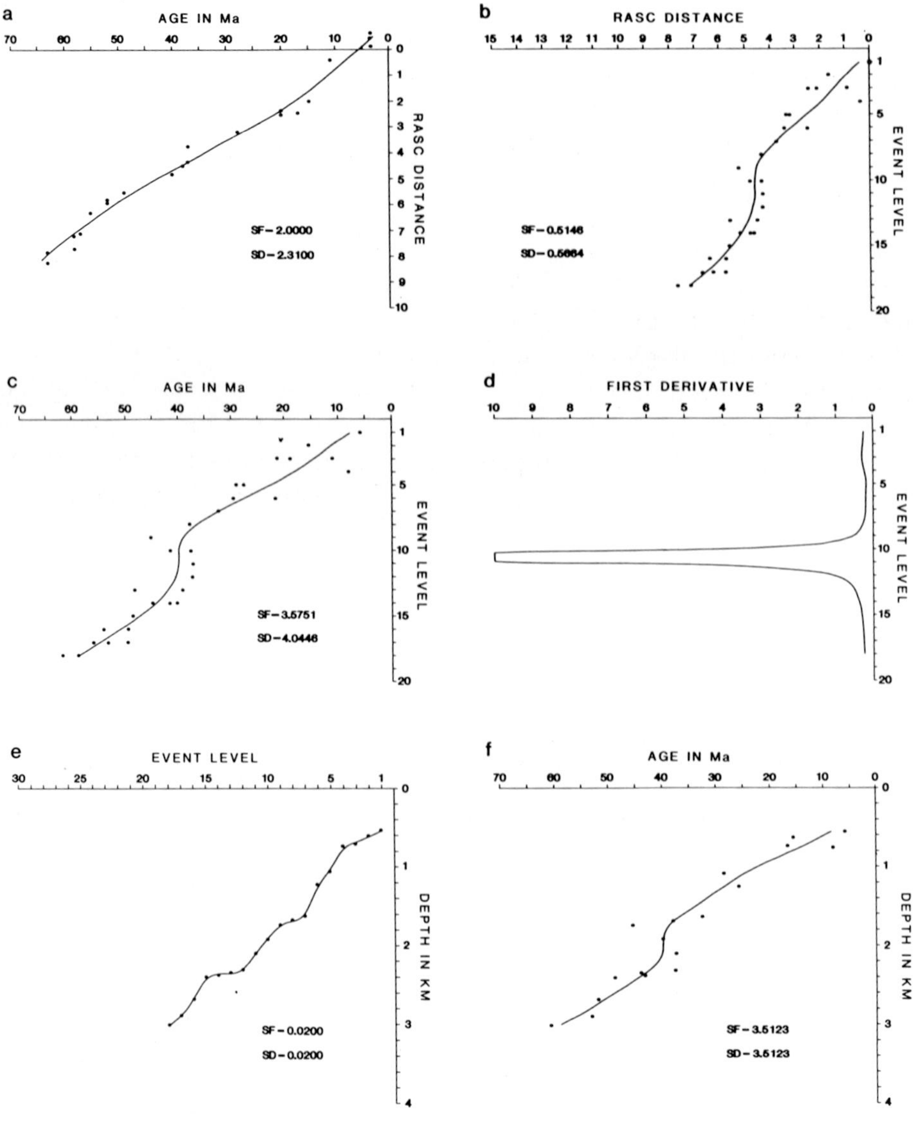

Fig. 9.9 Example of CASC displays for Indian Harbour well based on 7/2/4 RASC results for 21 wells. (a) Age-RASC distance relationship as derived from the 21 wells file. (b) Initial CASC plot for default smoothing factor. (c) Age-level plot for default SF. (d) First derivative of (c). (e) Level-depth plot. (f) Age-depth plot for default SF; spline-curve was fitted directly to the data, using irregularly spaced depths.

further analysis. As mentioned before, one of two different routes can be selected at the beginning of mainframe CASC. These consist of using either optimum sequence data or RASC distances for the events. In both subprograms, event levels for successive, non-coeval events are defined, as illustrated for Indian Harbour M-52, in the second column of Table 9.3. In the second subprogram, the RASC distances in a well are transformed into ages using the spline-curve fitted in Figure 9.9a (see last column in Table 9.3). The methods used in the two subprograms are identical, except that sequence position numbers instead of ages are used in the first subprogram. Only the option that uses the ages (in Ma) will be illustrated in detail here.

Mainframe CASC produces a number of successive plots. For each of these plots the user is required to answer one or more questions. The plot that comes after Figure 9.9a during a CASC session is shown in Figure 9.9b. It shows the RASC distances of Table 9.3 plotted against their event levels. Before this plot is actually shown on the Tektronix screen, the user is asked if he wishes to exercise the option of deleting anomalous events which are out of place with a probability of 99 percent according to the RASC normality test. Moreover, points can be deleted from Figure 9.10

TABLE 9.3

CASC input for Indian Harbour well; definition of 18 event levels; and transformation of RASC distances into ages using spline-curve in Fig. 9.9a.

Event No.	Event Level	RASC Distance	Age (Ma)	Event No.	Event Level	RASC Distance	Age (Ma)
10	1	0.000	5.6	34	12	4.311	37.0
18	2	1.647	15.2	263	13	4.518	38.9
15	3	2.442	21.0	-36	13	5.620	47.9
-20	3	2.108	18.6	29	14	4.809	41.4
-16	3	0.912	10.9	-40	14	4.645	39.9
17	4	0.391	7.9	-41	14	5.237	44.2
24	5	3.198	27.3	-42	14	5.189	44.4
-25	5	3.358	28.7	86	15	5.667	48.2
26	6	2.486	21.4	45	16	5.786	49.1
-27	6	3.418	29.2	-46	16	6.429	53.7
259	7	3.738	32.0	57	17	5.779	49.2
261	8	4.364	37.5	-54	17	6.738	55.6
30	9	5.251	44.9	-50	17	6.302	52.9
260	10	4.342	37.3	55	18	7.689	61.5
-32	10	4.302	41.2	-56	18	7.178	58.4
33	11	4.295	36.9	(59)			

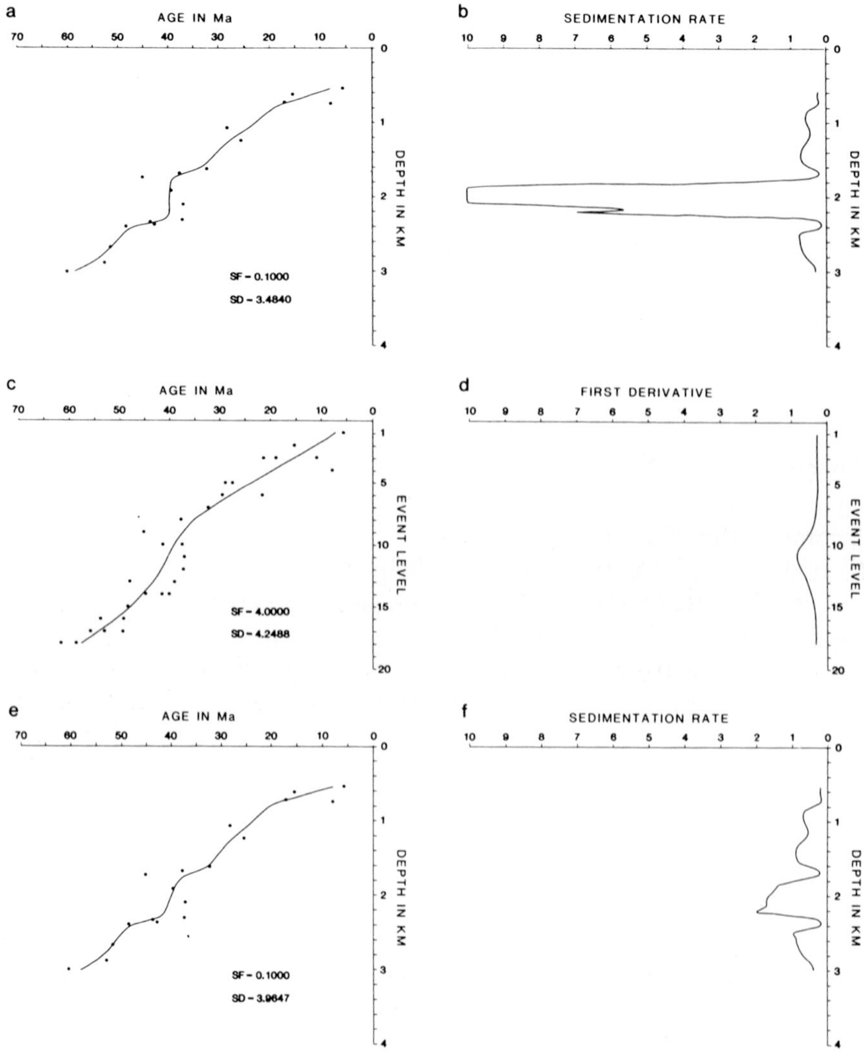

Fig. 9.10 Example of CASC displays for Indian Harbour well (continued from Fig. 9.9). (a) Spline-curve for small (default) SF fitted to combination of Figs. 9.9c and 9.9e; indirect method explained in Fig. 9.6 was used. (b) Sedimentation rate in 0.1 km/my (=first derivative of spline-curve in Fig. 9.10a multiplied by 10); local maximum and minimum are due to lack of smoothness of spline-curve as explained in text. (c) Age-level plot for SF=4.0 instead of default, used in Fig. 9.9c. (d) First derivative for Fig. 9.10c; magnitude of peak in Fig. 9.9d has been reduced. (e) Spline-curve for small (default) SF fitted to combination of Figs. 9.9e and 9.10c. (f) Sedimentation rate in 0.1 km/my corresponding to Fig. 9.10e.

itself by positioning the Tektronix cursor on top of them. No points were omitted in this example. Next a cubic spline-curve is fitted to the average RASC distance values in the third column of Table 9.4. First the user is shown the default smoothing factor (SF = 0.5146 for Fig. 9.9b) and asked if this value should be used. This default was obtained automatically, by fitting spline-curves with SF increasing from 0.0 until the first curve is

TABLE 9.4

Data used for fitting spline-curves in Indian Harbour well example shown in Figs. 9.9 to 9.11.

Event Level	Depth (m)	Average Distance	Average Age	Event Level	Depth (m)	Average Distance	Average Age
1	546	0.000	5.6	10	1912	4.572	39.3
2	619	1.647	15.2	11	2045	4.295	36.9
3	720	1.821	16.8	12	2305	4.311	37.0
4	747	0.391	7.9	13	2335	5.069	43.4
5	1067	3.278	28.0	14	2366	4.970	42.6
6	1232	2.952	25.3	15	2396	5.667	48.2
7	1616	3.738	32.0	16	2671	6.107	51.4
8	1674	4.364	37.5	17	2884	6.280	52.6
9	1732	5.251	44.9	18	3000	7.434	59.9

found for which the distance does not anywhere decrease with increasing depth. The default solution is shown in Figure 9.9b. The smoothing factor is the standard deviation of the differences between the 18 average RASC distances and the fitted spline curve. The standard deviation of residuals (=0.5664) representing differences between original RASC distances and fitted spline-curve is also given in Figure 9.9b. It is noted that this value is only slightly less than $\sigma = 0.7$ representing the theoretical standard deviation along the RASC scale.

Figure 9.9c shows a new default result, obtained after replacing RASC distance by age. It is possible to inspect the first derivative dx/dy of this graph (Fig. 9.9d). If the slope in the direction of increasing age for the curve in Figure 9.9c exceeds 10, its values are not displayed in Figure 9.9d. Because the default yields the first monotonically increasing spline-curve, normally at least one interval with very high sedimentation rate is introduced with this option. By increasing SF, the user can remove artificially high sedimentation rates.

Figure 9.9e shows the relationship between depth and event level with fitted spline-curve for SF = 0.02. It passes almost exactly through the observed values. After display of this plot, the CASC user has the option of either using this spline-curve in conjunction with the age-event level plot of Figure 9.9b, or of by-passing the indirect procedure by directly fitting a curve to the event-depth diagram in which event levels have been replaced by their depths. The default result for the direct method is shown in Figure 9.9f.

The result obtained by following the indirect method is shown in Figure 9.10a for small SF (=0.1). The first derivative corresponding to Figure 9.10a is given in Figure 9.10b. The irregularity between 2.2 and

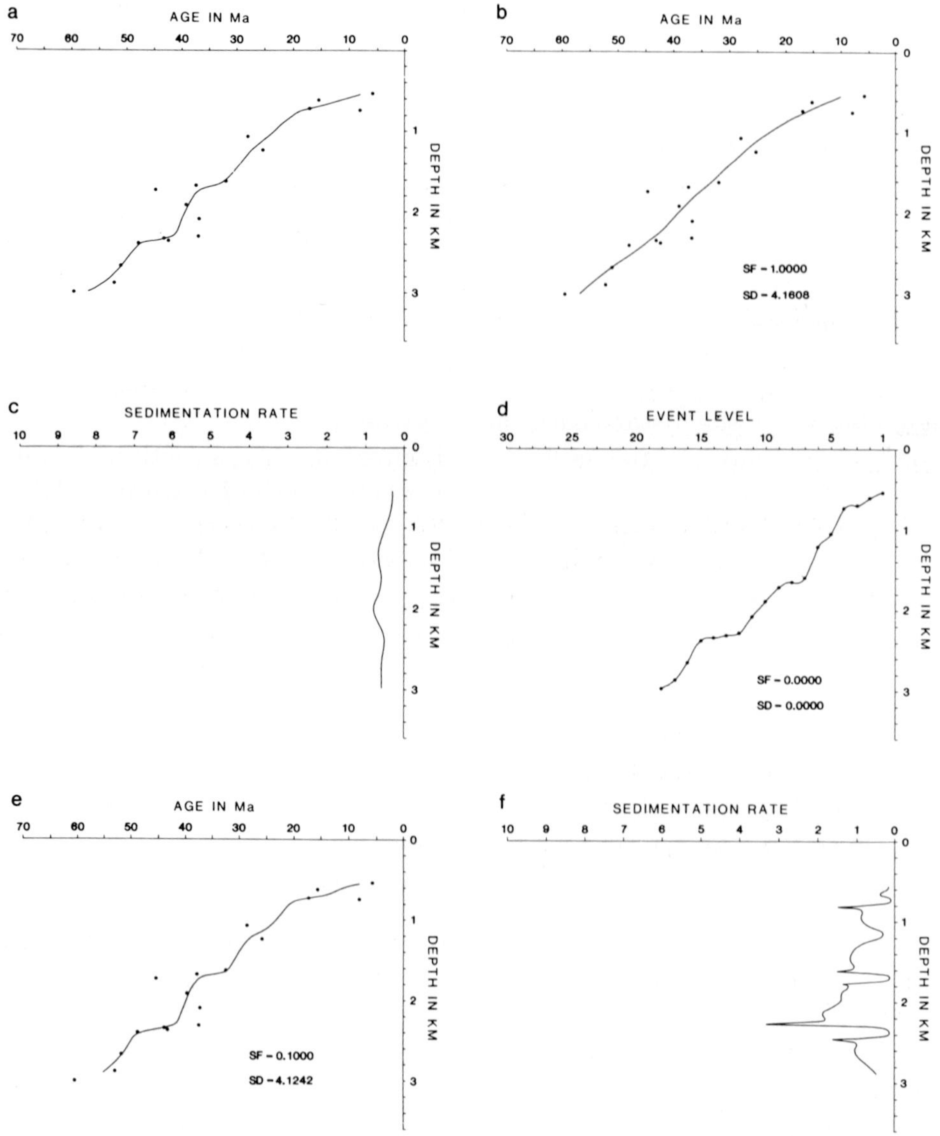

Fig. 9.11 Example of CASC displays for Indian Harbour well (continued from Figs. 9.9 and 9.10. (a) Unsmoothed combination of Figs. 9.9e and 9.10c; note similarity with spline-curve in Fig. 9.9e for SF=0.1. (b) Curve of Fig. 9.11a smoothed with SF=0.1. (c) Sedimentation rate in 0.1 km/my corresponding to Fig. 9.11b. (d) Level-depth plot for SF=0.0. (e) Spline-curve for small (default) SF fitted to combination of Figs. 9.10c and 9.11d; note similarity with spline-curve in Fig. 9.10e. (f) Sedimentation rate in 0.1 km/my corresponding to Fig. 9.11e; local maxima and minima are due to lack of smoothness of spline-curve as explained in text.

2.3 km in this diagram is due to lack of precision of the approximations, used in the indirect method, to obtain new values on the spline curve (see Figs. 9.6c and 9.6d). The regular spacing along the depth scale, used for this purpose in mainframe CASC, is 50 m. Consequently, irregularities due to lack of precision will not extend for more than 100 m along the depth scale.

Figures 9.10c to 9.10f show new results for the indirect method, obtained after changing the value of SF from the default (SF = 3.58 in Fig. 9.9c) to SF = 4.0 in Figure 9.10c. During a CASC session, the user is shown the unsmoothed values of $f\{g^{-1}(\underline{z})\}$ (cf. Fig. 9.6d) connected by straight lines. An example of the latter type of display is Figure 9.11a which originally appeared during the CASC session just before Figure 9.10e where SF = 0.1. It is not possible to see differences between the curves of Figures 9.10e and 9.11a. However, when SF is enlarged to 1.0, the smoother curve in Figure 9.11b is generated from Figure 9.11a. The rate of sedimentation for Figure 9.11b is shown in Figure 9.11c.

Figure 9.11d represents the depth versus event level curve that replaces Figure 9.9e, when SF = 0.00 instead of SF = 0.02 is selected. The difference between these two curves is small and when the curve of Figure 9.11d is combined with that of Figure 9.10c, the resulting plot (Fig. 9.11e) does not differ significantly from Figure 9.10e. However, the first derivative of Figure 9.11e which is shown in Figure 9.11f differs significantly from Figure 9.10f. It shows many 50 m irregularities, which are due to lack of precision as also discussed before (see Fig. 9.10b). In general, the final event-depth curve is less sensitive than the sedimentation rate curve to small changes in the choice of smoothing factors for successive curves during an interactive CASC session.

As a final example, Figure 9.12 contains various CASC displays for Adolphus D-50. Table 9.5 shows the input for this CASC run which consists of the partial DAT file for Adolphus D-50 (see Table 4.8) combined with output from a 5/2/3 RASC run on 24 wells. As shown in Figure 9.12a there are as many as 39 events on 28 levels in this well so that there is good control in the vertical direction. The first derivative of the spline-curve for SF = 2.2 (Fig. 9.12a) remains fairly constant. It has its largest value at event level 14 (see Fig. 9.12b). This indicates that place where the spline-curve in Figure 9.12a has its steepest dip. The pattern of Figure 9.12c suggests that rate of sedimentation was above average between events 6 and 7 and also between events 21 and 22. These two maxima also occur in Figure 9.12e which is the first derivative of the event-depth curve (Fig. 9.12d), obtained by combining the spline-curves of Figures 9.12a and 9.12c with one another using the indirect method. The smaller peak in Figure 9.12e, which occurs at a depth of about 1600m, represents the place (level 14) where the curve of Figure 9.12a has its steepest dip. The same

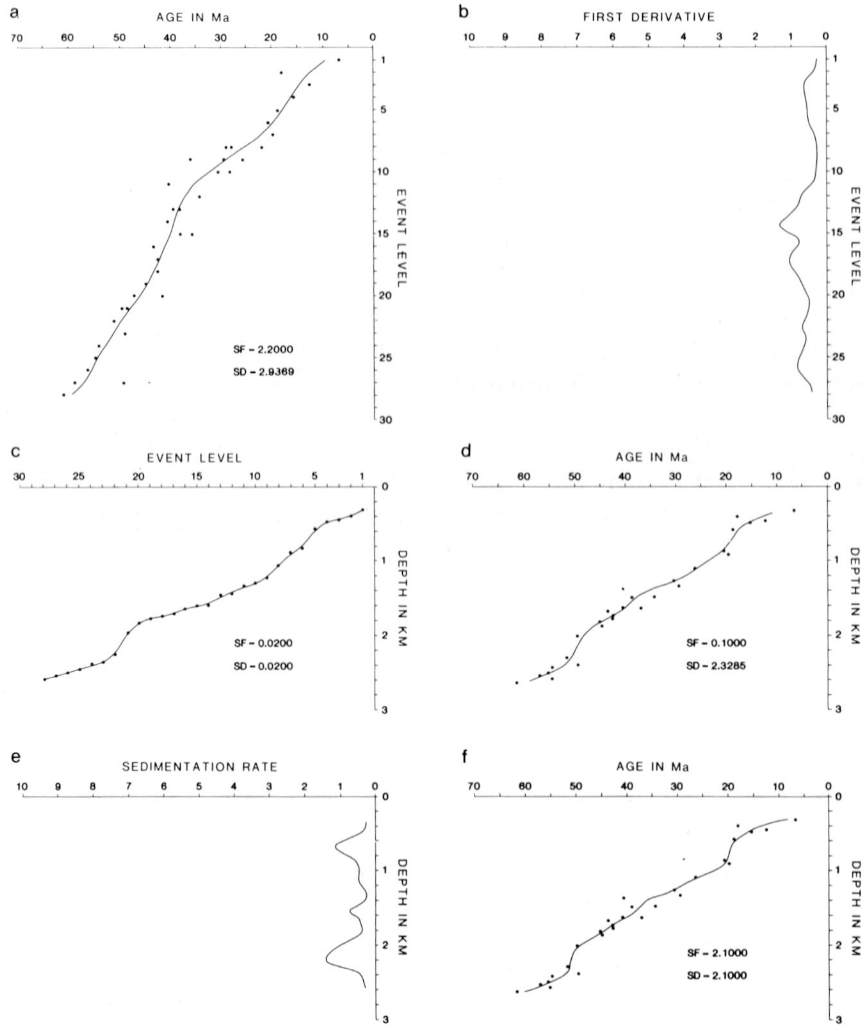

Fig. 9.12 Example of CASC displays for Adolphus well (5/2/3 RASC results using 24 wells). (a) Age-level plot for SF = 2.2. (b) First derivative corresponding to (a); note small peak near level 14. (c) Event level-depth plot; note relatively steep slopes at depths near 0.7 km and 2.2 km, respectively. (d) Spline-curve with small (default) SF fitted to combination of (a) and (c). (e) Sedimentation rate in 0.1 km/my corresponding to (d); two relatively high peaks correspond to steeper slopes in (c); intermediate small peak corresponds to highest first derivative in (b). (f) Event-depth spline-curve fitted directly to the data using irregularly spaced depths; note similarity with spline-curve of (d); direct method yields poorer results than indirect method when one or more intervals between successive ages are much larger than average, due to high sedimentation rate or relative lack of microfossils.

TABLE 9.5

Information for Adolphus D-50 well used for CASC experiments of Figs. 9.12 to 9.14; ID are identification numbers of foraminifers (cf. Tables 4.7 and 4.8); rank gives position of event in scaled optimum sequence; age was derived from RASC distance; level refers to successive samples taken at different depths.

ID	Rank	Age (m.y.)	Level	Depth (km)	ID	Rank	Age (m.y.)	Level	Depth (km)
10	3.0	6.634	1	0.318	147	23.0	35.561	15	1.622
71	8.0	17.695	2	0.400	−260	27.0	37.942	15	1.622
16	5.0	12.354	3	0.435	60	34.0	43.353	16	1.662
18	7.0	15.312	4	0.482	32	33.0	42.457	17	1.731
20	11.0	18.557	5	0.574	40	38.0	42.477	18	1.767
201	12.0	20.393	6	0.854	30	42.0	44.916	19	1.804
26	9.0	19.465	7	0.903	49	43.0	47.339	20	1.860
15	13.0	21.542	8	1.086	−29	31.0	41.523	20	1.860
−81	18.0	28.791	8	1.086	90	46.0	48.912	21	2.996
−69	16.0	27.683	8	1.086	−37	49.0	49.778	21	2.996
24	17.0	25.444	9	1.250	93	48.0	51.331	22	2.285
−33	24.0	35.964	9	1.250	36	47.0	49.189	23	2.383
−202	19.0	29.272	9	1.250	164	55.0	54.274	24	2.414
259	21.0	30.452	10	1.323	50	56.0	55.003	25	2.487
−25	20.0	28.015	10	1.323	−230	51.0	55.046	25	2.487
263	28.0	40.263	11	1.360	54	58.0	56.640	26	2.525
82	22.0	34.115	12	1.470	57	45.0	49.428	27	2.567
85	29.0	39.358	13	1.479	−56	59.0	59.009	27	2.567
−261	26.0	38.057	13	1.479	55	60.0	61.158	28	2.622
203	37.0	40.481	14	1.616					

three intervals with relatively steep slopes in the event-depth spline-curve can be observed in Figure 9.12f, which resulted from applying the direct method to the age-depth values. Without further statistical analysis or corroboration from other wells drilled in the immediate vicinity it is not possible to decide with certainty whether or not small fluctuations in the rate of sedimentation, as shown in Figure 9.12e, are significant. Increased smoothing in the event-depth diagram (Fig. 9.12d) would change the pattern of Figure 9.12e much more drastically than the pattern of Figure 9.12d itself.

As was illustrated in more detail for Indian Harbour M-52 in Figures 9.9 to 9.11, minor smoothing in Figure 9.12a for Adolphus D-50 would, in a multi-well comparison, only slightly change the position of isochrons in Adolphus D-50. However, the widths of the error bars are proportional to rate of sedimentation in both local and global error bar estimation and these widths would change drastically if smoothing is increased. This is because rate of sediment accumulation does depend strongly on choice of smoothing factors.

9.4 Statistical selection of optimum spline-curves

In the preceding sections, extensive use of smoothing splines has been discussed. In this respect, quantitative stratigraphy follows a general trend in computer-based statistics where smoothing splines have become widely employed after their invention (Reinsch, 1967, 1971; Schoenberg, 1964) in the late 1960's. The book by de Boor (1978) provides an introduction to spline-fitting with computer programs in FORTRAN. For comprehensive reviews of splines in statistics, see Eubank (1988) and Wegman and Wright (1983). The approach to spline-fitting taken in mainframe CASC is that the user decides in a subjective way on a best smoothing factor. It can be assumed that the latter lies somewhere between the "default" value which is based on the law of superposition of strata (age always increases in the stratigraphically downward direction) and the straight line which represents the smoothest possible spline. An additional guideline is provided by the value of the standard deviation (= 0.7071) originally selected for events along the linear scale in the RASC model. If all events in a well would occur at different levels, this value (0.7071) would represent a good choice for the smoothing factor in diagrams with RASC distance plotted along the horizontal axis. This guideline applies only if all events, approximately, have equal standard deviations as assumed in the scaling model.

The basic idea of the smoothing spline was explained in Section 3.11. It was pointed out that S representing the sum of standardized residuals in Equation (3.23) is distributed as chi-squared. This result is derived from statistical theory for the distribution of the variance s^2 (see e.g. Hald, 1957, p. 278) which has mean $E(s^2) = \sigma^2$ and variance $Var(s^2) = 2\sigma^4/f$ where $f = n$-1. Setting $S = ns^2/\sigma^2$, it follows that $E(S) = n$ and $Var(S) = 2f$. Thus the preceding interval extends from one standard deviation below the mean ($=n$) to one standard deviation above it. This idea has led users of smoothing splines to the choice $S = n$ ("Reinsch's suggestion", see e.g. Wahba, 1975). Because the smoothing factor is defined as $SF = (S/n)^{\frac{1}{2}}$, the use of Reinsch's suggestion is equivalent to setting $SF = 1.0$ if all values of $s(y_i)$ are known. This is in fact the method of spline-fitting previously used for constructing geological time scales (see Section 3.11) and in modified RASC (Chapter 8).

9.5 Cross-validation method

Wold (1974) conducted a number of computer simulation experiments for finding the optimum smoothing factor. In these experiments for which all data had known standard deviations, setting SF = 0.84 to SF = 0.97 provided better results than SF = 1.00. This illustrates that, even if good estimates of $s(y_i)$ are available, it is not known exactly which SF is optimum. For this reason, Wahba (1975) introduced the method of cross-validation for experimentally finding the best smoothing factor. This method has the additional advantage that it can be used even when estimates of $s(y_i)$ are not available as in most applications of CASC.

Suppose that y_i and s_i represent observed and fitted values, respectively, and that residuals are written as $R_i = y_i - s_i$ $(i = 1, 2, ..., n)$. Then:

$$\mathrm{SF} = \left[\frac{1}{n}\sum_{i=1}^{n} R_i^2\right]^{\frac{1}{2}} \tag{9.3}$$

In cross-validation, separate spline-curves are evaluated for m different smoothing factors. Let s_{ijk} $(i = 1, 2, ..., n; j = 2, 3, ..., n\text{-}1;$ and $k = 1, 2, ..., m)$ represent the ith value on a spline-curve for the kth smoothing factor fitted to a reduced data set of size $(n\text{-}1)$ obtained after deleting the value y_j. Then, a cross-validation value CV_k can be defined as:

$$\mathrm{CV}_k = \frac{1}{n-2}\sum_{i=2}^{n-1}\left(y_i - s_{iik}\right)^2 \tag{9.4}$$

It is noted that this sum is based on $(n\text{-}2)$ instead of n comparisons because the first and last values, s_{11k} and s_{nnk} $(k=1, 2, ..., m)$, are not available. The optimum smoothing factor has the lowest value of CV_k. In general, many spline-curves must be fitted before a satisfactory solution to the problem of optimizing SF is obtained. However, in our type of application the number of spline-curves to be fitted is not too large. For example, if $m = 30$ and $n = 22$, the total number of separately fitted spline-curves required for optimizing SF amounts to 600. Biostratigraphic datasets for single wells often have $n < 30$ and m can be kept small by establishing a range in which the optimum smoothing factor should fall before cross-validation is applied. This range extends from its minimum value corresponding to the law of superposition of strata (no decrease in age in the stratigraphically downward direction) to its maximum value corresponding to the SF value of the best-fitting straight line. The minimum and maximum values themselves are possible solutions for the optimum smoothing factor. For the preceding two reasons, cross-validation generally yields good results requiring relatively little computing time in biostratigraphic applications. It is not necessary to reduce the amount of computing time further by adopting one of the approximation methods known as "generalized cross-validation" (Craven and Wahba, 1979; Golub, Heath and Wahba, 1979; Utreras, 1981; Silverman, 1984).

The method of cross-validation is part of CASC 2 in micro-RASC (see Chapter 10). Table 9.5 showed CASC input for the Adolphus D-50 well. There are 39 ages occurring on 28 separate levels. In the CASC run previously applied to this dataset (see Fig. 9.12a), a spline-curve with SF = 2.2 was fitted to 28 values, after averaging ages on the same level. Table 9.6 shows results obtained by cross-validation applied to this example. The first step for cross-validation in CASC 2 consists of calculating the range for the optimum smoothing factor. The default spline had SF_{min} = 1.82 and the best-fitting straight line gave SF_{max} = 3.25, thus providing a range within which the optimum smoothing factor should fall. In CASC 2 this range is divided into 10 equal

TABLE 9.6

CASC 2 output for Adolphus D-50 (age-level plot, cf. Fig. 9.13a); smoothing factors SF_k range from 1.8158 for $k=1$ (first spline-curve satisfying law of superposition of strata) to 3.2519 for $k=11$ (best fitting straight line); optimum smoothing factor has lowest cross-validation value CV_k.

k	SF_k	CV_k
1	1.8158	13.2344
2	1.9594	12.1967
3	2.1030	11.3395
4	2.2466	10.4125
5	2.3902	9.4796
6	2.5338	9.0214
7	2.6775	9.2514
8	2.8211	9.7706
9	2.9647	10.3320
10	3.1083	10.9710
11	3.2519	11.3223

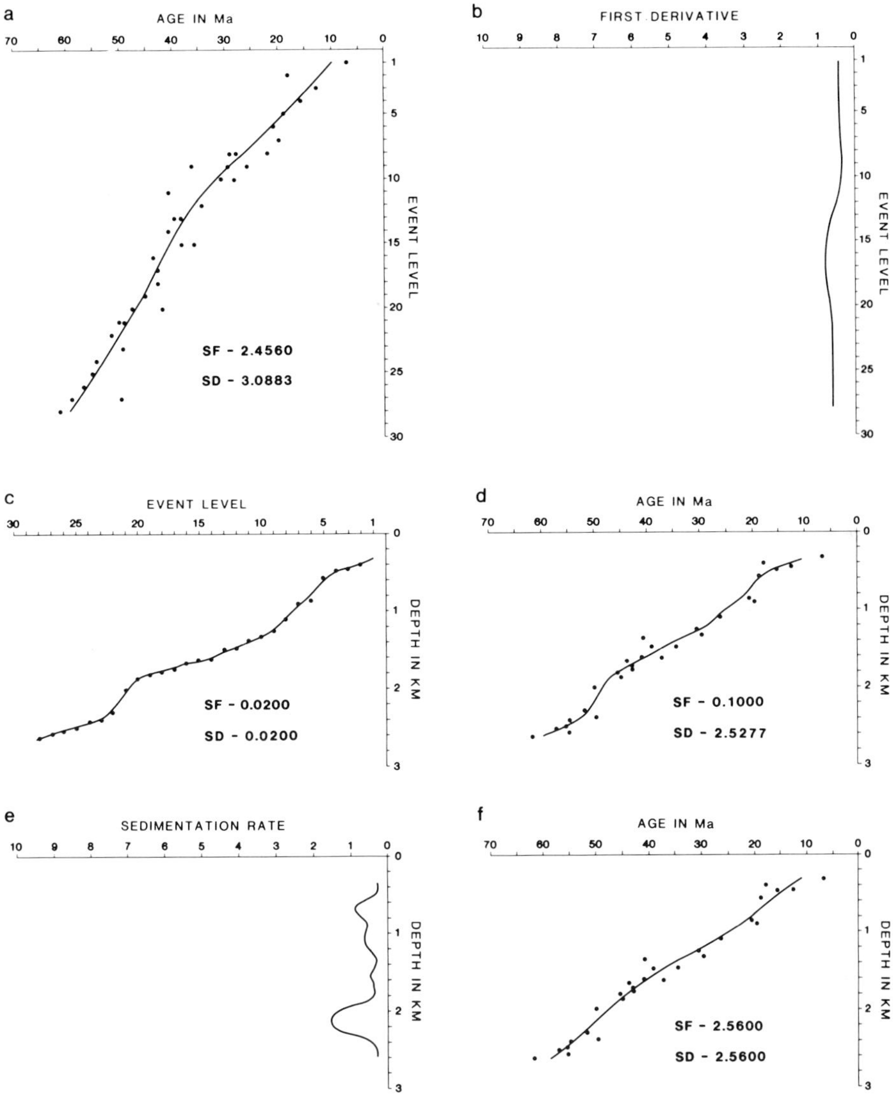

Fig. 9.13 Analysis on example of Fig. 9.12 for Adolphus D-50 repeated using optimum smoothing factors obtained by cross-validation for spline-curves in Figs. 9.13a and 9.13f. Largest differences in fitted curves occur in Figs. 9.13b (cf. 9.12b) and 9.13f (cf. 9.12f). For further explanation see text.

intervals and initially a cross-validation value is computed for each of the 11 equally spaced smoothing factors belonging to the range.

In Table 9.6, the minimum CV value occurs at SF = 2.53. A slightly improved estimate of SF = 2.546 was obtaining after zooming in on the vicinity of the initial minimum with a narrower range. Figure 9.13a shows the spline-curve for this optimum smoothing factor. It is only slightly smoother than the "subjective" spline-curve for SF = 2.2 shown in Figure 9.12e. However, the first derivative of the curve for SF = 2.546 (see Fig. 9.13b) is considerably smoother than its counterpart for SF = 2.2 in Figure 9.12b. The optimum age-level spline was combined with the same level-depth spline as before to give a new event-depth curve (Fig. 9.13d). This new spline closely resembles the previous result (see Fig. 9.12d). The new sedimentation rate (Fig. 9.13e) differs only slightly from the old one in that the central and smallest of the three maxima in Figure 9.12e has disappeared.

Finally, Figure 9.13f shows the spline-curve for SF = 2.56 representing the optimum smoothing factor obtained by cross-validation using irregularly spaced depth data. This new curve (Fig. 9.13f) based on the "direct" method is considerably smoother than its counterpart for SF = 2.1 in Figure 9.12f. It is also much smoother than the new event-depth curve (Fig. 9.13d) obtained by optimizing SF for the indirect method. Although the optimum smoothing factors for the indirect and direct method are nearly equal to one another, their corresponding splines turn out to be very different. It will be shown in the next section that the spline of Figure 9.13d which is based on the indirect method is better than the one shown in Figure 9.13e which is based on the direct method.

9.6 Jackknife method

Quenouille (1956) introduced the idea of splitting a sample of size n (for independent and identically distributed random variables) into g groups of size h ($n = gh$), analyzing the data in such a way that (1) bias would be reduced, and (2) a variance estimate would become available, for an estimator $\hat{\theta}$ of a parameter θ based on the sample of size n. Let $\hat{\theta}_i$ ($i = 1, 2, ..., g$) represent the same estimator based on the i th reduced data set of size (n-h). Then

$$\bar{\theta}_i = g\hat{\theta} - (g-1)\hat{\theta}_i \tag{9.5}$$

can be defined, leading to the jackknife estimator

$$\overline{\theta} = \sum_{i=1}^{g} \overline{\theta}_i / g \tag{9.6}$$

Tukey (1958) proposed that the g so-called pseudovalues $\overline{\theta}_i$ could be treated as approximately independent and identically distributed random variables in many situations. The statistic

$$\frac{g\,(\overline{\theta} - \theta)}{\sum (\overline{\theta}_i - \overline{\theta})^2 / (g-1)} \tag{9.7}$$

then should have on approximate t-distribution with (g-1) degrees of freedom. This would constitute a key statistic for robust confidence interval estimation (cf. Miller, 1974). Considerable research has gone into verifying and, by means of some counterexamples disverifying, the usefulness of the jackknife for robust variance estimation. For a review, see Efron (1982). The approach provides best results for ungrouped data, i.e. if $g = n$ ($h = 1$). Wold (1974) reported that good results were obtained by him applying the jackknife method for estimating confidence intervals of parameters of spline-functions fitted to data.

Keeping the notation previously used for cross-validation, one can define n pseudovalues for a spline-curve as

$$q_{ij} = (n-2)\,s_i - (n-3)\,s_{ij} \tag{9.8}$$

The subscript k has been dropped because a single value of SF is used in each jackknife experiment. The pseudovalues lead to the jackknife values q_i and their standard deviations $\hat{\sigma}(q_i)$:

$$q_i = \frac{1}{n-2} \sum_{j=2}^{n-1} q_{ij}; \quad \hat{\sigma}(q_i) = \left[\frac{\sum_{j=2}^{n-1} (q_{ij} - q_i)^2}{(n-2)(n-3)} \right]^{\frac{1}{2}} \tag{9.9}$$

Consequently:

$$q_i - E(q_i) = t(n-3)\,\hat{\sigma}(q_i) \tag{9.10}$$

Setting $t(n\text{-}3) = 2$, this leads to the approximate 95 percent confidence interval

$$q_i \pm 2\hat{\sigma}(q_i)$$

Jackknife values can be obtained for all four coefficients which determine a cubic curve for each the $(n\text{-}1)$ intervals between successive values x_i and x_{i+1} $(i = 1, 2, ..., n\text{-}1)$. Use of all coefficients results in a jackknife spline which interpolates between these successive values.

For example, Table 9.7 shows the values s_i, q_i and $s(q_i)$ for the spline-curve of Figure 9.13a. The corresponding jackknife spline based on complete sets of four coefficients is shown in Figure 9.14a together with 95 percent confidence intervals for the values q_i. Comparison of the values $s(q_i)$ in Table 9.7 indicates that spline and jackknife spline for SF = 2.456 are close to one another. Nearly all standard deviations $s(q_i)$ are less than SF. Only from level 10 to 14, the $s(q_i)$ values are relatively large as can also be seen in Figure 9.14a. It would be possible to transfer the error bars of Figure 9.14a to the data points in Figure 9.13d, and to project them along the depth scale by one of the methods illustrated in Figure 9.7. Instead of expressing the uncertainty of the observed events with respect to their most likely positions, these new error bars would give the uncertainty of the estimated ages themselves.

The jackknife method provides a valuable new tool for investigating the validity of splines fitted by different methods. This will be demonstrated by a comparison of the four jackknife splines shown in Figure 9.14. Figure 9.14b is the jackknife spline for SF = 2.56 corresponding to the spline of Figure 9.13f. This jackknife is for the spline with optimum smoothing factor in the situation of irregularly spaced depth data. It is fairly close to its counterpart with subjectively selected SF = 2.1 (Fig. 9.12f) except near the top and bottom of the section where the spline of Figure 9.13b would imply higher sedimentation rates. In fact, the age is decreasing with depth in the lowest part of the section thus violating the law of superposition of strata. This suggests that the method of fitting a cubic spline with irregularly spaced data may not give satisfactory results (also see experiments described in Section 3.6). Figure 9.14c with SF = 2.2 corresponds to the spline of Figure 9.12a; and Figure 9.14d with SF = 2.1 to that of Figure 9.12f. The jackknife spline of Figure 9.14c does not closely resemble the spline of Figure 9.12a. The jackknife

TABLE 9.7

CASC 2 output for Adolphus D-50 (age-level plot, cf. Figs. 9.13a and 9.14a); the values s_i are situated on the spline-curve with optimum smoothing factor (SF = 2.456); the values q_i with standard deviations $\hat{\sigma}(q_i)$ belong to the corresponding jackknife spline.

Level (i)	s_i	q_i	$\hat{\sigma}(q_i)$
1	9.3954	6.3569	1.8090
2	11.6993	9.0325	1.5035
3	13.9446	12.0333	1.3183
4	16.1968	15.2569	1.2735
5	18.5145	18.5826	1.2765
6	20.9281	21.9121	1.2336
7	23.4611	25.0837	1.1596
8	26.1030	27.9331	1.3708
9	28.7515	30.5142	2.0336
10	31.2879	32.9016	2.7904
11	33.6144	35.1614	3.3494
12	35.6327	37.4329	3.4880
13	37.3720	39.5062	3.2675
14	38.8722	41.2575	2.7789
15	40.2034	42.6240	2.1674
16	41.4596	43.5825	1.6073
17	42.6775	44.3336	1.2058
18	43.9157	45.0106	1.0935
19	45.2303	45.6888	1.1591
20	46.6426	46.4389	1.2063
21	48.1466	47.3135	1.1650
22	49.6923	48.4075	1.1132
23	51.2528	49.7226	1.0849
24	52.8331	51.1675	1.0637
25	54.4094	52.7214	1.1012
26	55.9815	54.3178	1.2085
27	57.5750	55.8434	1.3380
28	59.2145	57.2140	1.4501

spline for SF = 2.2 is not as smooth as the spline for SF = 2.2 that was originally selected in a subjective manner.

Although there is little difference between the splines of Figures 9.12a and 9.13a, the difference between the corresponding jackknife splines of these patterns suggests that only the spline for the optimum smoothing factor (SF = 2.456) is satisfactory. The jackknife spline of Figure 9.14c also violates the law of superposition in several places. Even more severe violations of this law can be observed in Figure 9.14d where the jackknife spline dips in the wrong direction in several places. These

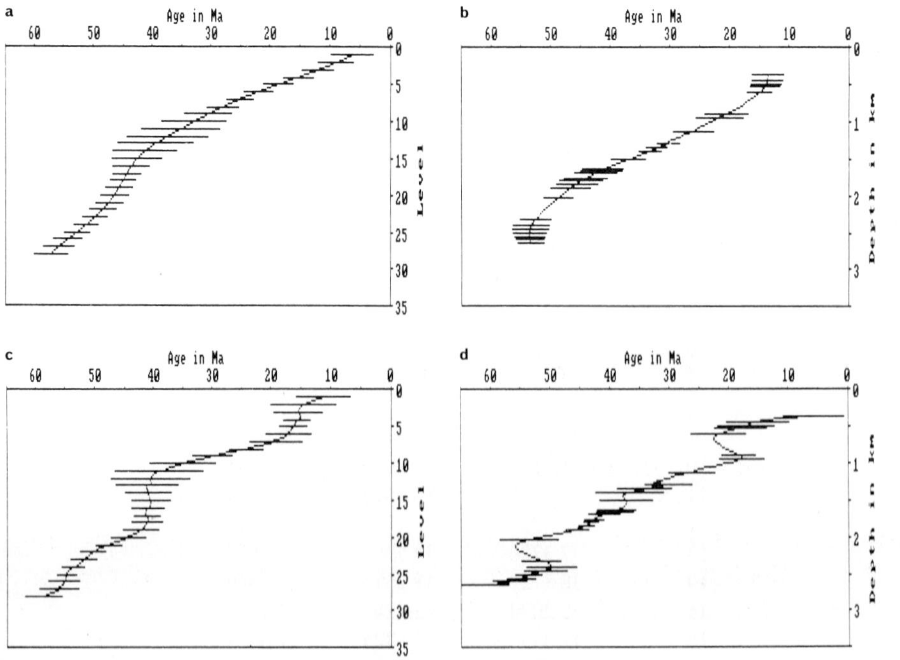

Fig. 9.14 Jackknife spline-curves with approximate 95% confidence limits for Adolphus D-50 results previously shown in : a. Fig. 9.13a; b. Fig. 9.13f; c. Fig. 9.12a; d. Fig. 9.12f. The optimum smoothing factor patterns of Figs. 9.13a and 9.13f are relatively closely approximated by their jackknife splines, contrary to the subjectively selected spline-curves of Figs. 9.12a and 9.12f. The latter two jackknife splines (Figs. 9.14c and 9.14d) show violations of the law of superposition of strata. In general, the indirect method (Figs. 9.14a and 9.14c) yields results which are superior to those of the direct method (Figs. 9.14b and 9.14d).

are characterized by lack of data along the depth axis due to relatively high sedimentation rates. Although the subjectively derived event-depth curve for SF = 2.1 (Fig. 9.12e) is relatively close to the "optimum" event-depth curve shown in Figure 9.13d, it obviously could not be duplicated by its jackknife estimator. This confirms that it may be dangerous to fit splines to irregularly spaced data. The results of Figure 9.14 clearly demonstrate that the indirect method of constructing event-depth curves illustrated in Figure 9.6 is to be preferred to the direct method. The discrepancy between the patterns of Figures 9.13d and 9.13f also can be explained now. Although the pattern of Figure 9.13f is for an optimum smoothing factor and was reasonably well duplicated by its jackknife spline (Fig 9.14b), the irregular spacing of control points along the depth axis resulted in a pattern which is too smooth in comparison with the pattern (Fig. 9.13d) obtained by the indirect method.

It is noted that the experiments with the CASC 2 computer program described in this section made use of the method of assigning equal weights to ages for all levels. It is possible in CASC 2 to assign more weight to level values based on more than a single age. This alternative procedure was applied but it led to results which are not markedly different from those described in this section.

9.7 Computer simulation experiment for event-depth spline fitting with error analysis

During development of the RASC/CASC procedures, three criteria of method evaluation were employed: (1) the method should have a firm stratigraphic foundation; (2) it should be logically coherent from a mathematical-statistical point of view; and (3) the computer programs should be efficient as well as user-friendly.

The first aim (1) is promoted by systematic comparison of RASC zonations independently obtained and correlations with subjective results, and by evaluation of computer outputs by stratigraphers. Computer simulation experiments are helpful in (2) and (3). Results for one such experiment are displayed in Figure 9.15 and Tables 9.8 and 9.9. A theoretical age-depth curve $t = 9.155 + 0.685x + 0.165x^2 - 0.0005x^3$ (unit of $x = 100$m) is shown in Figure 9.15a. Twenty-one random normal numbers E_{0i} (zero mean, unit variance) of regularly spaced points (labelled $i = 1, 2, ..., 21$ in Table 9.9) were added in order to simulate observations, y_i, of biostratigraphic events in a hypothetical Cenozoic basin. In Reinsch – De Boor spline-fitting, the smoothing factor (SF) fully determines the shape of the best-fitting spline-curve. In CASC 2 (see Chapter 10), SF is input and, as elsewhere in this book, has the following meaning: the square of SF is equal to the averaged squared deviation $(y_i\text{-}s_i)^2$ between observations y_i and their corresponding values s_i on the spline-curve s.

An optimum value of SF can be found by cross-validation (see Section 9.5) as illustrated in Table 9.8. In general, $SF_{min} \leq SF_{opt} \leq SF_{max}$ where the minimum smoothing factor (SF_{min}) represents the first-fitted smoothing spline curve whose age increases monotonically with depth.

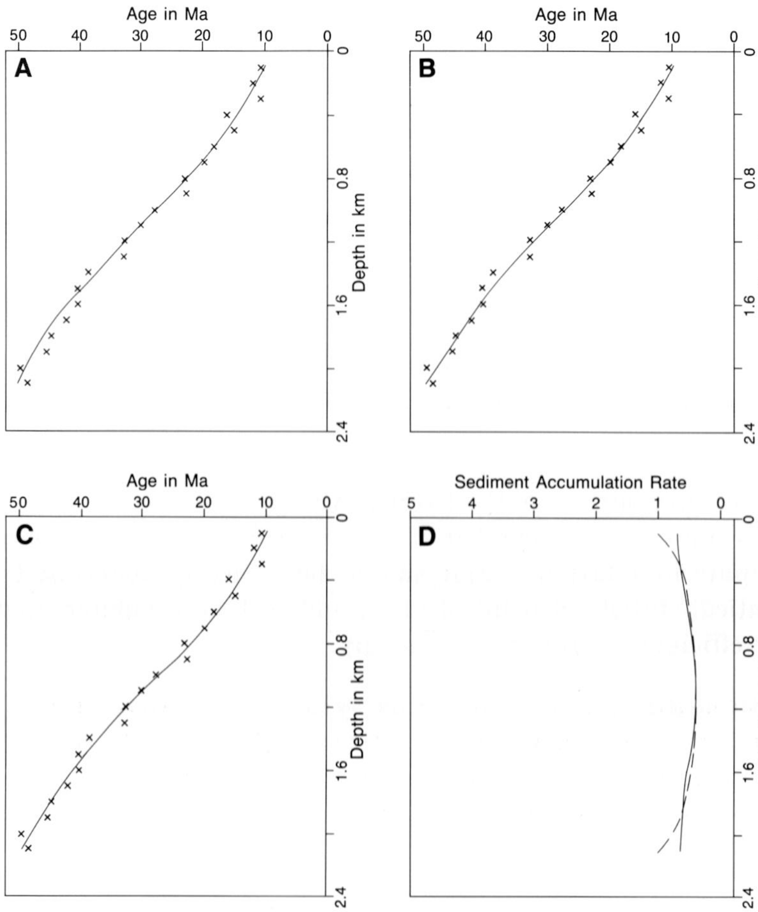

Fig. 9.15 Computer simulation experiment. Random normal deviates were added to theoretical curve (A). Cross-validated smoothing spline (B) was approximated by its jackknife estimate (C). First derivative of spline-curve (B) gave sediment accumulation rate curve (D, solid line) which is compared with first derivative of theoretical curve (D, broken line).

The maximum smoothing factor (SF_{max}) corresponds to the best-fitting straight line. The optimum smoothing factor (SF_{opt}) has minimum cross-validation value within the interval between SF_{min} and SF_{max}. The spline-fitting routine in CASC 2 is a modified version of De Boor's (1978) FORTRAN program which uses the secant method, returning a smoothing factor that is slightly greater than an input SF. For this reason, 1.170 was selected from Table 9.8 as input for the cross-validation smoothing spline with SF = 1.174 shown in Figure 9.15b. Obviously s in Figure 9.15b is close to t in Figure 9.15a. In practice, t is not known and one would like to

TABLE 9.8

CASC 2 output for computer simulation experiment of Fig. 9.15. See Table 9.6 for explanation of column headings.

k	SF_k	CV_k
1	0.828	3.607
2	0.897	3.314
3	0.967	2.968
4	1.036	2.565
5	1.106	2.219
6	1.175	2.131
7	1.245	2.192
8	1.314	2.290
9	1.384	2.395
10	1.453	2.445
11	1.523	2.448

construct confidence intervals on s to evaluate the difference between s and (unknown) t, and to check residuals (y_i-s_i) for possible outliers. This problem has been studied by Wahba (1983) and Eubank (1984) who have shown that the column vector S of values s_i are related to column vector Y for y_i by $S = HY$ where the hat matrix H has properties similar to the hat matrix in regression analysis. Several methods have been developed for obtaining H in explicit form. This allows estimation of the following two variance-covariance matrices: Var (s_i-t_i) $= \sigma^2 H$, and Var (y_i-s_i) $= \sigma^2 (I\text{-}H)$ where I is the identity matrix.

In this section, the following procedure is used for obtaining the diagonal elements h_{ii} of H. S was closely approximated by its jackknife spline-curve q with values q_i at the observation points (see Fig. 9.15c and Table 9.9). The jackknife method provides variances $s^2(q_i)$ of the values q_i which are approximately equal to $\sigma^2 h^2{}_{ii}$ being the diagonal elements Var (s_i-t_i). The smoothing factor SF = 1.174 estimates σ. The diagonal elements of Var (y_i-s_i) can also be estimated because $q_i \simeq s_i$. These are written as $s^2(E_{1i})$ in Table 9.9. Approximately 95% and 99% confidence intervals are obtained by multiplying $s(E_{1i})$ by 2 and 3, respectively. As expected, no outliers are indicated for the random normal numbers used in the computer simulation experiment. Because $s(E_{1i})$ applies to observations used for estimating the spline-curve s, Table 9.9 also shows residuals $E_{2i} = y_{2i} - q_i$ for new observations y_{2i} obtained by adding 21 other random numbers to t_i. These new observations have wider confidence belts with widths controlled by $s(E_{2i}) = \mathrm{SF}\ \surd(1+h_{ii}{}^2)$, also shown in Table 9.9. This second type of confidence belt would, for example, apply to test suspected outliers not used for calculating the smoothing spline.

The three standard deviations $s(q_i)$, $s(E_{1i})$ and $s(E_{2i})$ can be projected onto the depth axis by using the sediment accumulation rate curve (Fig. 9.15d). It is noted that this type of curve is more difficult to estimate than the event-depth curve as indicated by larger discrepancies in Figure 9.15d.

TABLE 9.9

Random normal deviates (E_{0i}) were added to theoretical values (t_i) on curve of Fig. 9.15a to give observed values y_i. Cross-validated smoothing spline values (s_i) on curve of Fig. 9.15b were approximated by their jackknife estimates (q_i) of which standard deviations $s(q_i)$ could be computed. Standard deviations s (E_{1i}) and s (E_{2i}) are for residuals of data used (E_{1i}) and not used (E_{2i}) for estimating q_i, respectively.

i	t_i	E_{0i}	y_i	s_i	q_i	$s(q_i)$	E_{1i}	$s(E_{1i})$	E_{2i}	$s(E_{2i})$
1	10.00	0.54	10.54	9.69	9.79	1.02	0.75	0.59	-0.48	1.55
2	11.15	0.63	11.77	11.14	11.23	0.62	0.54	1.00	-0.57	1.33
3	12.56	-1.99	10.57	12.66	12.64	0.57	-2.08	1.02	-0.91	1.31
4	14.22	1.66	15.87	14.29	14.03	0.71	1.85	0.97	-1.10	1.37
5	16.08	-1.22	14.86	16.05	15.59	0.84	-0.73	0.82	-0.46	1.44
6	18.13	-0.04	18.08	17.97	17.39	0.90	0.69	0.76	1.40	1.48
7	20.32	-0.56	19.76	20.03	19.46	0.90	0.30	0.75	1.12	1.48
8	22.63	0.36	23.00	22.24	21.82	0.87	1.17	0.79	1.62	1.46
9	25.04	-2.45	22.59	24.58	24.40	0.78	-1.81	0.88	1.71	1.41
10	27.50	0.16	27.67	27.04	27.04	0.59	0.62	1.02	-0.15	1.31
11	30.00	-0.15	29.85	29.55	29.73	0.45	0.12	1.08	0.73	1.26
12	32.50	0.13	32.63	32.04	32.39	0.51	0.24	1.06	-1.51	1.28
13	34.96	-2.33	32.63	34.47	34.90	0.70	-2.27	0.94	0.42	1.37
14	37.37	1.30	38.66	36.81	37.17	0.91	1.49	0.75	-0.56	1.48
15	39.68	0.75	40.43	38.96	39.32	0.94	1.21	0.70	1.72	1.51
16	41.87	-1.51	40.36	40.93	41.36	0.79	-1.00	0.87	-0.85	1.41
17	43.92	-1.74	42.18	42.78	43.21	0.59	-1.04	1.02	1.47	1.31
18	45.79	-1.10	44.68	44.54	44.87	0.45	-0.19	1.08	1.82	1.26
19	47.44	-2.04	45.40	46.26	46.42	0.44	-1.02	1.09	0.76	1.25
20	48.86	0.67	49.52	47.93	47.95	0.55	1.57	1.04	0.43	1.29
21	50.00	-1.52	48.48	49.55	49.58	0.78	-1.10	0.88	0.26	1.41

9.8 Regional application of RASC and CASC

The geological use of the RASC/CASC method and its value in sedimentary basin analysis will be illustrated by means of examples drawn from exploration micropaleontology. The first examples are based on the original distribution of 168 Cenozoic Foraminifera in 21 Grand Banks and Labrador wells (cf. Fig. 6.2), and of 116 taxa of Mesozoic foraminifers in 16 Grand Banks wells. The latter databank was largely prepared by Williamson (1987). Later examples will deal with the North Sea Basin and integration of different families of microfossils (foraminifers; dinoflagellates; spores and pollen). The Cenozoic databank used for running CASC on the Labrador Shelf of Grand Banks is the same as that previously used for RASC (Chapters 4-6), with the one difference that in the Bjarni H-81 well, taxa 54 and 55 (*Gavelinella beccariiformis* zone, Paleocene) have been added (cf. Table 4.8). Both taxa were observed at a depth of 6660ft.

The discussion in the remainder of this chapter deals with the following questions: (1) what is the stratigraphical meaning of RASC/CASC-type of correlation?; and (2) what is the degree of confidence, expressed in depth and in linear time units of CASC correlation, and to what extent is the error bar useful for geological interpretation?

In order to find answers to these questions, conventional or subjective and more objective CASC-type correlation of wells are compared using three related stratigraphic criteria: (a) selected zone markers; (b) assemblage zones; and (c) isochrons in Ma. In all examples, the underlying biostratigraphic zonations were defined with the RASC method, but in principle CASC can be applied to non-RASC zonations based on biostratigraphic and other stratigraphic events.

Firstly, ten selected zone markers were traced through six wells. Starting point was the Cenozoic scaled optimum sequence (Fig. 6.2, 7/2/4 run for 21 wells). For the interactive spline fitting of the bivariate plots, all CASC defaults were accepted, unless otherwise specified. The first CASC default is the smoothing factor (SF) that defines the spline-curve for which an increase in position or depth along one axis does not anywhere correpond to a decrease in position (or time) along the other axis. This default is obtained by means of an algorithm that calculates spline-curve fits with SF increasing or decreasing according to a binary search method.

The default satisfies the condition that the observed sedimentation rate is never negative.

In three wells the mainframe CASC cursor option was used to delete aberrantly positioned events: in Hibernia P-15, one point was deleted on event level 12; in Bonavista C-99, three points were deleted (on levels 4, 11 and 15), and SF for the events versus depth graph was changed from its default (=0.02) to 0.15; and in Snorri J-90, one point was deleted on level 6.

The results of the CASC multi-well comparison are shown in Table 9.10, listing both the observed and the most likely depths of the ten selected Paleocene through Miocene zone markers 50 (56), 90, 32, 29, 261 (260), 259, 24, 26(15), 18 and 16. In two wells substitute taxa were correlated rather than the three designated events 50, 261 and 26. The substitutes 56, 260 and 15 are neighbors of the original events in the optimum sequence.

In most instances, the observed and the most likely depth values are within half the length of the error bar (68% probability) around the most

TABLE 9.10

Observed (above line) and most likely depth (in m) of ten Eocene through Miocene zone markers in six wells. The fossil numbers are the RASC dictionary numbers. Results are based on optimum sequence CASC (21 wells; $k_c=7$, $m_{c1}=2$, $m_{c2}=4$); * means that at that site substitute fossils (neighbors in the optimum sequence) were used.

		Hibernia P-15	Adolphus D-50	Bonavista C-99	I. Harbour M-52	Bjarni H-81	Snorri J-90
Ceratobulimina contraria	- 16	310	485	887	750	872	1262
			350 ± 85	1130 ±381	571 ±132		
Spiroplectammina carinata	- 18	275	512	2030	649	1085	1701
		334 ± 65	481 ± 78	1681 ±363	673 ±126	1025 ± 18	1533 ±167
Uvigerina dumblei	- 26(15)	550	933	2377	1261		
		619 ±330	726 ±228	2059 ±162	786 ±155*	1094 ± 14	1756 ± 42
Turrilina alsatica	- 24	1075	1280	2377	2097	1298	
		1000 ± 75	1164 ±115	2372 ±13	1344 ±532	1307 ± 31	1910 ± 29
Ammodiscus latus	- 259	1125	1153	2316	1646		
		1083 ± 45	1291 ± 71	2571 ±135	1655 ±147	1490 ± 27	1983 ± 26
Haplophragmoides walteri	- 261(260)	1185	1509	3078	1704	1634	213
		1176 ± 40	1471 ±171	2889 ±215	1864 ±404*	1577 ± 16	2067 ± 22*
Cyclammina amplectens	- 29	1195	1890	3109	2396	1634	
			1721 ± 95	3099 ± 17	2212 ±489	1635 ± 14	
Ammosphaeroidina sp. 1	- 32	1125	1761	3386	1935	1695	2112
			1767 ± 62	3144 ± 99	2275 ±266	1653 ± 15	2155 ± 27
Acarinina densa	- 90		2062	3478			
			2293 ±249	3423 ± 96	2528 ±497	1763 ± 8	2651 ± 59
Subbotina patagonica	- 50(56)		2517		2914	2009	2932
			2501 ± 66		2861 ±327*	1812 ± 15	2798 ± 51*

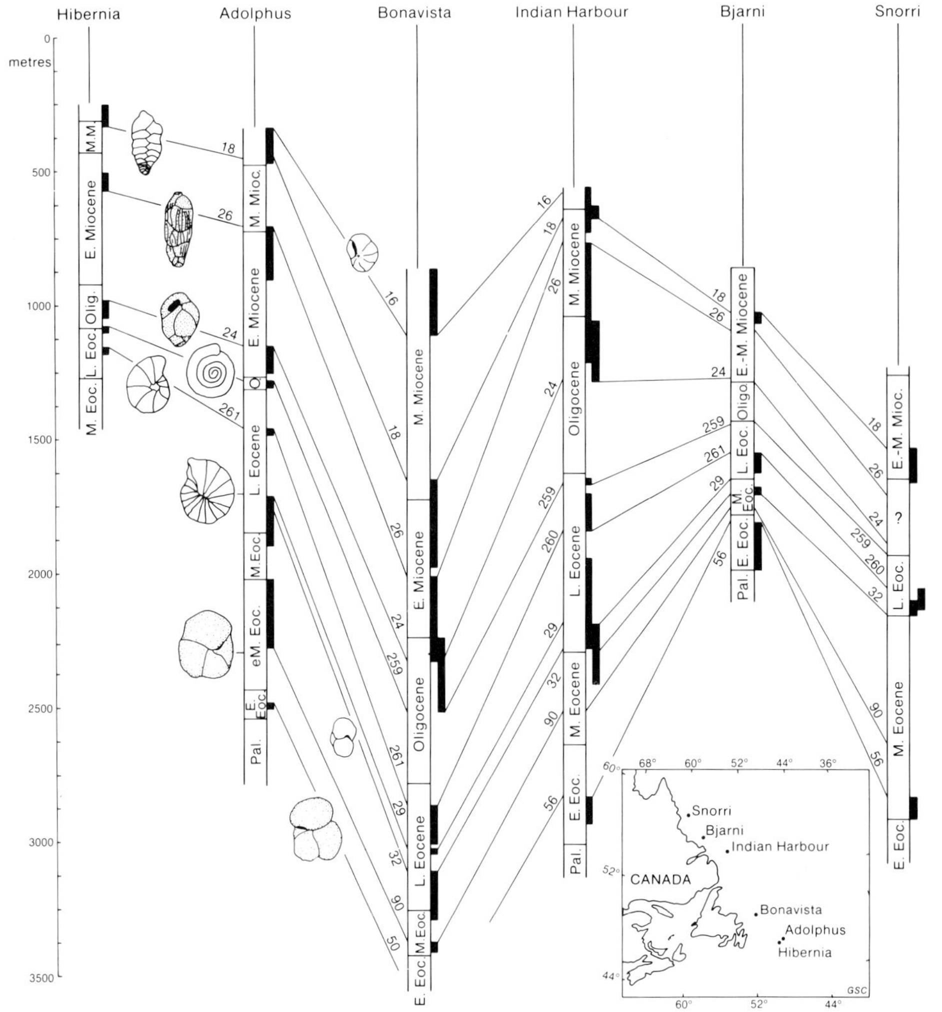

Fig. 9.16 Tracing of ten foraminiferal events through six wells, using the CASC (optimum sequence) method to calculate the most likely depths. Black bars show the deviations of these depths from the observed depths. The chronostratigraphic segmentation is based on observed depths only.

likely value. As pointed out in the previous section, the actual precision of the estimated depth of an event in a well is probably greater than that indicated by the local error bar for single event positions along the spline curve. Also, the local error bar at any depth is initially calculated over the time interval along the (scaled) optimum sequence scale (y), as defined by twice the standard devition (SD) in that (y) direction. It is directly proportional to the fitted average sedimentation rate for each point (cf. Fig. 9.7).

Figure 9.16 graphically correlates the ten events through the six wells. The conventional chronostratigraphic segmentation, which is shown for comparison, only uses the observed depth of events. The new, most likely, zone marker depths would lead to slight up or down adjustments of the age boundaries. It could be assumed that such a change might violate stratigraphic boundaries as adjusted for major lithology changes as determined from well logs. However, using sonic and gamma logs, no evidence for this was found. In the Snorri well there is no direct micropaleontological evidence for the presence of events 259, 24 and 26, associated with Oligocene-Early Miocene strata, although the CASC method predicts their likely depth in this well. These depths are not unreasonable given that Oligocene strata were thought to be present at that depth, based on palynology.

The conversion of the scaled optimum sequence to a (local) biochronology enables the stratigrapher familiar with CASC to trace isochrons in the same way as zones were traced. The procedure begins with the designation of numerical ages in Ma to those events in the scaled optimum sequence for which literature-based ages are available. In all, 23 events were dated this way, as explained below. The time scale is that of Berggren et al. (1985). The regional use of the standard planktonic zonation is as in Gradstein et al. (1985) who followed Gradstein and Srivastava (1980), and Gradstein and Agterberg (1982):

(1) 63 Ma – events 253 and 61 – *Subbotina triloculinoides* and *S. pseudobulloides* – two rare events that mark approximately the end of Danian time.

(2) 58 Ma – event 55 – *Gavelinella beccariiformis*; occurs up to standard zone P5 (Tjalsma and Lohman, 1983), which fits well with its disappearance in the Adolphus well together with *Aragonia velascoensis* (Paleocene) and below the appearance of *Pseudohastigerina* (post P5).

(3) 57 Ma – event 194 – *Planorotalites chapmani*; disappears in standard zone P6. Specimens are often transitional between *P. chapmani* and *Pseudohastigerina*. The latter is thought to appear at the boundary of P5 and P6, or ± 57 Ma ago.

(4) 55 Ma – event 50 – *Subbotina patagonica*; is frequent in the Ypresian of Belgium (Muller and Willems, 1981), in the Moe Clay of Denmark and in the Lower Eocene of the North Sea and Labrador Sea. The end of the *S. patagonica* peak occurs at the boundary of NP11/NP12, which

coincides with the boundary of the *Morozovella formosa formosa* Zone, at the time of Anomaly 24, just after 55 Ma.

(5) 52 Ma – event 93 – *Acarinina broedermanni*; the species has its top well below *A. densa*, probably in the *A. pentacamerata – Hantkenina aragonensis* Zone, near the Early-Middle Eocene boundary at 52 Ma. In some RASC runs, *A. broedermanni* falls between Early and Middle Eocene zones.

(6) 49 Ma – event 90 – *Acarinina densa*; this is the time of the optimum climatic warming in the Labrador Sea, in early Middle Eocene time. Less common at this time are *A. senni*, *A.* aff. *pentacamerata*, *A.* aff. *broedermanni*, *Morozovella caucasica*, *M. spinulosa*, and *M.* aff. *aragonensis*. The event probably falls in the *Hantkenina aragonensis – Globigerinatheka subconglobata* Zone at Anomaly 21 time or 52-46 Ma (average 49 Ma).

(7) 40 Ma – event 29 – *Cyclammina amplectens*; in RASC runs this event falls below *Turborotalia pomeroli* and *Globigerina yeguaensis* and above *Acarinina densa*. In Poland its peak occurrence is so-called Middle Eocene; it is less frequent in upper Eocene strata (Gradstein and Berggren, 1981). The event was tentatively placed at 40 Ma.

(8) 38 Ma – event 85 – *Pseudodhastigerina micra*; same reasoning as for *Turborotalia pomeroli* (see below), but often disappears in slighty older beds, as also shown in the scaled optimum sequence (Fig. 6.2).

(9) 37 Ma – event 33 – *Turborotalia pomeroli*; co-occurs in southern wells with *Subbotina linaperta*, *Globigerina yeguaensis* and *Pseudohastigerina micra*, of the *Turborotalia cerroazulensis* Zone, late Late Eocene. The top was placed just below the inferred Eocene/Oligocene boundary.

(10) 28 Ma – event 24 – *Turrilina alsatica*; the top of this distinctive Oligocene taxon roughly equates with the top of the Boom Clay in Belgium and the top of the *Globorotalia opima opima* Zone, at ± 28 Ma.

(11) ~20 Ma – event 26 – *Uvigerina dumblei*; slightly older than *Globigerina praebulloides*.

(12) 20 Ma – event 137 – *Globigerinoides primordius trilobus*; rare Early Miocene event.

(13) 17 Ma – event 15 – *Globigerina praebulloides*; disappears locally with *Sphaeroidinella seminula* and with or just below *Globorotalia scitula praescitula*, which may equate with the *G. fohsi peripheroronda* Zone, early Middle Miocene of Scotian Shelf wells (Gradstein and Agterberg, 1982). The RASC run 7/2/4 indicated an average disappearance in the *Uvigerina dumblei* zone. Its local extinction was placed between 14 and 20 Ma (average 17 Ma).

(14) 15 Ma – event 179 – *Globorotalia scitula praescitula*; probably occurs in the late Early to early Middle Miocene warming event, as observed from the northern incursion of warmer water planktonic taxa.

(15) 3.5 Ma – events 266, 4, 269 and 5 – Both *Globorotalia puncticulata*, *G. inflata*, *G. crassaformis*, and *Neogloboquadrina atlantica* are thought to disappear with the onset of major glaciation in the Labrador Sea, dated at approximately 3.5 Ma.

Four other events occur at or near significant breaks in the 5-2-3 and 7-2-4 scaling solutions for 21 and 24 wells. These breaks were equated with zonal boundaries and series breaks as follows:

(16) 58 Ma – event 56 – *Glomospira corona*; Paleocene-Eocene boundary on (upper) continental margin wells.

(17) 52 Ma – event 57 – *Spiroplectammina spectabilis* LCO; Early-Middle Eocene boundary; LCO = Last Common Occurrence.

(18) 37 Ma – event 259 – *Ammodiscus latus;* Eocene – Oligocene boundary.

(19) 11 Ma – event 17 – *Asterigerina gurichi*; Middle-Late Miocene boundary.

Figure 9.9a was a plot of the ages of the previously listed events in a RASC distance scale (21 wells, 7/2/4 run) versus linear time scale. Smoothing of the spline-curve function diminishes some of the uncertainty in subjective assignment. The spline function now can be used to convert the RASC distance scale into an age scale.

Next, the question can be asked what is the most likely depth in the wells of the principal boundaries between RASC zones, expressed in Ma. Gradstein and Agterberg (1985) have traced the boundaries between the successive Cenozoic RASC zones (Fig. 6.2), which are close approximations to the boundaries between Paleocene and Eocene (~56 Ma), Early Eocene and early Middle Eocene (~52 Ma), early Middle Eocene and Middle Eocene (~49 Ma), Middle and Late Eocene (~44 Ma), late Eocene and Oligocene (~36 Ma) Oligocene and Miocene (~24 Ma), Early and Middle Miocene (~16 Ma) and top of Middle Miocene (~12 Ma).

Table 9.11 lists most likely and subjectively determined (as far as known) depths for these isochrons. The same results are plotted in Figure 9.17, with the wells arranged latitudinally (48° - 58°N). The CASC depths are from a batch run that accepted all SF defaults. Although it yields more crude results than interactive runs, this procedure takes less time and the actual depth estimates are not influenced much. As explained in the previous section, the choice of SF has much more influence on the average rate of sedimentation and hence on the error estimation, than on the actual depth of the isochrons. In a few instances, default smoothing yielded unacceptably steep spline fits, and the local

TABLE 9.11

Observed (above line) and most likely depth (in m) of the 56, 52, 49, 44, 36, 24, 16 and 12 Ma isochrons in 10 wells on the Grand Banks and Labrador Shelf. Results are based on scaled optimum sequence or distance – CASC (21 wells; $k_c = 7$, $m_{c1} = 2$, $m_{c2} = 4$).

			Hibernia P-15	Fl. Foam I-13	Adolphus D-50	Bonavista C-99	Dominion O-23	I. Harbour M-52	Gudrid H-55	Bjarni H-81	Snorri J-90	Karlsefni H-13
12 Ma	Middle Miocene	Observed	330	576	485	585	859	676	613	871	1261	
		Most likely	262 ± 42	382 ± 92	385 ± 44	521 ± 45	986 ± 46	638 ± 82	528 ± 65	952 ± 53	1288 ± 88	
16 Ma	Early Miocene	Observed	460	713	762		1767				1618	
		Most likely	344 ± 77	502 ±114	508 ± 56	584 ± 84	1636 ± 56	723 ± 68	619 ±117	1019 ± 15	1546 ±159	
24 Ma	Oligocene	Observed	960	960	1310	1335	2320	1097	1066	1298		2060
		Most likely	813 ± 71	809 ±147	935 ± 94	1041 ±306	2254 ± 78	1072 ±164	1025 ±241	1041 ± 20	1826 ± 43	1977 ± 40
36 Ma	Late Eocene	Observed	1125	1289	1353	1719	2804	1645	1505	1481	1975	2060
		Most likely	1128 ± 41	2398 ± 98	1364 ± 46	1732 ±223	3027 ±113	1668 ± 65	1719 ±231	1524 ± 18	2041 ± 27	2137 ±118
44 Ma	Middle Eocene	Observed	1315	1585	1890	2307	2161	2377		1695	2164	
		Most likely	1216	1501 ±133	2324 ±298	3274 ±184	2359 ± 50	1961 ±138	1670 ± 32	2189 ± 68	2784 ± 49	
49 Ma	Early Middle Eocene	Observed		1615	2026	2438	3265		1862		2316	2791
		Most likely		1687	2311 ±133	2643 ±173	3426 ±208	2509 ±250	2091 ± 44	1779 ±178	2636 ±100	2949 ±179
52 Ma	Early Eocene	Observed		1763	2469	2764	3478	2700	2118	1847	2926	2944
		Most likely		1796	2467 ± 71	2730	3564 ±20	2726 ±225	2149 ±127	1824 ± 24	2770 ± 73	3220
56 Ma	Paleocene	Observed			2569			3030	2395	2029	2993	3828
		Most likely			2561 ± 29			2919 ±118	2324 ±179	2000	2995	3663 ±251
	Cretaceous		1400	2002	2667	3118	3660	3194	2400			

error bar estimate was deleted. In one well, Karlsefni H-13, both foraminifers and palynomorphs agree on the absence of Oligocene beds (*Turrilina alsatica* Zone). Batch CASC calculates a thin Oligocene interval (24-36 Ma). Above the Eocene, the well has only a few data points and results are crude.

The local error estimates of the most likely depths for the isochrons are within 1 to 10% of the actual depth values, and more frequently 2 to 5%. In about ten cases the subjectively assigned depths for the zonal boundaries as converted to isochrons are outside the 68% confidence limits (±1 SD). For geological interpretations, it should be borne in mind that the error in most likely depth is an upper limit, and the SD is probably smaller by a factor that, amongst others, is related to the number of observations per spline-curve, as explained earlier. Palynologically determined depths for these stage boundaries often are outside the depth interval (most likely depth ± 1 SD), calculated by CASC. The errors in this independent biostratigraphic correlation are unknown, but the comparison suggests that multiple biostratigraphy uncertainties exceed the CASC-type of errors using one fossil discipline only. The conclusion may be drawn that the CASC program is able to predict reliable and objective well to well isochrons. The error expression, that remains vague in conventional, subjective correlation schemes, is conservatively large when one fossil discipline only is used.

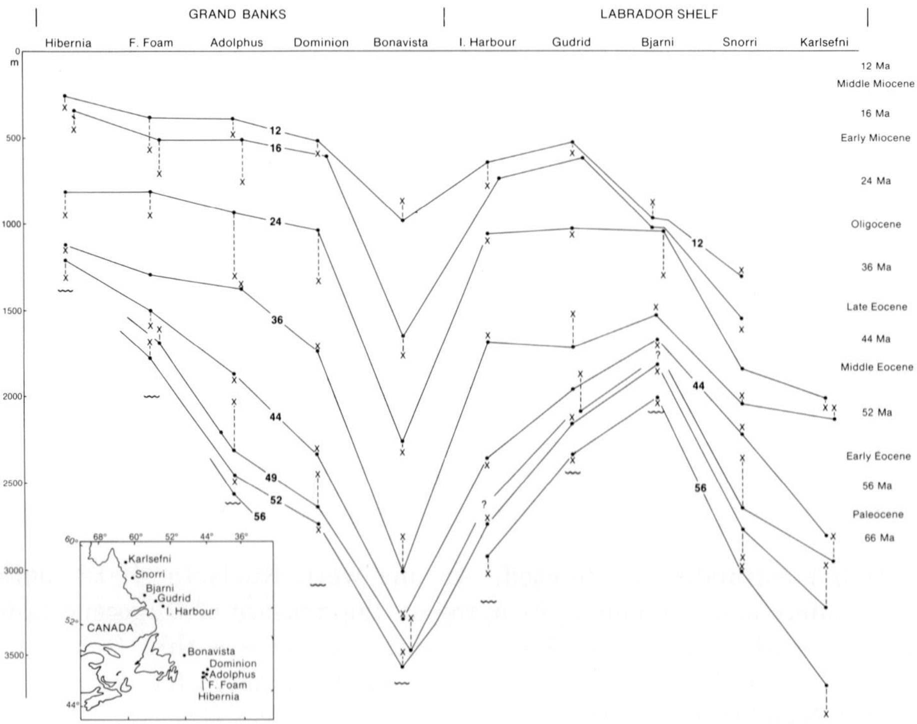

Fig. 9.17 Correlation of 8 Cenozoic isochrons, according to their most likely depths in 10 wells on the Grand Banks and Labrador Shelf. The depths were computed by means of the RASC-CASC method explained in the text. Subjective estimates for the depths of these isochrons are shown with x.

9.9 Application of RASC and CASC to the Hibernia Oilfield

Williamson (1987) has investigated the application of ranking, scaling and correlation in time to the Mesozoic microfossil record recovered from the Grand Banks, particularly for the area centered around the Hibernia oilfield. Figure 1.1 illustrated the cumulative frequency distribution of the highest occurrences of 116 taxa of foraminifers in 13 deep wells based on Williamson's original data (Fig. 9.18). This dataset later was enlarged to encompass the Upper Jurassic and Lower Cretaceous microfossil record of up to 25 wells. Comparison between Williamson's original zonation using thresholds of $k_c = 4$ and $m_c = m_{c1} = m_{c2} = 3$ which leaves 54 events, and subsequent runs of $k_c = 7$, $m_{c1} = 1$ and $m_{c2} = 4$ (or 5) using 18 to 25 wells, gave close to 50 events with virtually

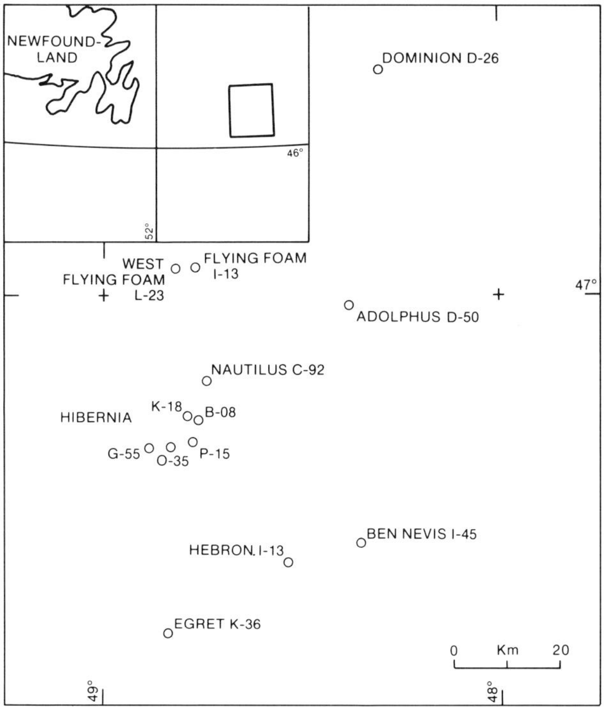

Fig. 9.18 Locations of 13 boreholes used by Williamson (1987) for RASC/CASC application on northern Grand Banks.

the same zonation. For this reason, the concise account of Williamson (1987) is followed with minor changes, using his original illustrations.

Figure 9.19 is the scaled optimum sequence with chronostratigraphically useful average interval zones highlighted through shading. Based on the original RASC run with 54 events, to which 9 unique events were added for (further) chronostratigraphic calibration, eleven RASC zones stand out, numbered from XI through I, Kimmeridgian-Cenomanian. This zonation considerably expands stratigraphic resolution previously available. There is good correspondence of the average position of the disappearance levels of the taxa in the wells and the upper part of stratigraphic ranges reported in the literature. Some longer ranging taxa of the literature, on the Grand Banks have relatively short ranges, as is the case with *L. nodosa* (no. 10) and *D. gradata* (no. 111). *N. varsoviensis* (no. 64) was not previously reported so young. The tight clustering of events in the Albian zones III and II reflects considerable uncertainty on their exact disappearance

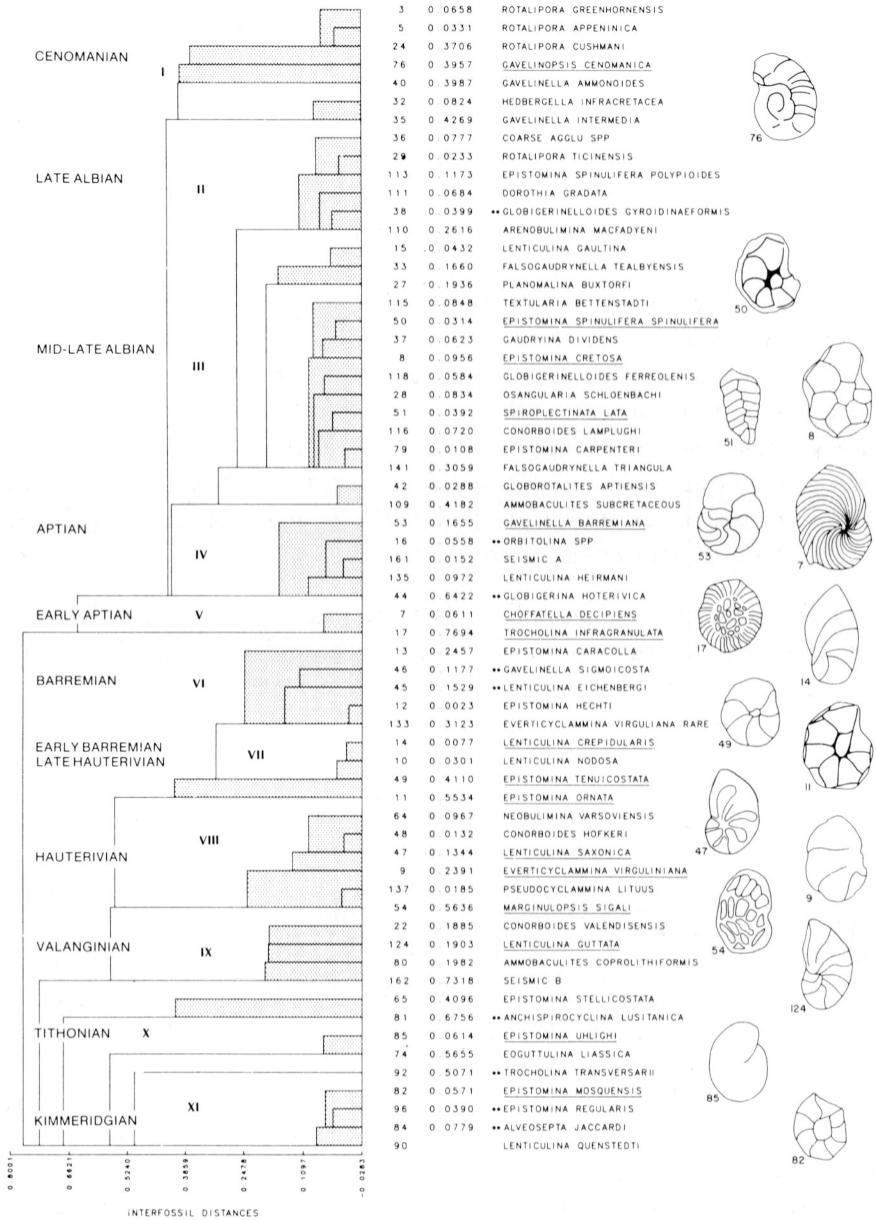

Fig. 9.19 Williamson's (1987) eleven-fold average interval zonation, using ranking and scaling for the Upper Jurassic and Lower Cretaceous foraminiferal record, northern Grand Banks. Asterisks indicate unique events.

levels. For example, *P. buxtorfi* is considered to disappear later than *R. ticinensis*, but in the zonation the order is reversed. It turns out that *R. ticinensis* is rare which leads to poor sampling for the event and that the more common *P. buxtorfi* in the wells is associated with other taxa of "older but less certain" stratigraphic position. Zones II and III, therefore, reveal strong overlap in age.

As reported earlier, large distances between successive interval zones in the scaled optimum sequence are caused by major sedimentary changes or breaks that separate the majority of events below from those occurring above it. Figure 9.19 shows large interfossil distances between zones X (Tithonian) and IX (Valanginian) and between zones VI (Barremian) and IV (Aptian). The lower of the two breaks is mainly due to the nonmarine or very shallow marine facies probably of Berriasian age, which has a paucity of shelly microfossils. This break also may be associated with a condensed limestone sequence (seismic marker A), and may be related to changes in sea level also observed in Portugal. The younger of the two large breaks is the so-called pre-Aptian unconformity associated with RASC zone V, below seismic marker B.

Because events in the RASC zonation are present in at least 4 wells, well to well correlation is relatively easy. Williamson (1987) executed both a subjective (manual) and an automated correlation. In the former exercise, boundaries of zones were placed with reference to the order of events in each well, and any event that clearly did not fit in, or accompanied a group of events it was not associated with in the scaled optimum sequence of Figure 9.19, was given less weight or ignored.

The quantitative correlation framework is based on the scaled optimum sequence. The upper part of Figure 9.20 shows depth values of RASC zones in each well. The numbers above each boundary are from a subjective interpretation of RASC results. Numbers underlying each boundary are the most likely depth for each zone calculated, using CASC. Numbers in parentheses are the local error bar estimates in meters below and above the CASC depth. As may be seen, the subjective zone depths generally are within the error ranges calculated. Although there is no easy choice between right or wrong in geological correlation, the close match of the two types of correlation is a means of model verification.

The next step is to convert the scaled optimum sequence of Figure 9.19 to a local time scale, using the ages in millions of years of

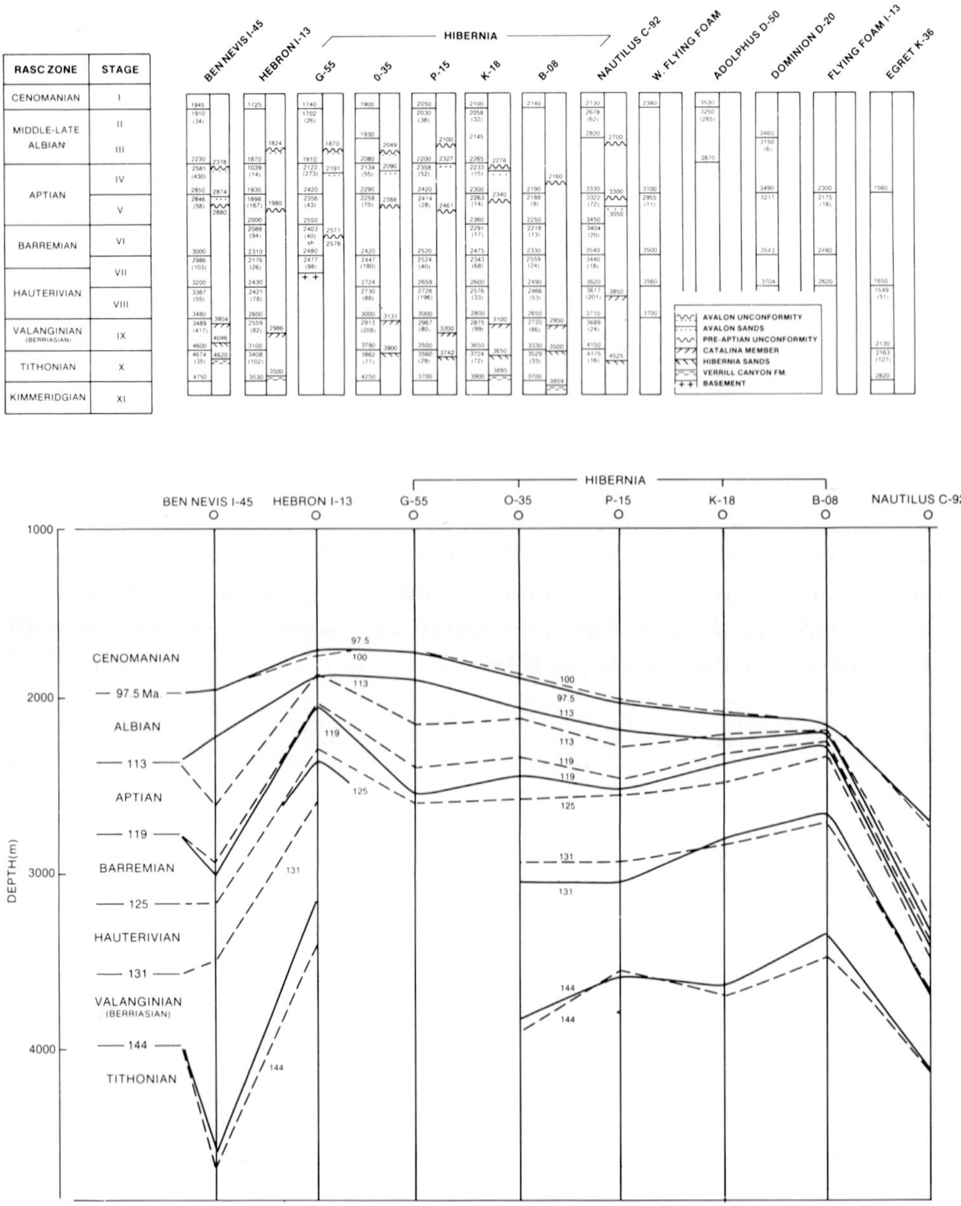

Fig. 9.20 **Upper part:** Depth values of RASC zones in northern Grand Banks wells. Numbers above each boundary are based on subjective interpretation. Below each boundary are most likely depths using the CASC method with error bars in meters in parentheses. **Lower part:** Comparison of subjective (solid line) and most likely (dashed) depths for Cretaceous isochrons in northern Grand Banks wells (after Williamson, 1987).

several good marker events. Each (CASC) age versus depth plot per well was executed with isochron boundaries for zones and the result is displayed in the lower half of Figure 9.20. The dashed lines are based on the CASC method and the solid lines are a subjective interpretation. An advantage of the CASC type of interpolation method is that it can be used for isochron cross-sections at for example 1 m.y. intervals. Such cross-sections as constructed by Williamson (cf. Williamson and Agterberg, 1990) have realistic geological properties and are of use in relating seismic cross-sections to geochronologic results and in detection of hiatuses in one or more wells. This type of application considerably enhances the role of biochronology in regional basin studies.

So far, the examples of automated correlation involved Cenozoic foraminiferal events, zones and isochrons based on the RASC zonation in Labrador and Grand Banks wells, and Early Cretaceous isochrons based on the RASC zonation in the Hibernia oilfield, off Newfoundland (Williamson, 1987). Previous analysis based on subjective age-depth data consistently confirmed results obtained by the CASC model. The error for the most likely depths of the correlation lines rarely exceeds 10%; it commonly is 2 to 5%. CASC-type of age/depth data offer the potential for significant contributions to analytical error analysis in tectonic subsidence and sedimentation calculations. RASC and CASC make subsidence analysis more objective and accurate, and easier to perform by non-paleontologists.

The procedure used to derive the objective schemes depicted in Figures 9.19 and 9.20 involved several steps. As has been seen, a prerequisite for the derivation of the correlation scheme is the successful application of the quantitative RASC program. This information was made use of by CASC together with additional information in DEP files that introduced recorded depth values (in meters) of each event in each well and age estimates of selected taxa (in Ma) from the scaled optimum sequence. The end result was a sequence of objectively derived isochrons plus standard errors.

As pointed out in Williamson and Agterberg (1990), the application of CASC to the foraminiferal data set of the Hibernia area enables a more precise chronological framework within which to consider the relationships of particular sandstone bodies, especially those bodies of economic interest. Early Cretaceous and Late Jurassic sedimentation in the study area resulted in the accumulation of a thick sandstone-shale sequence in a fluvio-deltaic setting which includes the Hibernia "Giant" oil field. Precise determination of the temporal interrelationships of the economically important sandstone sequences in the Hibernia area are depicted on Figure 9.20 (upper part). The Avalon sandstone member represents the youngest reservoir unit in this area and is thought to represent shoreline sand deposits (McKenzie, 1981). This sandstone lies within RASC zone IV and is closely associated with the 115 Ma isochron (mid-Late Aptian). CASC isochrons 105-115 Ma are "missing" or extremely condensed in some wells; for instance, in Hibernia B-08. Figure 9.20 shows how the chronologic position of this sand body fluctuates indicating a degree of diachroneity. The main Hibernia sand is markedly associated with RASC zones IX and X (Fig. 9.19) and isochrons 141-148 Ma (Fig. 9.20); i.e. from the data examined in this study the Hibernia sand sequence seems to straddle the Jurassic-Cretaceous boundary.

The results and discussions of applications of RASC and CASC in this and previous sections have a demonstrable reproducibility and furthermore allow experimentation of results using different threshold levels. Thus detailed interpretive scrutiny of results and the steps required to obtain them is possible. In addition, the methods allow development of final interpretations that allow easier communication in a scientific way to fellow workers. Biostratigraphers then are able to express numerically the uncertainty accompanying their zonation and correlation schemes. Other benefits such as the ability to deal with ever expanding databases, graphic display and data input and retrieval also are of significance. Of greater implication, however, is the potential contribution to basin history analysis. The following two examples serve to illustrate this point in more detail.

Burial history or subsidence curves can be derived and backstripped by computer to investigate the relative effects of sedimentary loading, eustacy, paleobathmetry and tectonics upon the geohistory of an area. Previously, the regional time scale or rather the biozonation which provides the chronostratigraphic control had been derived through conventional methods. The quantitative RASC and CASC approach with accompanying error analysis allows timing constraints to be imposed on basin modelling. In this way, the procedures allow important testing of hypotheses. Figure 9.21 shows this procedure as applied to Hibernia O-35. The burial history curve of Hibernia O-35 was derived using the program BURSUB described in Gradstein et al. (1989). The minimum and maximum age value as derived from error analysis of CASC was entered

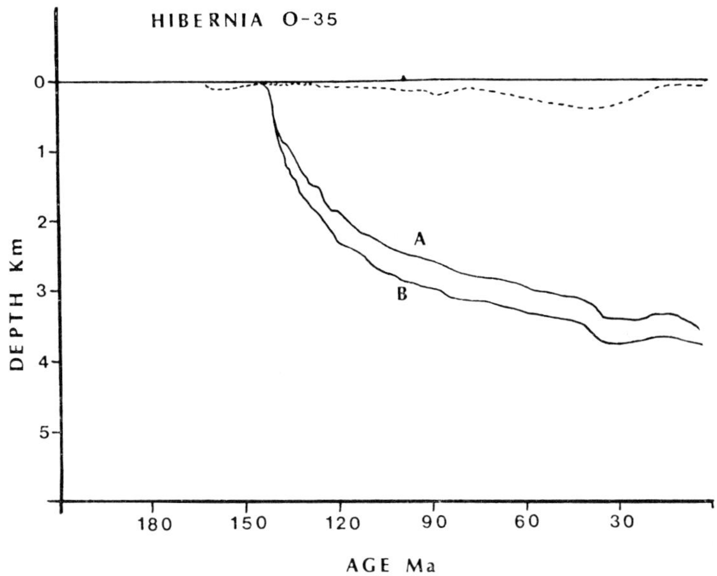

Fig. 9.21 Burial history of Hibernia O-35 well accounting for CASC derived error limits. Curve A has minimum associated error, Curve B has maximum error.

as input into the program producing the two observed subsidence curves shown. Such an approach provides an error envelope of burial curves within which maturity calculations can be made which would help determine the effect of chronology on the timing of peak generation and expulsion of hydrocarbons.

Another application stems from an idea of Van Hinte (1984) who describes the construction of synthetic seismic sections from biostratigraphic deata (the term synthetic isogram is perhaps more appropriate). The theory is quite simple and assumes that a seismic section "is an image of time stratigraphic depositional patterns" (van Hinte, 1984); and further: "...seismic section is a record of chronostratigraphic depositional and structural patterns and not a record of time transgressive lithostratigraphy" (Vail and Mitchum, 1979). Assuming that biostratigraphic correlation reflects these natural time stratigraphic markers, the correlation of routinely produced CASC isochrons (in 1 million year intervals) between suitable sections should mimic seismic sections and allow reconstruction of the general geometry of depositional sequences.

Such an isogram approach would not only enable a better integration of seismic sections and paleontological data (i.e. to determine if and where the two do not key together) but will also allow the use of seismic terminology of toplap, downlap, offlap and concordance with paleontologically derived schemes (with improved communication through a common glossary of terms). Similarly, Van Hinte (1984) believes that improved regional age calibrations of seismic sections would be apparent as would be the improvement of the correlation of regional seismostratigraphy to Vail and Mitchum's (1979) Global Cycle Charts and enhancing the understanding of the eustatics causing these changes in depositional styles (Van Hinte, 1984). The "isogram" between the wells Flying Foam and West Flying Foam (Fig. 9.22) resembles seismic sections between these wells and successfully predicted missing sections. Such correspondence is testament to the predictive ability of the overall model.

9.10 Application of CASC to Palmer's database

The application of RASC to Palmer's database for the Riley Formation in central Texas was discussed in Section 8.4. Table 8.13, showed normality test output for the Morgan Creek, White Creek, and

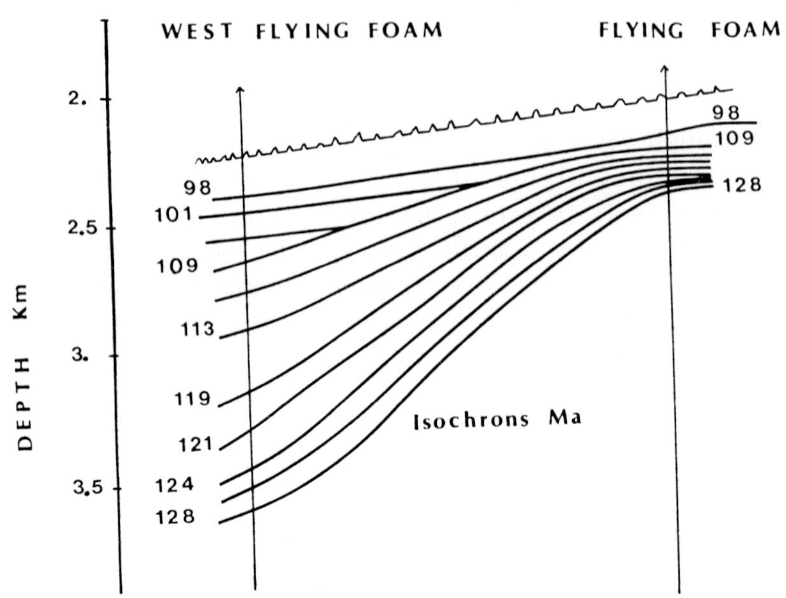

Fig. 9.22 Isochron correlation between West Flying Foam and Flying Foam wells showing unconformity and interpolations between known stratigraphic sections.

Pontotoc sections. This information was combined with distances in feet from base of section for each sample in order to create three DEP files for use in CASC. RASC distance-depth curves were constructed with "depth" measured in the stratigraphically downward direction taking new reference points at 600, 800 and 700 ft above base of section for the Morgan Creek, White Creek and Pontotoc sections, respectively. This will permit representation of CASC automated correlation lines together with Palmer's (1955) original zones and Shaw's (1964) Riley Composite Standard (R.S.T.) units for these three sections. The latter units were constructed by Shaw in order to project his final composite standard back onto each of the sections analyzed as a basis for biostratigraphic correlation and isopach mapping of time intervals (Shaw, 1964, pp. 314-316). The indirect method (called "cross-plots" in the CASC 2 module of micro-RASC, see Chapter 10) was used for spline-curve fitting. Figure 9.23 shows the 30 RASC distances of Table 8.13 for the Morgan Creek section plotted against the 18 event levels in this section. Table 9.12

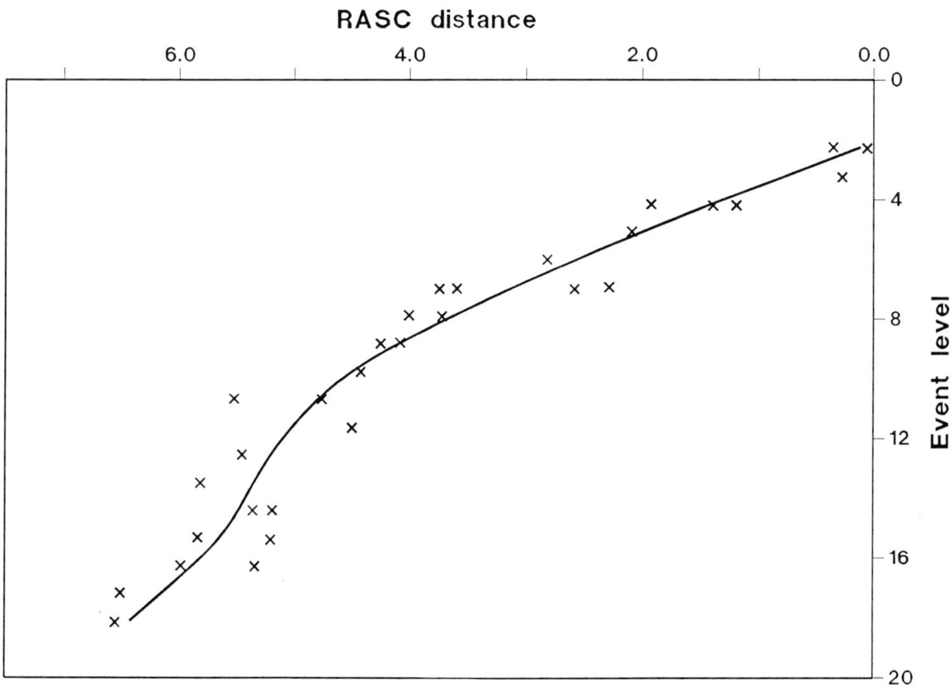

Fig. 9.23 RASC distance-event level plot for Morgan Creek section. Spline-curve is for optimum (cross-validation) smoothing factor SF = 0.382.

shows the results of cross-validation for selecting the best smoothing factor.

The smoothing factor (SF = 0.38) derived from Table 9.12 is nearly half as small as the standard deviation (= 0.71) tacitly assumed in the ordinary RASC method. This corroborates Shaw's assumption that the Morgan Creek section has better than average biostratigraphic information. The optimum smoothing factor (SF = 0.37) for the White Creek section which also is considered as better than average, is nearly the same. (Shaw selected the Morgan Creek and White Creek sections as the first two sections to be plotted against one another in the composite standard method). Cross-validation applied to the Pontotoc section, considered to have the poorest biostratigraphic information by Shaw, gave SF = 0.61.

Figure 9.23 shows the best-fitting spline-curve with SF = 0.38. It is interesting to compare this diagram with Figure 8.12 constructed for the Morgan Creek section during application of modified RASC (see Section 8.9). The spline-curve of Figure 9.23 is more realistic. Although modified RASC as a method has the capability of assigning different weights to different stratigraphic events, it is not possible to consider that the

TABLE 9.12

Smoothing factor (SF) and cross-validation value (CV) for RASC distance versus event level plot of Morgan Creek section (Fig. 9.23). **A.** Minimum and maximum SF values correspond to first monotonically increasing spline and best-fitting straight line, respectively. **B.** Zooming in on window provided optimum value SF = 0.38.

(A) SF	CV	(B) SF	CV
0.364	0.2132	0.360	0.2187
0.392	0.2049	0.370	0.2076
0.420	0.2251	0.380	0.2023
0.448	0.2499	0.390	0.2041
0.476	0.2767	0.400	0.2094
0.504	0.3050	0.410	0.2171
0.532	0.3351	0.420	0.2245
0.560	0.3661	0.430	0.2333
0.588	0.3999	0.440	0.2425
0.616	0.4309	0.450	0.2519
0.435	0.4514	0.460	0.2612

biostratigraphic information in some sections may be better than in others. In this respect, CASC has the advantage because each section can be analyzed separately in this technique and, on the average, the deviations between points and fitted curves then will be less in sections with better biostratigraphic information.

The curve of Figure 9.23 was combined with its line of observation (depth-versus-level curve) to produce the RASC distance-depth plot of Figure 9.24A. The lowest event (LO *Kormagnostus simplex*) with RASC distance equal to 6.55 in the Morgan Creek Section is not shown in this diagram which was redrafted from CASC 2 output. (The fitted curve does not extend to 6.55 because some information was lost at the edges due to use of cross-plots). Figures 9.24B and C show similar plots for the White Creek and Pontotoc sections. The standard deviations for the three curves are equal to 0.39, 0.36 and 0.61, respectively, and nearly equal to the optimum smoothing factors (see before). The three fitted curves become steeper in the downward stratigraphic direction reflecting higher sedimentation rate (cf. Shaw, 1964). Figure 9.24 can be used to determine the probable depths of specific RASC distances in the three sections for automated stratigraphic correlation. Figure 9.25 shows the results of the CASC comparison together with Palmer's zones and Shaw's R.S.T. values. The modified local error bars (± 1 SD) shown for RASC distances 2.0, 5.0, and 6.0 illustrate that the uncertainty increases in the downward direction due to the higher sedimentation rate. The three sets of lines of correlation agree closely with one another near the tops of the sections where biostratigraphic control is relatively good. It is noted that the lines for Palmer's zones were drawn through the locations of the collections with the highest stratigraphic position classified as belonging to a particular zone by Palmer (1955).

Shaw (1964) already had pointed out the good correspondence between his R.S.T. values and Palmer's zonation. In this section, it has been shown that essentially the same results were obtained automatically by using the RASC/CASC approach. Although the highest occurrences of individual fossils in the composite standard occur above the RASC distances estimated for the same fossils, while the reverse holds true for the lowest occurrences, this does not result in a systematic discrepancy during correlation (cf. Chapter 8). In fact, when there is considerable uncertainty in the biostratigraphic information (e.g. cuttings in wells during exploration), it is preferable to base the lines of correlation on average stratigraphic events instead of on subjective "total" ranges. This

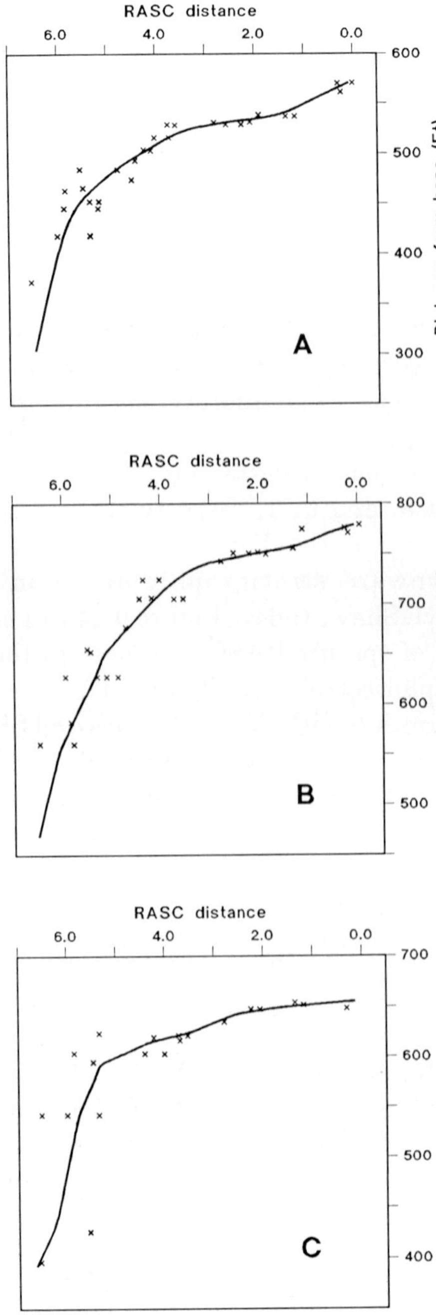

Fig. 9.24 Spline-curves for positions of RASC distance values in three sections obtained by means of indirect method. **A.** Morgan Creek section. Curve of Fig. 9.23 was combined with curve for positions of event levels according to method of Fig. 9.6. Second (cf. Fig. 9.6b) and third (cf. Fig. 9.6e) smoothing factors used were equal to 0.02 and 0.2, respectively. Final standard deviation of deviations from curve is SD = 0.390. **B.** White Creek section (SD = 0.357). **C.** Pontotoc section (SD = 0.615).

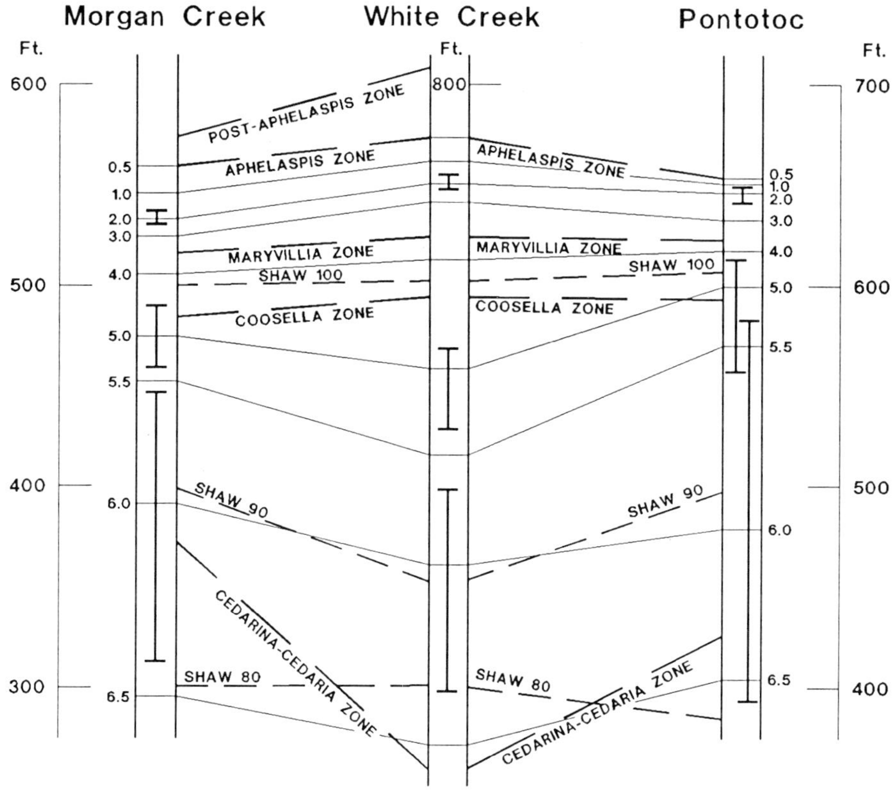

Fig. 9.25 Stratigraphic correlation of three sections by 3 methods. Palmer's (1955) zones and Shaw's (1964) R.S.T. value correlation lines are superimposed on CASC results using spline-curves of Fig. 9.24. Modified error bars extend one standard deviation on either side of probable positions for RASC distance values equal to 2.0, 5.0 and 6.0, respectively. The uncertainty of the correlation lines increases in the stratigraphically downward direction due to higher sedimentation rate.

is because, statistically speaking, the estimated average highest or lowest occurrence for a taxon is closer to its population value than the highest or lowest occurrence on the range chart can be to its population value (cf. Chapter 2).

9.11 Benthic foraminiferal zonation, central North Sea

The ranking and scaling method was used by Gradstein et al. (1988) and Agterberg and Gradstein (1988) to propose a benthic foraminiferal zonation for the Cenozoic deep marine deposits of the Central and Viking

Grabens, North Sea. Although CASC applications have not yet been published, this case-history study is interesting because it involves integration of biostratigraphic and lithostratigraphic information, seismic stratigraphy and correlation of Cenozoic hiatuses across the Atlantic Ocean.

Following the widespread deposition of Danian chalk, south of about 60°N, the North Sea Basin underwent rapid subsidence (Sclater and Christie, 1980; Gradstein and Berggren, 1981; Wood, 1981). As a result, terrigenous clastic sediments in excess of 3 km thick accumulated in the

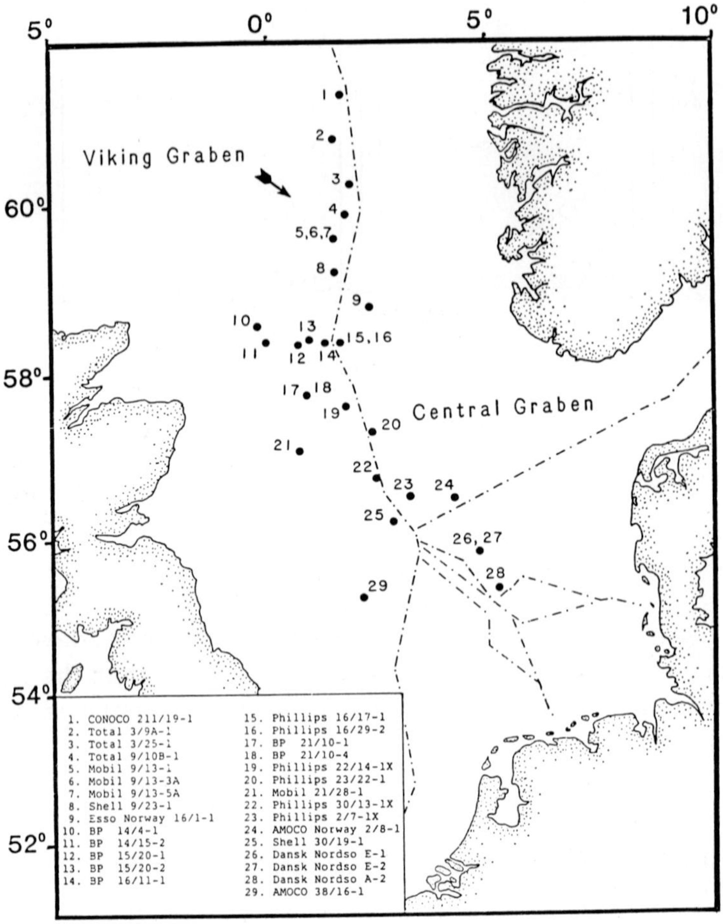

Fig. 9.26 Locations of 29 exploration wells, central North Sea.

central portion of the basin. Thickest sediments are found in the Central Graben, whereas the Viking Graben received between 2 and 3 km of sediment. Mudstones predominate, with deep marine clastic fans, like those of the Forties and Frigg oil fields developing during the early stage of Tertiary subsidence. In the Ekofisk area post-Danian olistostromes occur. By Middle Miocene time the North Sea trough had been filled, leaving a neritic environment with a predominantly calcareous benthic microfauna dominated by *Cassidulina, Elphidium*, *Fursenkoina*, and *Cibicidoides*.

The post-Danian, Paleocene through Early-Middle Miocene mudstones harbour a rich and diversified flysch-type agglutinated benthic fauna (Gradstein and Berggren, 1981), which includes over 60 taxa. Many benthic taxa show minor and some major inconsistencies in relative stratigraphic position of highest occurrence events as sampled in 29 wells (Fig. 9.26). Over 2000 cuttings samples, sidewall cores, and some core samples were analyzed, and the final analysis involves the tops of 147 benthics and relatively few planktonic taxa. The microfossil distribution data were augmented by the relative positions in the wells of physical log markers A through G as defined by A.C. Morton and R.B. Knox (personal communication, 1984).

A close look at the North Sea analytical data shows that the southern wells (blocks 21-38) contain more Oligocene-Miocene calcareous taxa, including several species of planktonics, than the northern wells which contain a more diversified Paleocene agglutinated record. This pattern of geographic differentiation was further confirmed using correspondence analysis (G.F. Bonham-Carter, personal communication, 1985). This method clarifies the spatial distribution of co-occurring taxa. There may be several reasons for this biogeographic trend, one of them being the fact that the principal deep water connection was to the north in the Norwegian Sea. The latter region does not have much of an indigenous planktonic record. Another reason is that the post-Danian, Late Paleocene-Eocene bathyal mudstone facies did not preserve much of a carbonate record, owing to diagenetic effects (Gradstein and Berggren, 1981). A third reason is climatic; apparently the transition from carbonate rich to carbonate poor rocks in Cenozoic time can be traced from south to north over the central North Sea (Ziegler, 1981). The biogeographic analysis indicates that for detailed regional studies two zonations are required, one emphasizing the northern Paleogene record and the other the southern Oliogocene-Miocene record. In this section, emphasis is on the generalized zonation which combines features from both the Central Graben and Viking Graben deep water troughs.

The generalized Cenozoic North Sea zonation uses the RASC thresholds $k_c = 8$ $m_{c1} = 1$ and $m_{c2} = 5$, which means that zonal taxa must occur in 8 or more out of 29 wells and each pair of taxa in the scaled optimum sequence in 5 or more wells. The threshold k_c reduces the original data set of 147 events to 49 (Fig. 9.27), including 8 planktonic and 25 agglutinated taxa and the log markers A-G of Knox and Morton as found in the majority of the wells studied. The dendrograms that display the interfossil distances between the ranked taxa (Fig. 9.27), are stable when RASC is run with $k_c = 9$ and 10 and $m_{c2} = 6$ and 7, which incorporate 45 and 41 taxa, respectively. In each situation the same zones are recognized.

In order to enhance the zonation with index taxa that are rare or other taxa that are thought to be potentially of such use, the RASC method allows introduction of special or unique events (UE) occurring in one or a few wells only. Twelve events were selected that occur in less than $k_c = 8$ wells, but are worth noting. These events are the highest occurrences of (from old to young) *Ammodiscus planus*, *Reticulophragmium garcilassoi*, *Bulimina trigonalis*, *Turrilina robertsi*, *Haplophragmoides* (aff.) *jarvisi*, *Adercotryma* sp. 1 (formally described as *Adercotryma agterbergi*, n.sp. by Gradstein and Kaminski, 1989), *Globigerinatheka index*, *Turrilina alsatica*, *Globigerina* ex gr. *officinalis*, *G. angustiumbilicata* and Neogene radiolarian flood. In the final RASC calculations, stratigraphic neighbors of these events are identified. A neighbor is a species that occurs in the scaled optimum sequence and also in the wells with the UE, and stratigraphically as close as possible to it. Each UE is positioned between these neighboring events in the scaled optimum sequence (cf. Section 6.8).

Eleven interval zones are recognized (Figs. 9.27 and 9.28), with the characteristic taxa listed stratigraphically in order of average

Fig. 9.27 Biozonation primarily based on agglutinated benthic foraminifers, Cenozoic, central North Sea. The scaled optimum sequence is for the average tops of 54 foraminifers and siliceous microfossils and physical log markers A-G in 29 wells. Dendrogram values are distances between events in relative time. Scaling is stratigraphically downward, in line with the study of the wells. The generalized 10-fold zonation is representative for the regional Cenozoic stratigraphy (see text). There are 11 unique events (=rare events) shown with * *. A shading pattern has been used to enhance the stratigraphically most useful parts of the dendrograms. The large interfossil distances at the top of the Danian, Late Selandian-Early Ypresian, Middle-Late Eocene, Late Oligocene-Early Miocene and Middle Miocene are sedimentary cycle boundaries (from Agterberg and Gradstein, 1988).

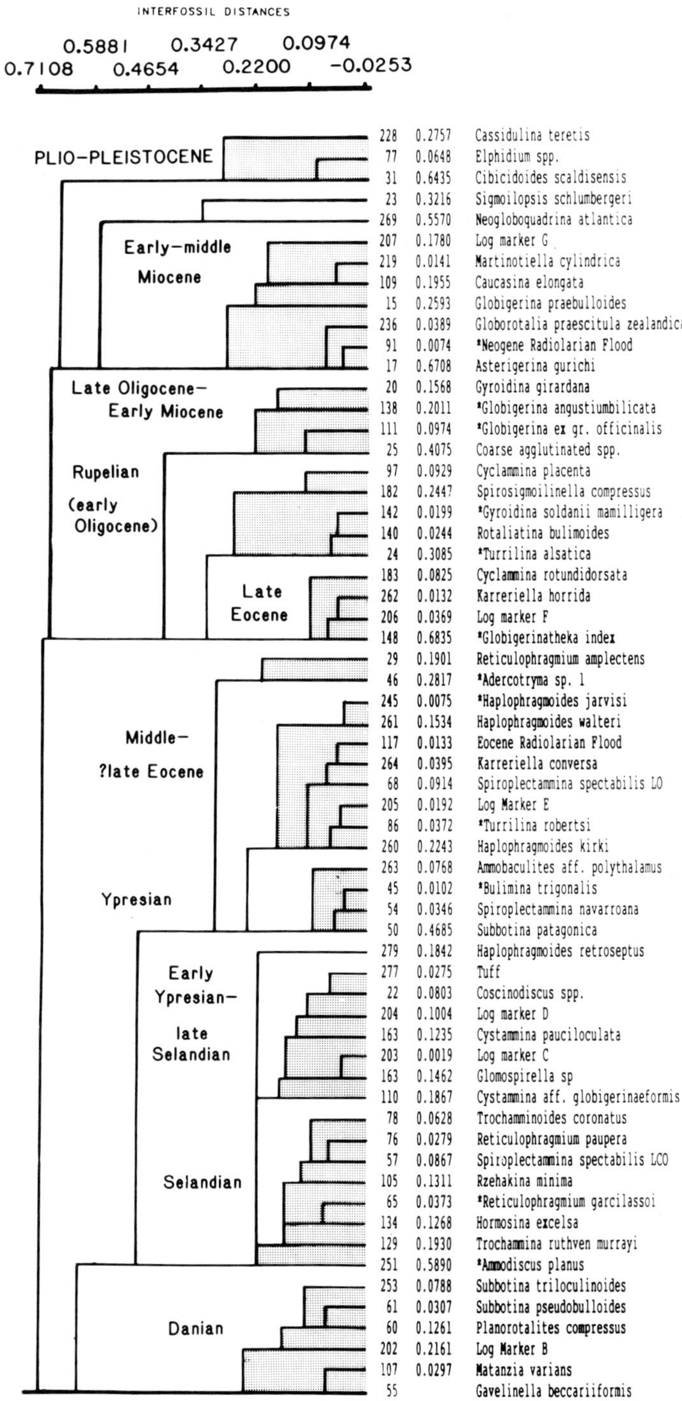
INTERFOSSIL DISTANCES
0.5881 0.3427 0.0974
0.7108 0.4654 0.2200 -0.0253
PLIO-PLEISTOCENE
Early-middle Miocene
Late Oligocene-Early Miocene
Rupelian (early Oligocene)
Late Eocene
Middle-?late Eocene
Ypresian
Early Ypresian-late Selandian
Selandian
Danian
228 0.2757 Cassidulina teretis
77 0.0648 Elphidium spp.
31 0.6435 Cibicidoides scaldisensis
23 0.3216 Sigmoilopsis schlumbergeri
269 0.5570 Neogloboquadrina atlantica
207 0.1780 Log marker G
219 0.0141 Martinotiella cylindrica
109 0.1955 Caucasina elongata
15 0.2593 Globigerina praebulloides
236 0.0389 Globorotalia praescitula zealandica
91 0.0074 *Neogene Radiolarian Flood
17 0.6708 Asterigerina gurichi
20 0.1568 Gyroidina girardana
138 0.2011 *Globigerina angustiumbilicata
111 0.0974 *Globigerina ex gr. officinalis
25 0.4075 Coarse agglutinated spp.
97 0.0929 Cyclammina placenta
182 0.2447 Spirosigmoilinella compressus
142 0.0199 *Gyroidina soldanii mamilligera
140 0.0244 Rotaliatina bulimoides
24 0.3085 *Turrilina alsatica
183 0.0825 Cyclammina rotundidorsata
262 0.0132 Karreriella horrida
206 0.0369 Log marker F
148 0.6835 *Globigerinatheka index
29 0.1901 Reticulophragmium amplectens
46 0.2817 *Adercotryma sp. 1
245 0.0075 *Haplophragmoides jarvisi
261 0.1534 Haplophragmoides walteri
117 0.0133 Eocene Radiolarian Flood
264 0.0395 Karreriella conversa
68 0.0914 Spiroplectammina spectabilis LO
205 0.0192 Log Marker E
86 0.0372 *Turrilina robertsi
260 0.2243 Haplophragmoides kirki
263 0.0768 Ammobaculites aff. polythalamus
45 0.0102 *Bulimina trigonalis
54 0.0346 Spiroplectammina navarroana
50 0.4685 Subbotina patagonica
279 0.1842 Haplophragmoides retroseptus
277 0.0275 Tuff
22 0.0803 Coscinodiscus spp.
204 0.1004 Log marker D
163 0.1235 Cystammina pauciloculata
203 0.0019 Log marker C
163 0.1462 Glomospirella sp
110 0.1867 Cystammina aff. globigerinaeformis
78 0.0628 Trochamminoides coronatus
76 0.0279 Reticulophragmium paupera
57 0.0867 Spiroplectammina spectabilis LCO
105 0.1311 Rzehakina minima
65 0.0373 *Reticulophragmium garcilassoi
134 0.1268 Hormosina excelsa
129 0.1930 Trochammina ruthven murrayi
251 0.5890 *Ammodiscus planus
253 0.0788 Subbotina triloculinoides
61 0.0307 Subbotina pseudobulloides
60 0.1261 Planorotalites compressus
202 0.2161 Log Marker B
107 0.0297 Matanzia varians
55 Gavelinella beccariiformis

disappearance. A more detailed zonation (which uses lower k_c and m_{c2} values) is possible for local correlation. Gradstein et al. (1988) have shown the approximate relation of the new zonation to several standard planktonic zones that can be recognized in the central North Sea, and to the regional foraminiferal zonation for the circum-central North Sea by King (1983).

Approximately 30 taxa of agglutinated benthic Foraminifera have distinct stratigraphic ranges in the central North Sea wells studied. A comprehensive summary for the ranges is given in the range chart of Agterberg and Gradstein (1988, pp. 21-26). So-called log markers are valuable in the numerical stratigraphic analysis of the central North Sea. These log markers are thought to be chronostratigraphic in nature and according to Morton and Knox (personal communication, 1984) correspond to the following approximate levels: marker G top Middle Miocene;

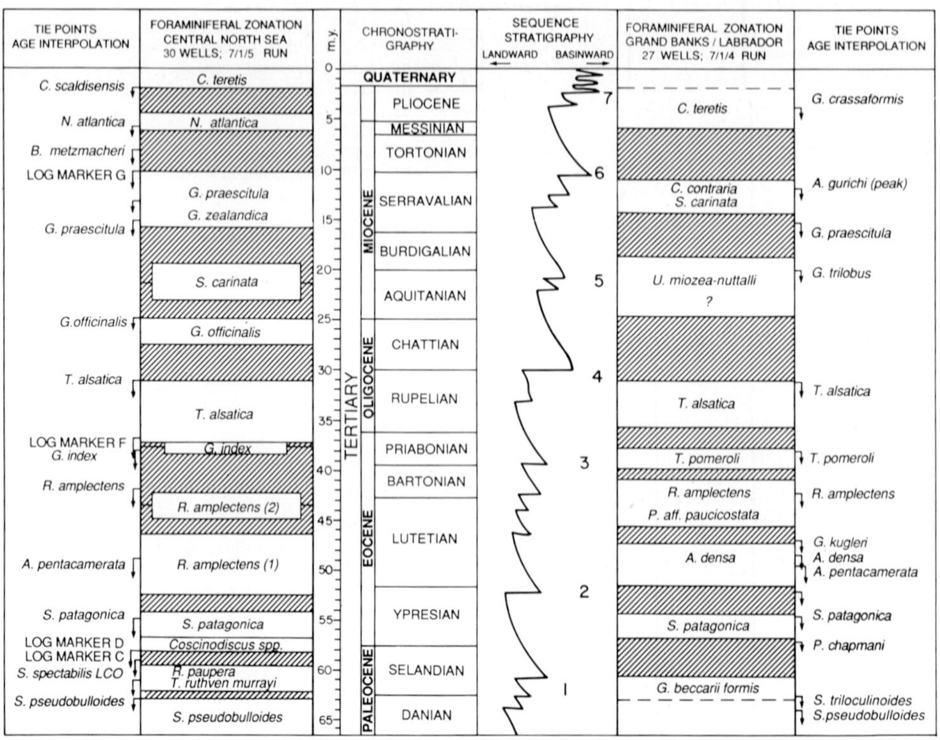

Fig. 9.28 Relation between global model for (seismic) sequences stratigraphy (Vail, Hardenbol and coworkers, pers. commun., 1986) and hiatuses based on scaling in time of the RASC zonations for the central North Sea shown in Fig. 9.27 and the Canadian Atlantic margin shown in Fig. 6.2. Age tiepoints for scaling are shown on each side of zonation in time. For explanation see text.

marker F top Upper Eocene; marker E top Lower Eocene; marker D top Sele Formation (or equivalent); marker C base Sele Formation (top Paleocene); marker B top Ekofisk Formation (top Lower Paleocene); and marker A top Cretaceous. The log picks were expected to vary slightly in stratigraphic position relative to the foraminiferal events in the wells and were treated as "fossil events" in the calculations. Figure 9.27 shows the calculated average stratigraphic position of these events. There is good agreement between the ages assigned by Knox and Morton to the log picks and the ages assigned to the accompanying zones.

Log marker A is always found at the level with *Globotruncana* below the Danian zone (not shown in Fig. 9.27). Log marker B on average is in the Danian, rather than at the top as suggested. Log markers C and D are in the *Coscinodiscus* zone that delineates the ash-series that straddles the Paleocene-Eocene boundary.

The top of log marker E is given as top of Lower Eocene in agreement with its average occurrence slightly above the *Subbotina patagonica* zone, Ypresian. The only serious exception to this average position was found to be in well 23/22-1 where E occurs with Danian planktonics. The latter may be reworked.

Log marker F occurs in the *Globigerinatheka index* interval, Upper Eocene. Log marker G fits well at the top of the *Globorotalia praescitula-G. zealandica* zone, Lower-Middle Miocene. An interesting observation is that the markers F and G are associated in Figure 9.27 with breaks in the scaled optimum sequence. These breaks are recognized by large interfossil distances between two events in adjacent zones. The large distance means that there is little or no cross-over in position between the events in the two successive zones. Such a situation is expected where there is a stratigraphic section missing between the zones or a sudden change in facies (may also be due to a hiatus). The latter is the case between the Danian and Selandian zones, where carbonates are replaced by mudstones and sands.

In computer runs without unique events such as *Globigerinatheka index*, log marker F falls exactly at the large interfossil distance between the *Rotaliatina bulimoides* zone and the underlying *Reticulophragmium amplectens* zone. Log marker F marks the Eocene-Oligocene boundary in the zonation and the large interfossil distance associated with it suggests a hiatus involving the uppermost Eocene to Lower Oligocene. In runs including unique events, the *Globigerinatheka* index-log marker F events are sandwiched between the *R. bulimoides* and *R. amplectens* zones, indicating that these events are closely tied to the position of a hiatus in many of the wells. Another break in the record is suggested at the base of the *Globorotalia praescitula-G. zealandica* zone, Oligocene-Miocene boundary.

Log marker G reflects an Upper Miocene hiatus in the wells in the central North Sea grabens. Few of the wells studied show unequivocal fossil evidence for an upper Miocene interval (based on rare planktonic foraminifers including *Neogloboquadrina atlantica*, left coiling and *N. acostaensis*), and no RASC zone stands out.

In order to further investigate the stratigraphic extent of these breaks or hiatuses in the central North Sea, a study was made of the most likely sequence of events in Figure 9.27, but now scaled in linear time. First, 16 North Sea age tiepoints were assumed for a subset of fossil events in the most likely (scaled) sequence with ages in million of years interpolated from recent geochronological literature. Details on the age assignments were given in Section 9.8 and in Gradstein et al. (1988). Each of these events also has a distance from the origin (top) in the scaled optimum sequence. Next, the best fit was calculated between this series of events scaled both in RASC units and in linear time using cubic spline fitting. The resulting function of age versus distance may be used to convert the RASC distances of all events in the scaled optimum sequence to millions of years. The result is shown in Table 9.13, which also gives the original tiepoints and their assumed ages. The spline-curve was only slightly smoothed and passes nearly through all points with SF (Smoothing Factor) of 0.37. As a result of this simple operation we can now stretch the North Sea zones can be stretched in linear time and with the lower and upper limits of the zones expressed in millions of years. This is shown in Figure 9.28 (left) which also gives again the input ages of the tiepoints.

The scaling in linear time operation enhances the detection of breaks or hiatuses that may occur in the central North Sea. Not unexpectedly, only about 50% of Cenozoic time is represented by zones and their representative sediments. For resolution and extent in time of the breaks it is useful to test this local time scale against similar ones based on other fossil groups, like nannofossils and dinoflagellates, but such data is not available at present. Individual error in age calibration does affect the position of zonal boundaries, but not the general trend. On a local scale one

TABLE 9.13

Interpolated ages of the events in the central North Sea zonation of Fig. 9.27, using cubic spline fitting for the age-RASC distance relationship of a subset of events (shown as *) for which age estimates (in parentheses) are available in the literature.

Fossil	RASC distance	Interpolated age (m.y.)	Fossil	RASC distance	Interpolated age (m.y.)
31 *	0.3408	1.9 (2)	117	6.0667	46.7
23	0.9841	4.2	68	6.1209	47.5
269 *	1.3056	6.1 (6)	264	6.1385	47.7
207 *	1.8527	10.9 (11)	205	6.2139	48.8
219	2.0405	12.8	86 *	6.2311	49.1 (49)
109	2.0548	12.9	260	6.2629	49.5
15 *	2.2503	14.9 (15)	50 *	6.5585	53.4 (55)
236 *	2.5099	17.2 (17)	263	6.6153	54.0
91	2.5529	17.6	54	6.6995	54.8
17	2.5561	17.6	45	7.0680	57.1
20	3.2268	22.7	279	7.1774	57.5
138	3.3777	23.7	277	7.3616	57.9
111 *	3.5783	24.9 (25)	204 *	7.4336	58.1 (58)
25	3.7001	25.5	22	7.4353	58.1
97	4.0920	27.7	136	7.5829	58.3
182	4.1952	28.5	110	8.0943	59.1
142	4.4267	30.9	203 *	8.1837	59.2 (59)
140 *	4.5548	32.1 (31)	163	8.2012	59.2
24	4.5939	32.6	134	8.2882	59.4
183	4.7408	34.6	78	8.4590	59.7
262	4.8159	35.6	76	8.5015	59.8
206 *	4.8528	36.0 (37)	105	8.5021	59.8
148 *	4.8713	36.2 (38)	57 *	8.5921	60.0 (60)
29 *	5.5634	40.7 (40)	65	8.6336	60.1
46	5.6672	41.6	129	9.0155	60.9
245	5.9375	44.9	251	9.0928	61.1
261	6.0104	45.9	253	9.7575	62.8
			61 *	9.8425	63.0 (63)

can expect improvements from more corroboration on the average disappearance in time of the events used as tiepoints. For example, it is assumed that *R. bulimoides* and *T. alsatica* disappear near 30 Ma, at the end of the Rupelian, but this needs verification in more well sites. Attention is drawn to the fact that the *G. index* extends over the Eocene-Oligocene boundary, although *G. index* itself is Eocene. This taxon was found reworked in Oligocene-Neogene deposits in several wells.

In order to emphasize breaks of a more general nature, the foraminiferal zonation using RASC on the record in 27 Labrador and Newfoundland offshore wells (see Fig. 6.27) was added to Figure 9.28, also stretching it in linear time. Several age tiepoints, like *S. pseudobulloides*

(63 Ma), *S. patagonica* (55 Ma), *T. robertsi* (49 Ma). *R. amplectens* (40 Ma), and *T. alsatica* (30 Ma) are in common with the central North Sea zonation. Again, large Eocene, Oligocene and Miocene hiatuses stand out.

Haq et al. (1987) have related a global seismic-sequence stratigraphy to chronostratigraphy. The sequences are composed of periods of offlap (basinward movement of the shoreline) and onlap (landwards movement of the shoreline). These sequences are thought to reflect global changes in sealevel. If rate of sealevel fall exceeds rate of basin subsidence, such events can exert considerable influence on shallow deep marine clastic or carbonate deposition. A relative shift seaward of the shoreline may disrupt sedimentation in shallow basins, and lead to a hiatus. In deeper water, more mass-flow sediments may occur causing local deposition or erosion. The sequences were adjusted to conform to the linear time scale used for the tiepoints (Berggren et al., 1985), and the North Sea and Canadian Atlantic margin zones and the seismic sequence stratigraphy were placed side by side (Fig. 9.28). Not unexpectedly, the more prominent basinward shifts of the shoreline, for convenience numbered 1 through 7, approximately coincide in time with breaks in the zonations. As discussed earlier, large breaks in the scaled optimum sequence of fossil events are likely to match hiatuses or sudden changes in facies. Major shifts in position of shorelines influence the sediment supply as well as erosion and can be expected to exert control over the sedimentary sequences in the Canadian offshore and Central and Viking Grabens. The latter, in turn, influence the zonal boundaries of fossil assemblages.

Shift 1 in the North Sea may have coincided with replacement of the Danian carbonates (*S. pseudobuloides* zone) by clastics (*R. paupera* – *T. ruthven murrayi* zone). Shift 2 also is seen on the Rockall and Grand Banks and may tie to a late Ypresian hiatus. Shifts 3 and 4 appear associated with breaks in the uppermost Eocene and Oligocene, which caused major disruptions in the fossil sequence both in Labrador and North Sea wells. It is not easy to explain why in the deep central North Sea a Late Eocene hiatus occurs. The mid-Oligocene shift 4 event appears to have affected the deeper North Sea less than the shallower beds offshore Canada. This is to be expected. Shift 5 does not match an Early Miocene hiatus but events 6 and 7 bracket a Late Miocene break. In general, as expected, the extent of the hiatuses and the presumed influence of sea level changes increases stratigraphically upward with decreasing rate of subsidence and sedimentation.

Further study, particularly geological, will clarify the relation between the global seismic-sequence stratigraphy and regional sedimentary and paleontological history. It should be emphasized that the RASC biochronology expresses average depositional sequence trends, not necessarily reflected in each single well section. This brief case history on the North Sea biostratigraphy and (local) biochronology has highlighted the use of numerical methods to advance the application of the fossil record in subsurface geology.

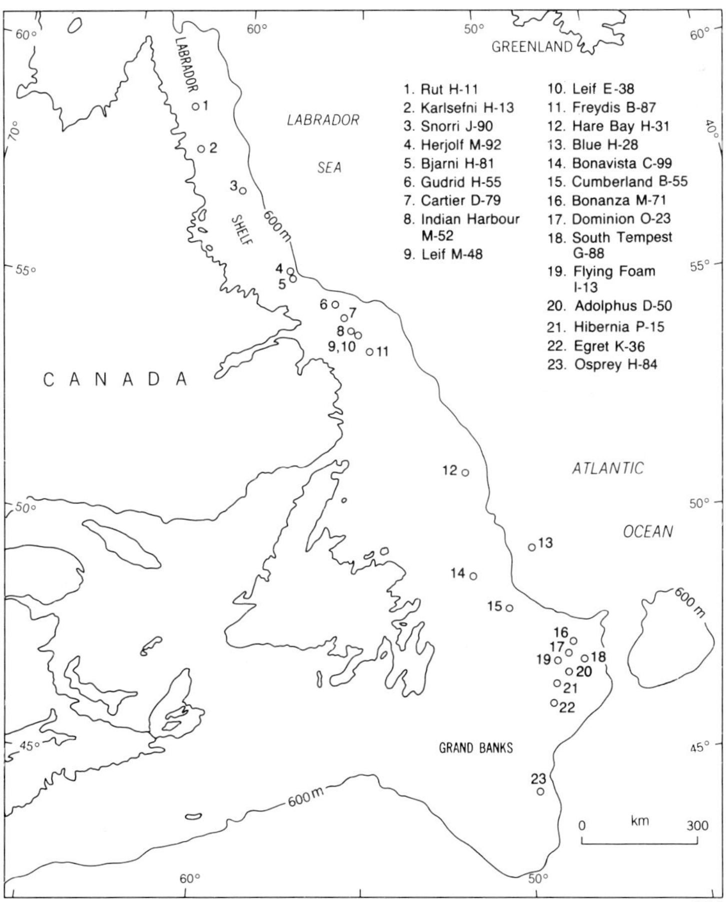

Fig. 9.29 Location map of the Labrador Shelf and Grand Banks wells used by D'Iorio and Agterberg (1989).

9.12 Integration of foraminiferal and dinoflagellate datasets, Labrador Shelf – Grand Banks

D'Iorio (1986, 1987, 1988) and D'Iorio and Agterberg (1989) have dealt with the problem of using RASC and CASC for combined analysis of microfossils belonging to different families. A RASC biozonation and CASC correlation lines between 23 wells on the Labrador Shelf and Grand Banks were based on the positions of highest occurrences of palynomorph or foraminiferal taxa in wells and on their positions in a regional biozonation model. The locations of these wells are shown in Figure 9.29.

The Cenozoic biozonation was established using the ranking and scaling method (RASC). The automated correlation technique of CASC provided an effective method of identifying patterns of sediment accumulation by tracing biozones through the wells of the study area. The

TABLE 9.14

Names and ages of biozones of Fig. 9.30 and list of boundary events used to trace RASC biozones in CASC multi-well comparison.

Zone	Age of Zone	Name of Marker Event
I	Paleocene	*Gavelinella beccariiformis*
II	Early Eocene	*Subbotina patagonica*
III	Early Middle Eocene	*Acarinina densa*
IV	Late Middle Eocene	*Plectofrondicularia aff. paucicostata*
V	Late Eocene	*Reticulophragmium amplectens*
VI	Late Eocene	*Turborotalia pomeroli*
VII	Oligocene	*Turrilina alsatica*
VIII	Late Oligocene to Early Miocene	*Uvigerina ex. gr. miozea-nuttali*
IX	Middle Miocene	*Spiroplectamina carinata*
X	Middle to Late Miocene	*Asterigerina gurichi*
XI	Pliocene-Pleistocene	*Cassidulina teretis*

Boundary	Event Number	Event Name
I - II	52	*Acarinina soldadoensis*
II - III	37	*Acarinina* aff. *pentacamerata*
III - IV	90	*Acarinina densa*
IV - V	29	*Reticulophragmium amplectens*
V - VI	263	*Ammobaculites aff. polythalamus*
VI - VII	259	*Ammodiscus latus*
VII - VIII	24	*Turrilina alsatica*
VIII - IX	21	*Guttulina problema*
IX - X	67	Scaphopod sp. 1
X - XI	17	*Asterigerina gurichi*

CASC estimated depths of Cenozoic epoch boundaries agree well with similar results determined from biostratigraphy. Slight differences between the two estimates reflected systematic deviations at the top and bottom of the scaled optimum sequence. These trends were consistent with results of a frequency distribution analysis of foraminiferal last occurrences by means of modified RASC. Histograms of the frequency distributions of events were found to be useful tools for identifying potential marker events. These results could be used to trace vertical migration of biostratigraphic events.

The construction and interpretation of the integrated biozonation model has been described by D'Iorio (1987). The eleven zones identified are named in Table 9.14. The discussion in this section will be restricted to automated correlation (CASC multi-well comparison). The scaled optimum sequence of Figure 9.30 was used as the correlation standard of this study. Preliminary regional age estimates of events for CASC were the same as those determined by D'Iorio (1987). The smoothing factors were optimized with the cross-validation technique. The smoothing factors which are equal to the standard deviations of residuals used in the CASC analysis of individual wells are listed in Table 9.15.

Labrador Shelf

Eleven wells were included in the Labrador Shelf group, the southernmost one being Freydis. RASC biozones were correlated between wells by tracing the depths of zone boundary events. These events were chosen from Figure 9.30 and are listed in Table 9.14. When an event is not found in a well, its expected depth was estimated from its RASC position. The depths of the zone boundaries are listed in Table 9.15 and plotted in Figure 9.31 (left side) for the Labrador Shelf wells.

The zone boundaries in the youngest or oldest parts of the wells may not always be shown because of either the scarcity of data points, or the specific shapes of the spline curves. The Bjarni, Cartier, Leif M-48 and Freydis wells show more closely spaced zone boundaries, probably indicating a lower sediment accumulation rate. This is in contrast with the northern wells, which appear to have greater sediment accumulation rates.

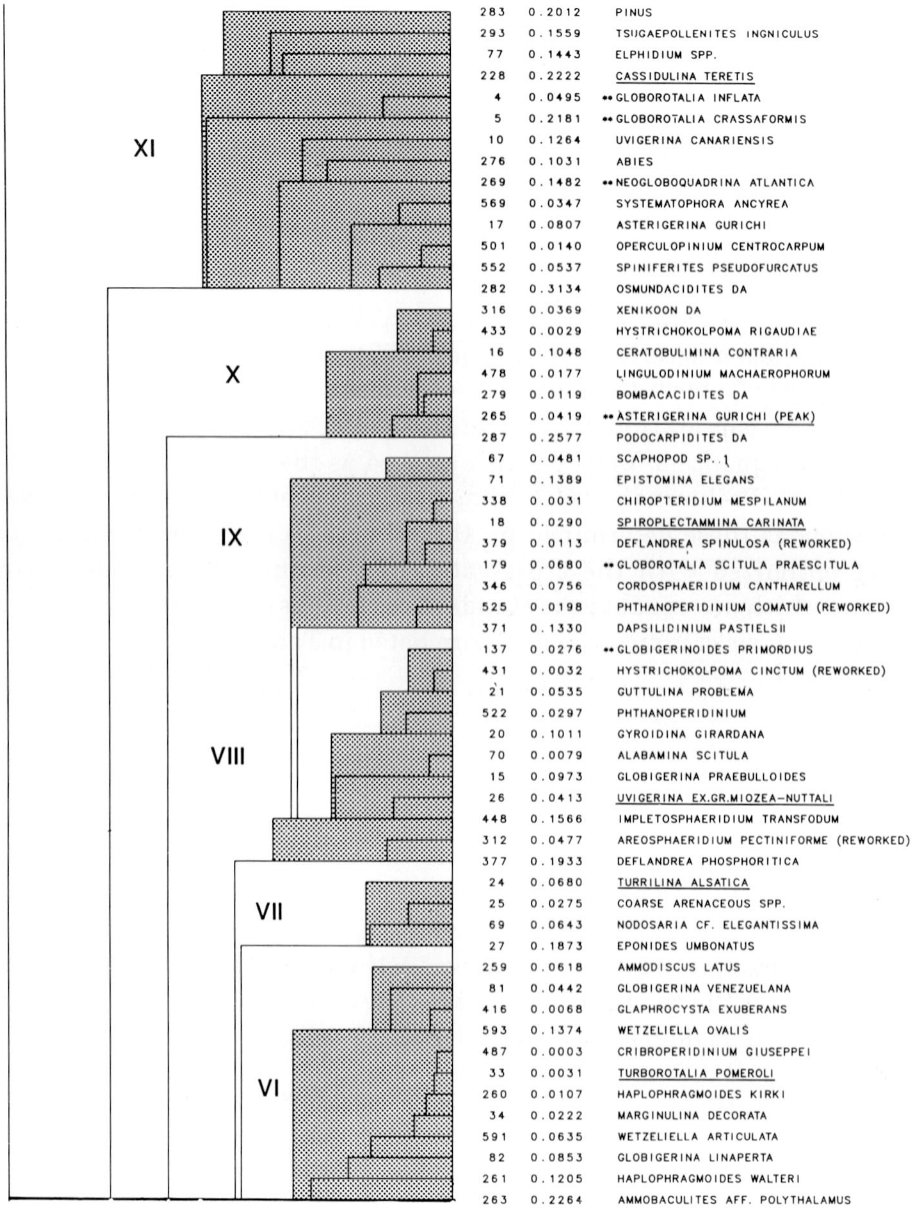

Fig. 9.30 Biozonation model of the Cenozoic of the Labrador Shelf and Grand Banks based on an integrated databank of foraminifers, dinoflagellates, and spores and pollen.

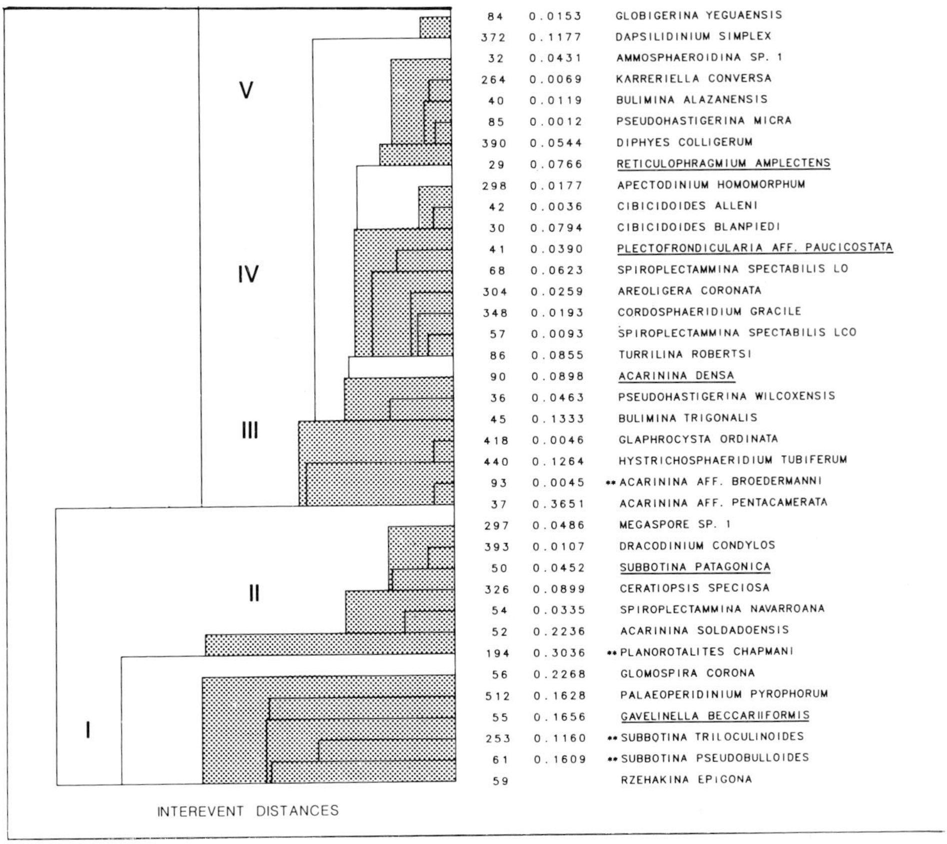

Fig. 9.30 (continued)

Zone boundaries reveal greater than average sediment accumulation rates in zone IX in the Karlsefni and Herjolf wells, zone VIII in the Gudrid well and zone VI in the Indian Harbour well. Low sediment accumulation rates or an unconformity would explain the mutual proximity of zone boundaries in Snorri from zones VI to VIII, and in Freydis from zones II to VII.

TABLE 9.15

CASC depths of biozone boundaries of Table 9.14. Errors are standard deviations.

Well Name	I - II	II - III	III - IV	IV - V	V - VI
Rut H-11				*3.16 ± 0.63	*3.00 ± 0.63
Karlsefni H-13	3.71 ± 0.90	*2.94 ± 0.20	*2.91 ± 0.15	2.76 ± 0.09	*2.74 ± 0.10
Snorri J-90	*2.95 ± 0.06	*2.53 ± 0.51	*2.48 ± 0.21	2.15 ± 0.05	2.14 ± 0.03
Herjolf M-92		*1.99 ± 0.19	*1.97 ± 0.18	1.76 ± 0.21	1.72 ± 0.22
Bjarni H-81	*1.99 ± 0.12	*1.83 ± 0.13	*1.82 ± 0.12	1.67 ± 0.08	*1.65 ± 0.08
Gudrid H-55	*2.37 ± 0.39	2.10 ± 0.16	2.08 ± 0.16	1.90 ± 0.11	*1.88 ± 0.09
Cartier D-70		*1.77 ± 0.08	*1.76 ± 0.08	1.55 ± 0.26	1.50 ± 0.17
Indian Harbour M-52	2.99 ± 0.07	2.53 ± 0.27	*2.47 ± 0.35	2.28 ± 0.32	2.16 ± 0.47
Leif M-48		*1.69 ± 0.07	*1.67 ± 0.09	1.59 ± 0.03	1.58 ± 0.04
Leif E-38					
Freydis B-87	*1.54 ± 0.10	*1.39 ± 0.12	*1.38 ± 0.11	1.28 ± 0.05	*1.27 ± 0.07
Hare Bay E-21	*3.06 ± 0.12	*2.88 ± 0.11	*2.86 ± 0.10	2.27 ± 0.67	2.18 ± 0.26
Blue H-28		4.74 ± 0.05	4.73 ± 0.05	4.43 ± 0.64	*4.29 ± 0.55
Bonavista C-99		*3.46 ± 0.13	3.44 ± 0.13	3.21 ± 0.37	3.12 ± 0.25
Cumberland B-55	3.58 ± 0.18	*3.31 ± 0.10	3.29 ± 0.12	2.89 ± 0.28	*2.81 ± 0.28
Bonanza M-71		3.31 ± 0.18	3.28 ± 0.18	2.90 ± 0.40	2.78 ± 0.63
Dominion O-23		2.50 ± 0.44	2.44 ± 0.29	1.94 ± 0.21	1.89 ± 0.22
South Tempest G-88		2.30 ± 0.11	2.27 ± 0.13	1.71 ± 0.25	1.59 ± 0.44
Flying Foam I-13		*1.96 ± 0.15	*1.94 ± 0.18	1.67 ± 0.16	1.62 ± 0.23
Adolphus D-50	*2.63 ± 0.10	2.28 ± 0.45	2.16 ± 0.46	1.78 ± 0.13	1.69 ± 0.34
Hibernia P-15				1.25 ± 0.19	1.20 ± 0.10
Egret K-36					
Osprey H-84				*0.76 ± 0.06	*0.75 ± 0.06

Well Name	VI - VII	VII - VIII	VIII - IX	IX - X	X - XI
Rut H-11	2.24 ± 0.08	*2.16 ± 0.07	*1.74 ± 0.52	*1.26 ± 0.51	0.76 ± 0.52
Karlsefni H-13	*2.31 ± 0.37	*2.15 ± 0.08	*2.00 ± 0.19	1.22 ± 1.79	*0.69 ± 0.11
Snorri J-90	*2.03 ± 0.14	*1.83 ± 0.12	1.73 ± 0.04	1.50 ± 0.68	*1.00 ± 0.97
Herjolf M-92	1.39 ± 0.15	*1.29 ± 0.12	*1.12 ± 0.28	0.60 ± 0.25	
Bjarni H-81	*1.42 ± 0.19	1.21 ± 0.34	0.87 ± 0.23	0.64 ± 0.58	
Gudrid H-55	1.78 ± 0.06	1.72 ± 0.11	0.70 ± 0.35	*0.58 ± 0.11	0.51 ± 0.10
Cartier D-70	1.31 ± 0.15	1.20 ± 0.12	0.99 ± 0.55	0.75 ± 0.15	*0.66 ± 0.09
Indian Harbour M-52	1.46 ± 0.36	1.18 ± 0.15	*0.96 ± 0.42	*0.65 ± 0.22	0.58 ± 0.04
Leif M-48	*1.20 ± 0.18	1.15 ± 0.08	1.01 ± 0.17	0.77 ± 0.20	*0.56 ± 0.36
Leif E-38			0.57 ± 0.80	0.45 ± 0.18	0.39 ± 0.06
Freydis B-87	*1.14 ± 0.10	*0.87 ± 0.51	0.67 ± 0.41	0.51 ± 0.19	
Hare Bay E-21	1.56 ± 0.22	1.35 ± 0.25	*1.14 ± 0.05	*1.08 ± 0.10	*0.81 ± 0.41
Blue H-28	*3.70 ± 0.15	*3.52 ± 0.21	*3.22 ± 0.31	*2.96 ± 0.29	*2.69 ± 0.35
Bonavista C-99	2.37 ± 0.15	1.95 ± 0.88	1.21 ± 0.49	*0.93 ± 0.13	0.82 ± 0.21
Cumberland B-55	2.22 ± 0.21	1.73 ± 0.90	*1.10 ± 0.37	*0.75 ± 0.16	0.63 ± 0.16
Bonanza M-71	1.62 ± 0.11	*1.52 ± 0.09	1.39 ± 0.08	*1.31 ± 0.08	1.23 ± 0.08
Dominion O-23	1.30 ± 0.33	1.05 ± 0.29	0.77 ± 0.54	*0.57 ± 0.12	0.44 ± 0.31
South Tempest G-88	1.29 ± 0.17	1.17 ± 0.13	*0.96 ± 0.13	*0.82 ± 0.05	0.76 ± 0.06
Flying Foam I-13	1.08 ± 0.32	0.86 ± 0.28	*0.61 ± 0.21	*0.42 ± 0.29	0.31 ± 0.10
Adolphus D-50	1.22 ± 0.17	0.97 ± 0.19	*0.54 ± 0.15	*0.46 ± 0.03	*0.38 ± 0.08
Hibernia P-15	0.99 ± 0.05	0.81 ± 0.35	*0.43 ± 0.29	*0.27 ± 0.07	
Egret K-36	*0.51 ± 0.05	0.48 ± 0.03	0.43 ± 0.08	*0.34 ± 0.15	
Osprey H-84	*0.62 ± 0.12	*0.54 ± 0.12	*0.43 ± 0.10	*0.37 ± 0.08	0.32 ± 0.09

*The event used as a boundary indicator was not observed

Grand Banks

The Grand Banks group consists of twelve wells, the northernmost one being Hare Bay. The Egret K-36 well is not included in the correlation chart because it is shallow and has a very condensed section.

The zone boundary events listed in Table 9.14 also were traced in the Grand Banks wells and plotted in Figure 9.31 (right side). The depth of the boundaries and their respective local error estimates are presented in Table 9.15.

A noticeable feature of Figure 9.31 (right side) is the thickening of zone VI in the Bonanza well, suggesting a higher sediment accumulation

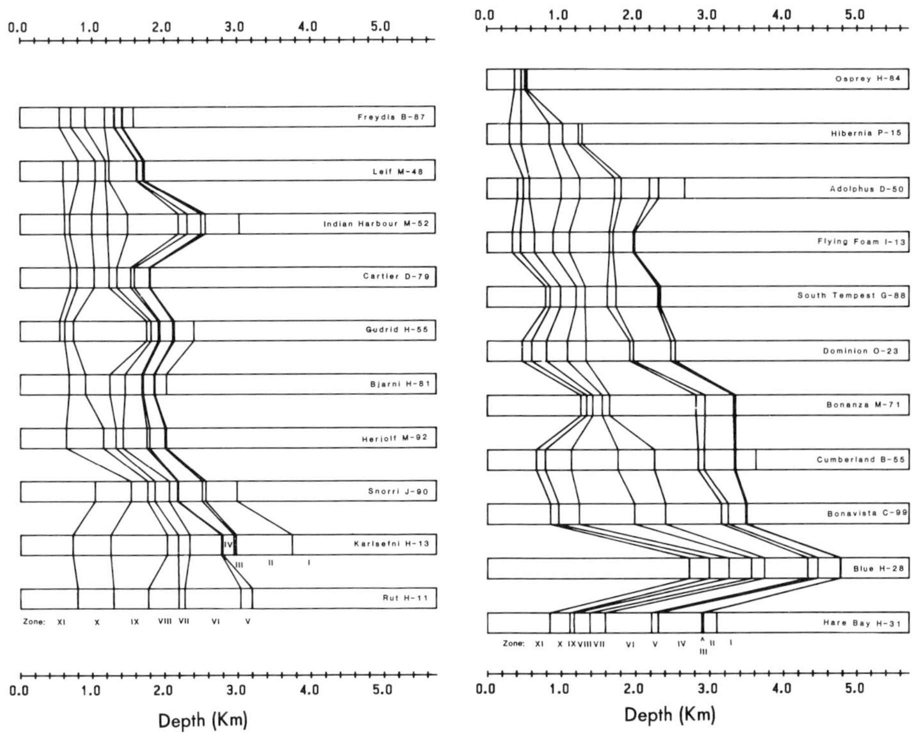

Fig. 9.31 Biozone correlation chart of the Labrador Shelf wells (left side) and the Grand Banks wells (right side). The zone boundaries are given in Table 9.14.

rate. The Osprey well exhibits relatively more closely spaced zone boundaries than other wells; this is presumably due to its more distant position from the terrigenous sediment supply (see Fig. 9.29). The Blue well shows all zones at greater depths to the sea floor.

D'Iorio (1986) has shown that, for Cenozoic quantitative stratigraphy of the Labrador Shelf - Grand Banks, combining different families of microfossils resulted in biozonations of improved resolution. The results described in this section illustrate that the RASC biozones can be traced between all wells of the study area and that the CASC method effectively identifies patterns of sediment accumulation rates. In D'Iorio and Agterberg (1989) it also was shown that the CASC depth estimates of Cenozoic age boundaries are consistent with the correponding boundary depths assigned in the Atlas of the Labrador Sea (Srivastava, Editor, 1986).

CHAPTER 10

COMPUTER PROGRAMS FOR RANKING, SCALING AND REGIONAL CORRELATION OF STRATIGRAPHIC EVENTS

10.1 Introduction

The RASC computer program for ranking and scaling of biostratigraphic events was originally written between 1978 and 1981 for mainframe computers. It was followed by the CASC program for correlation and scaling in time. In 1985, it became possible, after relatively minor modification, to compile the FORTRAN code of the RASC and CASC computer programs on IBM compatible microcomputers.

At present, several versions of these programs are in existence in different languages (primarily FORTRAN, C and BASIC). A brief history of the development of RASC and CASC with references is given at the end of this chapter. The existing programs are only slightly different from one another. As a rule, later versions are more user-friendly than earlier ones. The reader wishing to use RASC on a microcomputer (or mainframe) may obtain a copy of Program RASC (Ranking and Scaling), version 12, which at the time of writing (1990) is distributed free of charge by the Committee on Quantitative Stratigraphy (CQS). (Please send 360 KB floppy diskette to F.M. Gradstein, Chairman, CQS, Atlantic Geoscience Centre, Bedford Institute of Oceanography, Dartmouth, N.S., Canada, B2Y 4A2). This enhanced batch version of RASC in FORTRAN 77 by Agterberg et al. (1989) contains source code, executable (EXE) files and test data files. It can be executed on a PC with math co-processor. CASC is available as a mainframe program (Agterberg et al., 1985). Agterberg and Byron (1990) are preparing micro-RASC for release as a Geological Survey of Canada Open File.

The micro-RASC system consists of 12 separate program modules. It makes use of the characteristic features of microcomputers. Except for Module 1 which can be used to create new input files, each module reads one or more input files and creates one or more output files. This allows flexibility for program development because separate modules can be revised and replaced without changing the remainder of the system.

Micro-RASC contains slightly modified code previously published in the RASC (RAnking and SCaling) and CASC (Correlation And SCaling in time) computer programs for ranking, scaling and correlation of biostratigraphic events. This code has been supplemented by more recently developed algorithms including use of the jackknife method for estimating variances of cumulative RASC distances in scaling, modified RASC for frequency distribution analysis of stratigraphic events, and cross-validation to decide on optimum smoothing factors in spline-curve fitting for CASC. Micro-RASC can be used on any IBM compatible microcomputer with math co-processor and a FORTRAN compiler. This version of RASC and CASC is exclusively numeric using simple graphics programmed in FORTRAN 77.

The contents of the 12 modules is summarized in the next section with references to sections in earlier chapters where more details can be found. Important decisions to be made by the RASC user in order to create the parameter (PAR) file needed for running the modules are listed separately, in Section 10.3, together with those parameters that can be changed from their default values. These decisions are numbered as in micro-RASC. They are of a general nature in that any RASC user should consider the questions asked here. Parts of Module 1 (Data input) were previously published by Heller et al. (1985). Modules 2, 3, 4, 5 and 6 are equivalent to RASC version 12 (Agterberg et al., 1989). Modules 7 and 8 can be added to RASC version 12 relatively quickly, by using the procedures described in Chapter 7 (Jackknife scaling) and Chapter 8 (Modified RASC). An earlier version of Module 8 consisting of FORTRAN and BASIC programs was included in D'Iorio (1988). Earlier versions of Modules 9 and 11 were distributed on an informal basis under the names TSREG (Regional time scales) and SPLIN (Spline-curve fitting), respectively. Modules 10 and 12 emulate mainframe CASC.

It should be kept in mind that the purpose of the RASC computer programs is to order and correlate stratigraphic events. In most applications, the events are stratigraphically highest and lowest occurrences of (micro-)fossils, although peak occurrences and abrupt changes in relative abundance can be used equally well for correlation if these can be defined systematically. Lithostratigraphic, seismic and magnetostratigraphic events can be combined with biostratigraphic events. However, these other types of events may need special consideration; e.g. by defining them as marker horizons or by evaluating their relative uncertainty independently by means of modified RASC for frequency distribution analysis. Although the RASC computer programs provide automatic stratigraphic correlations, the user should remain in control of input and output, e.g. the input can be modified or amplified on the basis of new information provided in successive outputs.

10.2 Summary of contents of the 12 modules of micro-RASC

Module 1: DATA INPUT (cf. Section 4.2)

The RASC method requires as input a sequence (SEQ) file with coded sequences of stratigraphic events for individual sections, a dictionary with event names (DIC file), and a parameter (PAR) file with settings of switches and values of parameters. The CASC method requires depth (DEP) files for individual sections. Module 1 allows preparation of data (DAT) files from which SEQ files and preliminary DEP files are generated automatically. Examples of DAT file formats are:

(a) Depths (in feet or metres) followed by dictionary code numbers of events (feet will be converted into meters); and

(b) Fossil code numbers followed by depths of lowest and highest occurrences (DIC file for use in RASC then will be created with separate entries for lowest and highest occurrence of each fossil).

Module 2: PREPROCESSING (cf. Section 4.8)

Frequencies of events are determined as follows: (a) number of sections for each event; (b) number of events occurring in k sections and number of events occurring in k or more sections ($k=1, 2, ..., n$; n represents total number of sections). The threshold parameter k_c must be selected. Further analysis will be restricted to events that occur in at least k_c sections. Special rare events ("unique" events) which are to be re-inserted later in the biozonation, but which occur in fewer than k_c sections, should be identified. It is also possible to define marker horizons (e.g. seismic events or bentonite layers) which are not subject to biostratigraphic uncertainty.

Module 3: RANKING (cf. Sections 5.3 and 5.5)

The optimum sequence of events is determined by probabilistic ranking or "presorting" with or without the modified Hay method. The sequence obtained by presorting may be improved by sorting on the basis of superpositional relations ("above", "below", coeval) between pairs of events using the modified Hay method. Inconsistencies involving three or more events (cycles) will be identified (cf. Sections 5.7 and 5.8). The threshold parameter m_{c1} may be selected by changing its default value $m_{c1} = 1$. The modified Hay method will be applied only to pairs of events occurring in at least m_{c1} sections. It may not be possible to determine the relative order of two or more events. This type of uncertainty is expressed by means of the uncertainty range (Section 5.4) assigned to all events in the optimum sequence.

Module 4: SCALING (cf. Section 6.3)

The scaled optimum sequence of events is determined by estimating intervals between successive events in the optimum sequence previously obtained by ranking. This process usually involves minor reordering of the events. Final distances between successive events are clustered in a dendrogram which is useful as a regional biozonation. The threshold parameter m_{2c} must be selected. Scaling calculations are restricted to frequencies of order relations between pairs of events occurring in at least m_{c2} sections. Unweighted and weighted scaling can be performed. Further analysis (e.g. normality test) will be based on weighted scaling with the weights determined by frequencies of superpositional relations between events. Standard deviations of inter-event distances are provided. The cumulative RASC distance of each event is computed and added to the preliminary DEP files in order to create (complete) DEP files if CASC will be used.

Module 5: RANK EVALUATION (cf. Section 7.3)

The optimum sequence resulting from Module 3 or 4 can be used for construction of the occurrence table, "step model" and scattergrams. Events are shown for individual sections in the occurrence table. Their observed sequence in each section is compared with the optimum sequence in the step model. Penalty points are assigned for each position that an event is out of place in a section. Kendall's rank correlation coefficient can be computed from the total number of penalty points per section. The relative order of events in each section is compared to that in the optimum sequence in the scattergrams.

Module 6: NORMALITY TEST (cf. Sections 6.6 and 8.2)

The observed sequence of events in each section is compared to the scaled optimum sequence using cumulative RASC distances. Second-order differences are computed for events by comparing their observed positions to those of their neighbors in the stratigraphically upward and downward directions. Events that are out of place with probabilities greater than 95% and 99% are identified. A frequency distribution analysis of second-order differences is performed in order to evaluate: (a) autocorrelation of successive events in the scaled optimum sequence; and (b) overall frequencies of anomalous events occurring either too low (e.g. due to contamination during drilling), or too high (e.g. due to geological reworking) in the sections.

Module 7: JACKKNIFE SCALING (cf. Section 7.5)

In scaling, each estimated distance $\bar{D}_{ij}$ between successive events (i and j) is the average of a number of primary distance estimates $D_{ij.k}$ based on the superpositional relations of i and j with other events (k). By successively deleting individual events, and scaling of the reduced datasets, it is possible to obtain measures of precision of the estimates by means of the jackknife method which is non-parametric. This also results in jackknife estimates of the cumulative RASC distances of the events. Only if the latter estimates are close to the original cumulative RASC distances, as obtained by Module 4, their jackknife standard deviations can be used as standard deviations of the cumulative RASC distances.

Module 8: MODIFIED RASC (cf. Section 8.5)

The scaling method is based on transforming frequencies for superpositional relations between events into fractiles of the normal distribution in standard form. It is assumed that all events have the same variance for deviations between their regional mean positions and observed positions within individual sections. In modified RASC, the variances of the events can be different. They are estimated by means of an iterative procedure. Firstly, spline-curves are fitted to the events in common between the scaled optimum sequence and individual sections in order to project the regional mean positions onto the sections, and to collect all deviations for each event. Secondly, the variance of the deviations for each event is used for scaling which yields a new set of cumulative RASC distances. These two steps are repeated until approximate convergence is reached. Modified RASC allows identification of low-variance events which can be used as marker horizons. In addition to different event variances, this procedure provides frequency distributions of individual events which may be positively or negatively skewed. Maximum deviations can be used for constructing a conservative range chart in which the ranges are based on regional highest and lowest observed occurrences of fossils (cf. Sections 8.7 to 8.9).

Module 9: REGIONAL TIME SCALE (cf. Section 9.8)

The age (in millions of years) may be known for a subgroup of the stratigraphic events used in a regional RASC study. It may be possible to establish the relationship between age and cumulative RASC distance for this subgroup. This relationship, expressed as a spline-curve, can be used to transform all RASC distances into ages in linear time. These ages can be used to replace the cumulative RASC distances in the DEP files for CASC. Events can be weighted differently, either by using standard deviations resulting from Module 7, or by using subjective weights.

Module 10: CASC 1: EVENT-DEPTH CURVES (cf. Section 9.3)

The CASC (Correlation And SCaling in time) method consists of two steps: (a) construction of event-depth curves (this module or Module 11); and (b) multiwell comparison (Module 12). Module 10 closely resembles the age-depth curve-fitting part of the CASC mainframe computer program. Supplementary statistical techniques (cross-validation, jackknife spline-curve fitting) are given in Module 11 which amplifies Module 10. Input for CASC in Module 10 or 11 consists of DEP files for the sections to be studied. A spline-curve is fitted for each section. The dependent variable is (a) rank, (b) RASC distance, or (c) age in Ma; the independent variable is (a) relative event level, or (b) depth (in metric units). Because events generally are spaced irregularly along the depth-axis, the indirect method can be used for estimating the age-depth curve. This algorithm called "cross-plots" consists of fitting separate spline-curves for the age-level and depth-level relations. Elimination of level then gives an event-depth curve which usually is better than the one obtained by direct spline-curve fitting. Sediment accumulation (sedimentation) rate curves can be obtained from the first derivatives of the spline-curves.

Module 11: CASC 2: STATISTICAL ANALYSIS (cf. Section 9.4)

The shape of a spline-curve is to a large extent controlled by its smoothing factor (SF) representing the standard deviation of the differences between observed and fitted values. The law of superposition of strata requires that age never decreases in the stratigraphically downward direction. This provides an estimate of the minimum smoothing factor. The maximum smoothing factor correspond to the best-fitting straight line of least squares. The optimum smoothing factor has a value within the open interval bounded by these two extremes. Cross-validation is a method for estimating the optimum smoothing factor. The best-fitting spline-curve deviates from the unknown true age-depth curve. The error of the fitted spline-curve values can be estimated by using the jackknife method. The method for spline-curve fitting in Modules 10 and 11 is a modified version of De Boor's (1978) FORTRAN program. Module 12 allows two alternative techniques for spline-curve fitting. These are discrete cubic spline fitting (Duris, 1980) and a beam deformation analogue method (Hibbert, 1990), respectively.

Module 12: CASC 3: MULTI-WELL COMPARISON (cf. Section 9.8)

Probable depths of selected events or isochrons (e.g. multiples of 10 Ma) determined by means of the age-depth curves can be correlated between sections. This multi-well comparison is performed by means of a table in which the probable depths are accompanied by estimated 68% or 95% confidence intervals. Various types of confidence intervals can be obtained. These include the local and modified local error bars for deviations between observed depth of events and the probable depths used for correlation. Local and modified local error bars basically are error bars along the time axis which have been projected along the depth axis by assuming locally constant and variable rates of sediment accumulation, respectively.

10.3 List of decisions to be made by user of the RASC computer programs

During a complete micro-RASC session, the user can be asked 80 questions numbered separately in each of the twelve modules. The answer to each question is "yes" or "no". If the answer is "yes", the **switch** corresponding to the question is turned **on**. It is left **off** if the answer is "no" and a **default** decision would be made which is displayed on the monitor. The user then is given the chance to change "no" into "yes". Some questions are asked only if certain **conditions** are satisfied. Eleven questions are about a **parameter** with a default value that can be changed. The settings of the switches and the values of the parameters are entered in the PAR file needed to run the micro-RASC programs. At the beginning of each module, the user is asked if the switches are to be set for that module. If the answer is "no", an existing PAR file must be used.

Module 1: DATA INPUT

1.1 Do you wish to prepare a new dictionary?

Default: It will be assumed that you work with an existing dictionary.

1.2 Do entries represent stratigraphic events?

Default: It will be assumed that you wish to work with the highest and lowest occurrences of fossils.

1.3 Do you wish to make a HI and LO occurrences dictionary?

Default: It will be assumed that a HI and LO dictionary is in existence, and will not have to be created from a single entry dictionary.

1.4 Are you working in the stratigraphically downward direction?

Default: It will be assumed that you work in the stratigraphically upward direction.

1.5 Will you work with the depths of the samples?

Default: It will be assumed that you work with event levels along a relative depth scale.

1.6 Do you wish to enter rotary table height and water depth?

Condition: Switch 1.5 is on.

1.7 Are your depths metric?

Condition: Switch 1.5 is on.

Remark: If Switch 1.7 is turned on, the following supplementary question is asked: Are your depths in meters? If the answer to the supplementary question is "no", the user is asked to: Enter conversion factor from meters to the units of your depth (Example: if your depths are in kilometers, enter 1000).

Default: It is assumed that you work with feet. These will be automatically changed into metric units.

1.8 Do you want the depth files for use in CASC?

Default: It will be assumed that you will not wish to use CASC.

1.9 Do you wish to create preliminary depth files?

Default: It will be assumed that your depth files already exist.

1.10 Do you wish to create a new sequence (RASC input) file?

Default: It will be assumed that you work with an existing sequence file.

1.11 Do you wish to create a new data file?

Default: It will be assumed that you work with an existing data file.

1.12 Do you wish to subtract a constant from all dictionary numbers that are read in?

Parameter name: NSTART (Default value NSTART = 0).

Default: As usual, no changes are made in the dictionary numbers.

Module 2: PREPROCESSING

2.1 Do you wish to set the threshold parameter for minimum number of sections in which an event should occur?

Parameter name: IOCR (Default value: IOCR = 3)

Default: The minimum number of sections in which an event should occur is equal to 3.

2.2 Are you dealing with two separate groups of fossils which should have different threshold parameters?

Condition: Switch 1.12 is on.

Parameter name: IOCR2 (Default value: IOCR2 = 0)

Default: As usual, you wish to use a single threshold parameter for minimum number of sections.

2.3 Do you wish to define unique events? (i.e. special rare events that occur fewer than IOCR times)

2.4 Do you wish to define marker horizons?

2.5 Do you wish to see intermediate tabulations?

Default: Intermediate tabulations (e.g. recoded sequence data) will not be shown in the output.

Module 3: RANKING

3.1 Do you wish to perform presorting?

3.2 Do you wish to apply the modified Hay method?

3.3 Do you wish to set the threshold parameter for minimum number of sections in which a pair of events should occur?

Parameter name: CRIT1 (Default value: CRIT1 = 1.0)

Default: All frequencies will be used for the modified Hay method.

3.4 Do you wish to re-set the tolerance?

Parameter name: TOL (Default value: TOL = 0.0)

Default: As usual, the tolerance parameter is kept equal to zero.

3.5 Do you wish to change the maximum number of iterations?

Parameter name: ITER (Default value: ITER = 10,000)

Default: The maximum number of iterations allowed for the modified Hay method is 10,000.

3.6 Do you wish to see the cycling tabulations?

Default: The cycling tabulations will not be shown in the output.

3.7 Do you wish to see all intermediate tabulations?

Default: Intermediate tabulations (e.g., matrices with initial and reordered frequency scores) will not be shown in the output.

3.8 Do you wish to go on to the scaling module?

Default: RASC run will be terminated after ranking and input for Module 4 will not be created.

3.9 Do you wish to perform ranking evaluation?

Default: Input for ranking evaluation (Module 5) will not be created.

3.10 Do you wish to add ranking results to depth files for use in CASC?

Condition: Switch 1.9 is on.

Default: As usual, CASC will not be applied to ranking results.

3.11 Do you wish to re-insert unique events into the optimum sequence?

Default: Unique events will not be re-inserted into the optimum sequence.

Module 4: SCALING

4.1 Do you wish to set the threshold parameter for minimum number of sections in which a pair of events should occur?

Parameter name: CRIT2 (Default value: CRIT2 = 2.0)

Default: All frequencies for pairs occurring in two or more sections will be used for scaling.

4.2 Do you wish to change the truncation limit?

Parameter name: AAA (Default value: AAA = 0.95)

Default: Frequency of an event observed to occur above another event in all sections (containing both events) will be changed from 1.00 to 0.95.

4.3 Do you wish to delete scaling tables from output?

Default: Only dendrograms will be shown in the output.

4.4 Should long distances be suppressed during estimation?

Default: As usual, long distances will not be suppressed.

4.5 Should final reordering be applied?

4.6 Do you wish to apply scaling more than five times before accepting the final reordering results?

Condition: Switch 4.5 is on.

Parameter name: KKL (Default value: KKL = 5)

Default: Total number of interations during reordering is not allowed to exceed 5.

4.7 Do you wish to see intermediate tabulations?

Default: Intermediate tabulations (tables of fractiles) will not be shown in the output.

4.8 Do you wish to suppress re-insertion of unique events into the scaled optimum sequence?

Default: As usual, unique events will be re-inserted into the scaled optimum sequence.

4.9 Do you wish to perform rank evaluation?

Default: Rank evaluation (Module 5) of scaling results will not be performed.

4.10 Do you wish to perform the normality test?

Default: Normality test (Module 6) will not be performed.

4.11 Do you wish to perform jackknife scaling?

Default: Jackknife scaling (Module 7) will not be performed.

4.12 Do you wish to apply the modified RASC method?

Default: Modified RASC (Module 8) will not be performed.

4.13 Are you planning to construct a regional time scale using ages (in Ma) of selected events?

Default: Regional time scale (Module 9) will not be constructed.

4.14 Do you wish to add scaling results to depth files for use in CASC?

Condition: Switch 1.9 is on; Switch 3.10 is off.

Module 5: RANK EVALUATION

5.1 Do you wish to construct the occurrence table?

5.2 Do you wish to apply the step model?

5.3 Do you wish to see scattergrams for separate sections?

Module 6: NORMALITY TEST

6.1 Do you wish to see the detailed statistical analysis results? (i.e. study of autocorrelation based on second-order differences).

Module 7: JACKKNIFE SCALING

7.1 Do you wish to change the width of the window on the RASC scale?

Parameter name: WDW (Default value: WDW = 2.0)

Default: No use will be made of observed superposional relations between events that are above one another in the original scaled optimum sequence with a probability of approximately 95 percent.

7.2 Do you wish to use the jackknife standard deviations for construction of a regional time scale?

Condition: Switch 4.13 is on.

Module 8: MODIFIED RASC

8.1 Do you wish to perform more than three complete iterations?

Parameter name: KKM (Default value: KKM = 3)

Default: As usual, the cumulative RASC distance estimates will be refined three times using successive approximations of the event variances.

8.2 Do you wish to see frequency tables for separate events?

8.3 Do you wish to see plots of observed and calculated values for separate sections?

8.4 Do you wish to construct the range chart table?

8.5 Do you wish to save the event variances for weighting in CASC?

Condition: Switch 4.14 is on.

Module 9: REGIONAL TIME SCALE

9.1 Do you want to use automated version?

Condition: Switch 7.2 is on.

Default: You will choose your own smoothing factor for spline-curve fitting with age as the dependent variable.

9.2 Do you wish to define subjective weights in order to assign more or less influence to ages of events?

Default: All ages will have equal weights during spline-curve smoothing.

9.3 Do you want to substitute ages for RASC distances in depth files?

Condition: Switch 4.14 is on.

Default: CASC will be based on the RASC distances.

Module 10: CASC 1: EVENT-DEPTH CURVES

10.1 Are you using an optimum sequence with ranks only?

Condition: Switch 3.10 or Switch 4.14 is on.

Default: You are using the scaled optimum sequence supplemented by RASC distances or ages (in Ma).

10.2 If some events are observed to be coeval, do you wish to work with separate events at approximately the same event levels?

Default: Events observed to be coeval at a given level will be averaged.

10.3 Should each average for an event level be weighted according to the numbers of coeval events on which it is based?

Condition: Switch 10.1 is off.

10.4 Do your depth files contain standard deviations for separate events which are not equal to one another?

Condition: Switch 10.1 is off.

Default: All events will be weighted equally.

10.5 Are you using weights determined by means of modified RASC?

Condition: Switches 8.4 and 10.3 are on; Switch 10.1 is off.

10.6 Will you be performing a multi-well comparison?

Default: Age-depth results will not be saved for multi-well comparison.

10.7 Will you use the indirect method for estimating event-depth relations?

Default: The direct method will be used for estimation.

10.8 Do you want to study the first derivatives and sediment accumulation curves?

10.9 Do you wish to use defaults except for the age-level relation?

Condition: Switch 10.7 is on.

Default: You will have to select smoothing factors for the event-depth and age-interpolated depth relations in each section.

10.10 Do you wish to use the minimum smoothing factor and other defaults in all sections?

Condition: Switch 10.6 is on; Switch 10.7 is off.

Default: Sections will be analyzed separately one after another.

10.11 Do you wish to use plot axes defined during analysis of the first depth file later, for the other depth files?

Default: You can let the program define default plot axes or define new plot axes for any section.

10.12 Do you wish to perform detailed statistical analysis (e.g. cross-validation) for at least one of your sections?

Default: It will not be possible to use Module 11.

Remark: If Switch 10.12 is off, the next prompt asks for the name of the first depth file to be analyzed by means of Module 10.

Module 11: CASC 2: STATISTICAL ANALYSIS

11.1 Do you wish to use cross-validation?

Condition: Switch 10.12 is on and Module 11 has been activated.

Default: Optimum smoothing factor will be determined by autocorrelation method.

11.2 Do you wish to see additional tabulations (e.g. spline coefficients) in the output?

11.3 Do you wish to obtain the jackknife spline-curve?

11.4 Do you wish to use discrete cubic spline smoothing?

Default: As usual, a modification of De Boor's program for cubic spline smoothing will be used.

11.5 Do you wish to use the beam deformation analogue method for cubic spline smoothing?

Condition: Switch 10.2 is on; Switch 10.4 is off.

Default: As usual, a modification of De Boor's program for cubic spline smoothing will be used.

Remark: The next prompt asks for the name of the first depth file to be analyzed by means of Module 11.

Module 12: CASC 3: MULTI-WELL COMPARISON

12.1 Do you wish to specify the sections to be used for correlation?

Default: All sections analyzed by means of Module 10 or Module 11 will be used for correlation.

12.2 Do you wish to correlate selected events?

Default: Your correlation will be based on ages in millions of years.

12.3 Do you wish to compute probable positions of isochrons?

Condition: Switch 12.2 is off.

Default: Ages for correlation between sections will have to be selected individually.

12.4 Do you want modified local error bars?

Default: Local error bars will be given only.

12.5 Do you want approximate 95 per cent confidence intervals?

Default: Standard deviations will be used for the error bars (i.e., approximate 68 per cent confidence intervals will be given).

12.6 Do you wish to define a new t-value for the error bars?

Condition: Switch 12.5 is on.

Parameter name: TVALUE (Default value: TVALUE = 2.0)

Default: As usual, the approximation $t = 2.0$ for 95 per cent confidence intervals will be used.

12.7 Do you want statistical analysis results for spline-curve values and studentized residuals as well?

Default: As usual, statistical analysis will be restricted to deviations between observed and calculated values.

10.4 Brief history of the development of RASC and CASC

The basic ideas incorporated in the RASC RAnking and SCaling computer program originated during 1978 in collaboration with F.M. Gradstein (Bedford Institute of Oceanography in Dartmouth, Nova Scotia, Canada). For initial program development in FORTRAN IV, use was made of the Cyber 74 computer of the Department for Energy, Mines and Resources in Ottawa. Agterberg and Nel (1982a,b) published the ranking and scaling algorithms in the journal "Computers & Geosciences". Stratigraphic and statistical model verification with applications in exploration biostratigraphy in petroleum basins were given in Gradstein and Agterberg (1982).

During spring, 1979, an earlier version of the program was implemented by W.A. Burroughs on the DECSystem 10 of Syracuse University and tested by graduate students participating in a seminar on quantitative stratigraphic correlation. Their comments and discussions with J.C. Brower (Syracuse University) resulted in many improvements. The program also was implemented by K.G. Shih and A. Johnston at the Bedford Institute of Oceanography for demonstration in August, 1979, during the first meeting of the Canadian Working Group of the International Geological Correlation Programme (IGCP) Project 148 (Quantitative Stratigraphic Correlation Techniques). Suggestions were received during this workshop and later from participants including P.H. Doeven (PetroCanada, Calgary, Canada), L.E. Edwards (U.S. Geological Survey, Reston, Virginia, U.S.A.), P. Moore (Shell Resources Canada, Calgary, Canada), E.M. Oliver (Robertson Research, Calgary, Canada), and R.J. Price (Amoco Canada, Calgary, Canada). A number of results obtained by RASC were presented during the second meeting of the Canadian working group for ICGP Project 148 in Ottawa, February, 1980 (Agterberg and Gradstein, 1981). This included comprehensive scaling studies carried out in Ottawa by C.B. Hudson (University of South Carolina, Columbia, U.S.A.; see Hudson and Agterberg, 1982), and presentation of RASC output using DISSPLA by A. Jackson (Bedford Institute of Oceanography, Dartmouth, Canada). The version of RASC published in "Computers & Geosciences" was implemented by S. Briggs on the DECSystem 10 and IBM 370 computers of Syracuse University during spring, 1981.

An interactive version of mainframe RASC using a Tektronix 4014 terminal was prepared with the help of C.F. Chung (Geological Survey of Canada, Ottawa) and R. Lessard (University of Sherbrooke, Quebec) and used for demonstration during the Second International Quantitative Stratigraphy Short Course held during the Calgary 1982 meetings of the American Association of Petroleum Geologists (co-sponsored by IGCP Project 148, the Canadian Society of Petroleum Geologists and the University of Calgary). New implementations by oil companies including use on the UNIVAC 1108 by Shell Resources Canada in Calgary resulted in further suggestions for improvement. M. Heller and W.S. Gradstein (Consultants in Halifax, Nova Scotia) prepared a user guide for RASC which was released in 1983 as Geological Survey of Canada (GSC) Open File 922 (Heller et al., 1983). This Open File also contained revised

FORTRAN IV code for mainframe RASC (printout with examples and magnetic tape).

In 1982, with the help of J. Oliver (University of Ottawa), development of CASC (Correlation And SCaling in time) commenced in Ottawa (Agterberg and Gradstein, 1983). This interactive program was developed in FORTRAN IV using a Cyber 730 mainframe with Tektronix 4014 terminal. The program was demonstrated during the Third International Quantitative Stratigraphy Short Course, held in Dartmouth, Nova Scotia, October 1983 and the Seventh Meeting of the Canadian Working Group for IGCP Project 148 held in Ottawa, March 1984. The CASC program was released in 1985 as GSC Open File 1179 (Agterberg et al., 1985). Applications of CASC are described in Gradstein and Agterberg (1985), Williamson (1987) and D'Iorio and Agterberg (1989).

By 1985, microcomputer hardware and software had advanced to the stage that RASC could be run on IBM PC's and compatibles equipped with the 8087 Math Co-processor. S.N. Lew prepared a FORTRAN 77 version of RASC which, together with the revised user's manual, was published as GSC Open File 1203 (Heller et al., 1985). This program can be compiled and run on microcomputers. GSC Open Files 1179 and 1203 can be obtained from the Publications Office of the Geological Survey of Canada, 601 Booth Street, Ottawa K1A 0E8 (Each Open File consists of a manual and two 5.25-inch double-sided, double-density diskettes with IBM-PC readable code; cost $20.00 for OF 1179 and $25.00 for OF 1203). The DENO program (Jackson et al., 1984) serves to display dendrograms of scaled optimum sequences and the optimum sequences of stratigraphic events from RASC output by means of a CALCOMP plotter. It is written in the plotting language DISSPLA.

Alethic Software Incorporated (52 Parkhill Road, Halifax, Nova Scotia, B3P 1R5) has developed three computer programs in the language C for IBM personal computers (XT, AT, PS2) and compatibles (with math co-processor). Their GEOSCI-1 program is for data entry. It prepares sequence files and dictionaries for GEOSCI-2 which is a C version of RASC. Alethic's GEOSCI-3 program is a C version of CASC. These programs are marketed by Alethic and can be obtained at the address shown above or by phoning 902-423-9860.

Assisted by D. Gillis (Atlantic Geoscience Centre, Dartmouth, N.S.), F.M. Gradstein introduced output redirection to the code of RASC for IBM-PC. This feature improves its use on microcomputers that usually lack high-speed printers. This version (RASC 011) was later used during the Eighth International Quantitative Stratigraphy Shortcourse held at the Free University, Amsterdam, February 1989.

With the help of S.N. Lew (Geological Survey of Canada, Ottawa), a FORTRAN 77 program called SPLIN for microcomputers using IBM Graphics Development Toolkit was developed for spline-fitting of age-depth curves with cross-validation. Use was made of De Boor's (1978) FORTRAN programs. SPLIN was demonstrated during the Fifth International Quantitative Stratigraphy Short Course held in Aberdeen, Scotland, April 1986. For method and applications, see Agterberg and Gradstein (1988) and Gradstein et al. (1989). Discussions with M. Fearon (Consultant in Halifax, Nova Scotia) resulted in improvements of the spline-fitting algorithm.

SPLIN was combined with a microcomputer version of CASC with the help of J. Kirk (Informatics Applications Division, Energy, Mines and Resources Canada, Ottawa). This program (SPLIN2) was demonstrated during the Seventh International Quantitative Stratigraphy Short Course held at PETROBRAS, Rio de Janeiro, Brazil, November 1987. Modified RASC for frequency distribution analysis of biostratigraphic events (Agterberg and D'Iorio, in press) was developed in collaboration with M.A. D'Iorio (1988) whose doctoral dissertation contains FORTRAN 77 and BASIC programs for an earlier version of modified RASC.

Development of the micro-RASC computer programs was commenced in 1988 with the help of D. Byron at the Geological Survey of Canada in Ottawa, with contributions by P. Hibbert (Informatics Applications Division, Energy, Mines and Resources Canada, Ottawa). F.M. Gradstein provided valuable contributions by reviewing the blueprints several times with many comments. Micro-RASC will be available as a GSC Open File (Agterberg and Byron, 1990a,b).

During 1989, Z. Huang (Dalhousie University) added "help" files and enhanced output formats of the batch version of RASC (equivalent to Modules 2 to 6 of micro-RASC). This program (RASC, version 12) is available through the Committee on Quantitative Stratigraphy (see Section 10.1 on how to obtain it).

REFERENCES

Agterberg, F.P., 1974. Geomathematics. Elsevier, Amsterdam, 596 pp.

Agterberg, F.P., 1984. Binomial and trinomial methods in quantitative biostratigraphy. Computers and Geosciences, 10: 31-41.

Agterberg, F.P. (Editor), 1984. Theory, Application and Comparison of Stratigraphic Correlation Methods. Comput. Geosci., 10 (1): 1-186.

Agterberg, F.P., 1985. Normality testing and comparison of RASC to Unitary Associations Method. In: F.M. Gradstein et al., Quantitative Stratigraphy, UNESCO, Paris and Reidel, Dordrecht, pp. 243-262.

Agterberg, F.P., 1988. Quality of time scales – a statistical appraisal. In: D.F. Merriam (Editor), Current Trends in Geomathematics, Plenum, New York, pp. 57-103.

Agterberg, F.P. and Bonham-Carter, G.F. (Editors), 1990. Statistical applications in the Earth Sciences. Geological Survey of Canada Paper 89-9.

Agterberg, F.P. and Byron, D.N., 1990a. FORTRAN 77 microcomputer programs for ranking, scaling and regional correlation of stratigraphic events. In: F.P. Agterberg and G.F. Bonham-Carter (Editors), Statistical Applications in the Earth Sciences, Geol. Surv. Can. Paper 89-9.

Agterberg, F.P. and Byron, D.N., 1990b. Micro-RASC System of 12 FORTRAN 77 microcomputer programs for ranking, scaling and regional correlation of biostratigraphic events. Geological Survey of Canada Open File.

Agterberg, F.P. and D'Iorio, M.A., in press. Frequency distributions of highest occurrences of Cenozoic Foraminifera along the northwestern Atlantic Margin. Proceedings, 4th South American COGEODATA Symposium, held in Ouro Preto, Brazil, November 1987.

Agterberg, F.P. and Gradstein, F.M., 1981. Workshop on quantitative stratigraphic correlation. Math. Geol., 13(1): 81-91.

Agterberg, F.P. and Gradstein, F.M., 1983. Interactive system of computer programs for stratigraphic correlation. Current Research, Geological Survey of Canada, Paper 83-1A, pp. 83-87.

Agterberg, F.P., and Gradstein, F.M., 1988, Recent developments in quantitative stratigraphy. Earth-Science Reviews, 25(1): 1-73.

Agterberg, F.P. and Nel, L.D., 1982a. Algorithms for the ranking of stratigraphic events. Computers and Geosciences, 8: 69-90.

Agterberg, F.P. and Nel, L.D., 1982b. Algorithms for the scaling of stratigraphic events. Computers and Geosciences, 8: 163-189.

Agterberg, F.P. and Rao, S.N. (Editors), 1988. Recent Advances in Stratigraphic Correlation. Hindustan Publishing Corporation, Delhi, 192 pp.

Agterberg, F.P., Gradstein, F.M., Lew, S.N. and Thomas, F.C., 1985. Nine databases with applications of ranking and scaling of stratigraphic events. In: F.M. Gradstein et al., Quantitative Stratigraphy, UNESCO, Paris and Reidel, Dordrecht, pp. 473-564.

Agterberg, F.P., Gradstein, F.M. and Nazli, K., 1990. Correlation of Jurassic microfossil abundance data from the Tojeira sections, Portugal. In: F.P. Agterberg and G.F. Bonham-Carter (Editors), Statistical Applications in the Earth Sciences, Geol. Surv. Can. Paper 89-9.

Agterberg, F.P., Gradstein, F.M., Nel, L.D., Lew, S.N., Heller, M., Gradstein, W.S., D'Iorio, M.A., Gillis, D. and Huang, Z., 1989. Program RASC (Ranking and Scaling) version 12. Comm. Quantitative Stratigraphy, Bedford Inst. Oceanogr., Dartmouth, N.S., Canada.

Agterberg, F.P., Oliver, J., Lew, S.N., Gradstein, F.M. and Williamson, M.A., 1985. CASC Fortran IV interactive computer program for correlation and scaling in time of biostratigraphic events. Geological Survey of Canada Open File Report 1179.

Armstrong, R.L., 1978. Pre-Cenozoic Phanerozoic time scale. In: G.V. Cohee, M.G. Glaessner and H.D. Hedberg (Editors), Contributions to the Geologic Time Scale, Am. Assoc. Petroleum Geol., Studies in Geology, No. 6, pp. 73-91.

Barrell, J., 1917. Rhythms and the measurements of geologic time. Geol. Soc. Am., Bull., 28: 745-904.

Baumgartner, P.O., 1984. A Middle Jurassic-Early Cretaceous low-latitude radiolarian zonation based on Unitary Associations and age of Tethyan radiolarians. Eclogae Helv., 77: 729-837.

Baumgartner, P.O., 1987. Age and genesis of Tethyan Jurassic radiolarites. Eclogae Geol. Helv., 30: 831-879.

Berge, C., 1973. Graphes et Hypergraphes. Dunod, Paris, 516 pp.

Berger, W.H. and Heath, G.R., 1968. Vertical mixing in pelagic sediments. J. Marine Res., 26: 134-143.

Berggren, W.A., 1972. A Cenozoic time-scale, some implications for regional geology and paleogeography. Lethaia, 5: 195-215.

Berggren, W.A., Kent, D.V., Flynn, J.J. and Van Couvering, J.A., 1985. Cenozoic geochronology. Geol. Soc. Am. Bull., 96: 1402-1418.

Blank, R.G., 1979. Applications of probabilistic biostratigraphy to chronostratigraphy. J. Geol., 87: 647-670.

Blank, R.G., 1984. Comparison of two binomial models in probabilistic biostratigraphy. Computers and Geosciences, 10: 59-67.

Blank, R.G. and Ellis, C.H., 1982. The probable range concept applied to the biostratigraphy of marine microfossils. J. Geol., 90: 415-433.

Bliss, C.I., 1935. The calculation of the dosage-mortality curve. Ann. Appl. Biol., 22: 134-167.

Blow, W.H., 1969. Late middle Eocene to Recent planktonic foraminiferal biostratigraphy. In: P. Bronnimann and J.H.H. Renz (Editors), Proc. 1st International Conf. on Planktonic Microfossils, Geneva 1967, E.J. Brill, Leiden, pp. 339-378.

Bonham-Carter, G.F., Gradstein, F.M. and D'Iorio, M.A., 1986. Distribution of Cenozoic Foraminifera from the Northwestern Atlantic Margin analyzed by correspondence analysis. Computers and Geosciences, 12: 621-635.

Box, G.E.P. and Jenkins, G.M., 1976. Time Series Analysis: Forecasting and Control. Holden-Day, San Francisco, 575 pp.

Bramlette, M.N. and Sullivan, F.R., 1961. Coccolithophorids and related nannoplankton of the Early Tertiary in California. Micropal., 7: 129-188.

Brinkmann, R., 1929. Statistisch-biostratigraphische Untersuchungen an Mitteljurassischen Ammoniten: Uber Artbegriff und Stammesentwicklung. Abhandlungen der Gesellschaft der Wissenschaften zu Göttingen, Mathematisch-Physikalische Klasse, Neue Folge 13(3), 249 pp.

Brower, J.C., 1981. Quantitative biostratigraphy, 1830-1980. In: D.F. Merriam (Editor), Computer Applications in the Earth Sciences, Plenum, New York, pp. 63-103.

Brower, J.C., 1985a. Multivariate analysis of assemblage zones. In: F.M. Gradstein et al., Quantitative Stratigraphy, UNESCO, Paris and Reidel, Dordrecht, pp. 65-94

Brower, J.C., 1985b. Archaeological seriation of an original data matrix. In: F.M. Gradstein et al., Quantitative Stratigraphy, UNESCO, Paris and Reidel, Dordrecht, pp. 95-108.

Brower, J.C., 1990. A case study for comparison of some biostratigraphic techniques using Paleogene alveolinids from Slovenia and Istria. In: F.P. Agterberg and G.F. Bonham-Carter (Editors), Statistical Applications in the Earth Sciences, Geol. Surv. Can. Paper 89-9.

Brower, J.C., Millendorf, S.A. and Dyman, T.S., 1978. Quantification of assemblage zones based on multivariate analysis of weighted and unweighted data. Computers and Geosciences, 4: 221-227.

Brunk, H.D., 1960. Mathematical models for ranking from paired comparisons. J. Am. Stat. Assoc., 55: 503-520.

Burroughs, W.A. and Brower, J.C., 1982. SER, a FORTRAN program for the seriation of biostratigraphic data. Computers and Geosciences, 8: 137-148.

Buzas, M.A., Koch, C.F., Culver, S.J. and Sohl, N.F., 1982. On the distribution of species occurrence. Paleobiology, 8: 143-150.

Carinati, R., Marini, A. and Potenza, R.G., 1982. The mathematical formalization of the geological relations identifying the basic structure of a geological relations identifying the basic structure of a geological data bank. In: J.M. Cubitt and R.A. Reyment (Editors), Quantitative Stratigraphic Correlation, Wiley, Chichester, pp. 13-18.

Carr, P.F., Jones, B.G., Quinn, B.G. and Wright, A.J., 1984. Toward an objective Phanerozoic time scale. Geology, 12: 274-277.

Carré, B., 1979. Graphs and Networks. Clarendon Press, Oxford, 277 pp.

Cheetham, A.H. and Deboo, P.B., 1963. A numerical index for biostratigraphic zonation in the mid-Tertiary of the Eastern Gulf. Gulf Coast Association of Geological Societies, Transactions, 13: 139-147.

Christopher, R.A., 1978. Quantitative palynologic correlation of three Campanian and Maestrichtian sections (Upper Cretaceous) from the Atlantic coastal plain. Palynology, 2: 1-27.

Clark, R.M., 1989. A randomization test for the comparison of ordered sequences. Math. Geol., 21: 429-442.

Cowie, J.W. and Bassett, M.G. (Compilers), 1989. International Union of Geological Sciences, 1989 Global Stratigraphic chart. Supplement to Episodes, 12(2).

Cox, A.V. and Dalrymple, G.B., 1967. Statistical analysis of geomagnetic reversal data and the precision of potassium-argon dating. J. Geophysical Res., 72: 2603-2614.

Craven, P. and Wahba, G., 1979. Smoothing noisy data with spline functions. Numerische Mathematik, 31: 377-403.

Cross, T.A. (Editor), 1990. Quantitative Dynamic Stratigraphy. Prentice Hall, Englewood Cliffs, New Jersey, 625 pp.

Cubitt, J.C. (Editor), 1978. Quantitative Stratigraphic Correlation. Comput. Geosci., 4 (3): 215-318.

Cubitt, J.C. and Reyment, R.A. (Editors), 1982. Quantitative Stratigraphic Correlation. Wiley, Chichester, U.K.,320 pp.

Davaud, E., 1982. The automation of biochronological correlation. In: J.M. Cubitt and R.A. Reyment (Editors), Quantitative Stratigraphic Correlation, Wiley, Chichester, pp. 85-99.

Davaud, E. and Guex, J., 1978. Traitement analytique 'manuel' et algorithmique de problèmes complexes de correlations biochronologiques. Eclogae Geol. Helv., 71: 581-610.

David, H.A., 1988. The Method of Paired Comparisons (Second Edition). Oxford Univ. Press, New York, N.Y., 200 pp.

David, M., 1977. Geostatistical Ore Reserve Estimation. Elsevier, Amsterdam, 364 pp.

Davidson, R.R., 1970. On extending the Bradley-Terry model to accommodate ties in paired comparison experiments. J. Amer. Statist. Assoc., 65: 317-328.

Davis, J.C., 1986. Statistics and Data Analysis in Geology, 2nd Edition. Wiley, New York, N.Y., 646 pp.

De Boor, C., 1978. A Practical Guide to Splines. Springer Verlag, New York, 392 pp.

Dienes, I., 1974. General formulation of the correlation problem and its solution in two special situations. Math. Geol, 6: 73-81.

Dienes, I., 1982. Formalized Eocene stratigraphy of Dorog Basin, Transdanubia, Hungary, and related areas. In: J.M. Cubitt and R.A. Reyment (Editors), Quantitative Stratigraphic Correlation, Wiley, Chichester, pp. 19-42.

Dienes, I. and Mann, C.J., 1977. Mathematical formalization of stratigraphic terminology. Math. Geol., 9: 587-603.

D'Iorio, M.A., 1986. Integration of foraminiferal and dinoflagellate data sets in quantitative stratigraphy of the Grand Banks and Labrador Shelf. Bull. Canadian Petroleum Geology, 34: 277-283.

D'Iorio, M.A., 1987. Quantitative biostratigraphic analysis of the Cenozoic of 23 Canadian Atlantic offshore wells. The Compass, 64: 264-277.

D'Iorio, M.A., 1988. Quantitative biostratigraphic analysis of the Cenozoic of the Labrador Shelf and Grand Banks: Unpublished Ph.D. thesis, Univ. of Ottawa, 404 p.

D'Iorio, MA., 1990. Sensitivity of the RASC model to its critical probit value. In: F.P. Agterberg and G.F. Bonham-Carter (Editors), Statistical Applications in the Earth Sciences, Geol. Surv. Can. Paper 89-9.

D'Iorio, M.A. and Agterberg, F.P., 1989. Marker event identification technique and correlation of Cenozoic biozones on the Labrador Shelf and Grand Banks. Bull. Canadian Petroleum Geol., 37: 346-357.

Dixon, W.J. and Massey, F.J., 1957. Introduction to Statistical Analysis. McGraw-Hill, New York, N.Y., 488 pp.

Doeven, P.H., 1983. Cretaceous nannofossil stratigraphy and paleoecology of the Canadian Atlantic Margin. Bull. Geol. Surv. Can. no. 356, 70 pp.

Doeven, P.H., Gradstein, F.M., Jackson, A., Agterberg, F.P. and Nel, L.D., 1982. A quantitative nannofossil range chart. Micropal., 28: 85-92.

Doveton, J.H., 1986. Log analysis of Subsurface Geology Concepts and Computer Methods. Wiley, New York, N.Y., 273 p.

Drobne, K., 1977. Alvéolines Paléogènes de la Slovénie et de l'Istrie. Mém. Suisses Paleont., 99, 175 pp.

Drooger, C.W., 1974. The boundaries and limits of stratigraphy. Proc. Kon. Ned. Akad. Wet. Ser. 11B, 17: 159-176.

Duris, C.S., 1980. Algorithm 547, FORTRAN routines for discrete cubic spline interpolation and smoothing. ACM Transact. Math. Softw., 6: 92-103.

Edwards, L.E., 1978. Range charts and no-space graphs. Computers and Geosc., 4: 247-258.

Edwards, L.E., 1982. Numerical and semi-objective biostratigraphy: Review and predictions. Proc. 3rd North Am. Pal. Conv., Montreal, August 1982, 1: 147-152.

Edwards, L.E., 1984. Insights on why graphic correlation (Shaw's method) works. J. Geology, 92: 583-597.

Edwards, L.E., 1989. Supplemented graphic correlation: A powerful tool for paleontologists and nonpaleontologists. Soc. Econ. Paleontologists and Mineralogists, Research Reports, pp. 127-143.

Edwards, L.E. and Beaver, R.J., 1978. The use of paired comparison models in orders stratigraphic events. J. Math. Geol. 10: 261-272.

Efron, B., 1982. The Jackknife, the Bootstrap and Other Resampling Plans. Soc. for Industrial and Applied Mathematics, Philadelphia, Pennsylvania, 92 pp.

Eubank, R.L., 1984. The hat matrix for smoothing splines. Statist. and Prob. Letters, 2: 9-14.

Eubank, R.L., 1988. Spline Smoothing and Nonparametric Regression. Dekker, New York, N.Y., 438 pp.

Finney, D.J., 1971. Probit Analysis (3rd Edition). Cambridge Univ. Press, 333 pp.

Fisher, R.A. and Yates, F., 1964. Statistical Tables for Biological, Agricultural and Medical Research (6th Edition). Oliver and Boyd, Edinburgh, 146 pp.

Foster, N.H., 1966. Stratigraphic leak. Am. Assoc. Pet. Geol. Bull., 50: 2604-2606.

Fulkerson, D.R. and Gross, O.A., 1965. Incidence matrices and interval graphs. Pacific J. Math. 15: 835-855.

Gale, N.H., Beckinsale, R.D. and Wadge, A.J., 1980. Discussion of a paper by McKerrow, Lambert and Chamberlain on the Ordovician, Silurian and Devonian time scales. Earth Plan. Sc. L., 51: 9-17.

Gill, D. and Merriam, D.F. (Editors), 1979. Geomathematical and Petrophysical Studies in Sedimentology. Pergamon, Oxford, 266 pp.

Gilmore, P.C. and Hoffman, A.J., 1964. A characterization of comparability graphs and interval graphs. Can. J. Math. 6: 539-548.

Glenn, W.A. and David, H.A., 1960. Ties in paired-comparison experiments using a modified Thurtone-Mosteller model. Biometrics, 16: 86-109.

Golub, G.H., Heath, M. and Wahba, G., 1979. Generalized cross-validation as a method for choosing a good ridge parameter. Technometrics, 21: 215-223.

Gordon, A.D., 1982. An investigation of two sequence-comparison statistics. Austral. J. Statistics, 24: 332-342.

Gordon, A.D., Clark, A.M. and Thomson, R., 1988. The use of constraints in sequence slotting. In: E. Diday (Editor), Data Analysis and Informatics V, North Holland Publishing Co., Amsterdam, pp. 353-364.

Gradstein, F.M., 1984. On stratigraphic normality. Computers and Geosciences, 10: 43-57.

Gradstein, F.M., 1985. Ranking and scaling in exploration micropaleontology. In: F.M. Gradstein et al., Quantitative Stratigraphy, Unesco, Paris and Reidel, Dordrecht, pp. 109-160.

Gradstein, F.M. and Agterberg, F.P., 1982. Models of Cenozoic foraminiferal stratigraphy – Northwestern Atlantic Margin. In: J.M. Cubitt, and R.A. Reyment (Editors), Quantitative Stratigraphic Correlation, Wiley, Chichester, pp. 119-173.

Gradstein, F.M., and Agterberg, F.P., 1985. Quantitative correlation in exploration micropaleontology. In: F.M. Gradstein et al., Quantitative Stratigraphy, UNESCO, Paris and Reidel, Dordrecht, pp. 309-360.

Gradstein, F.M. and Berggren, W.A., 1981. Flysch-type agglutinated foraminifera and the Maestrichtian to Paleogene history of the Labrador and North Seas. Marine Micropal., 6: 211-268.

Gradstein, F.M. and Fearon, M., 1990. STRATCOR, a new method for biozonation and correlation with applications to exploration micropaleontology (Summary). In: F.P. Agterberg and G.F. Bonham-Carter (Editors), Statistical Applications in the Earth Sciences, Geol. Surv. Pap. 89-9.

Gradstein, F.M. and Kaminski, M.A., 1989. Taxonomy and biostratigraphy of new and emended species of Cenozoic deep-water agglutinated Foraminifera from the Labrador and North Seas. Micropal. 35: 72-92.

Gradstein, F.M. and Srivastava, S.P., 1980. Aspects of Cenozoic stratigraphy and paleogeography of the Labrador Sea and Baffin Bay. Palaeogeogr., Palaeoclimatol., Palaeoecol., 30: 261-295.

Gradstein, F.M. and Williams, G.L., 1976. Biostratigraphy of the Labrador Shelf, I. Geol. Surv. Canada Rept. 349, 40 pp.

Gradstein, F.M., Agterberg, F.P., Aubry, M.-P., Berggren, W.A., Flynn, J.J., Hewitt, R., Kent, D.V., Klitgord, K.D., Miller, K.G., Obradovitch, J., Ogg, J.G., Prothero, D.R. and Westerman, G.E.G., 1988. Sea level history. Science 241: 599-605.

Gradstein, F.M., Agterberg, F.P., Brower, J.C. and Schwarzacher, W.S., 1985. Quantitative Stratigraphy. Unesco, Paris, and Reidel, Dordrecht, 598 p.

Gradstein, F.M., Agterberg, F.P. and D'Iorio, M.A., 1990. Time in quantitative stratigraphy: In: T.A. Cross (Editor), Quantitative Dynamic Stratigraphy. Prentice-Hall, Englewood Cliffs, N.J., pp. 519-542.

Gradstein, F.M., Fearon, J.M. and Huang, Z., 1989. BURSUB and DEPOR version 3.50 - Two FORTRAN 77 programs for porosity and subsidence analysis. Geol. Surv. Can. Open File 1283.

Gradstein, F.M., Kaminsky, M. and Berggren, W.A., 1988. Cenozoic foraminiferal stratigraphy of the Central North Sea. In: F. Rogl and F.M. Gradstein (Editors), Proc. 2nd Agglutinated Foraminifera Workshop, Vienna, 1986, Abhandlungen der Geologischen Bundesanstalt, 41: 97-108.

Gradstein, F.M., Williams, G.L., Jenkins, W.A.M. and Ascoli, P., 1975. Mesozoic and Cenozoic stratigraphy of the Altantic continental margin, eastern Canada. In: G.T. Yorath et al. (Editors), Canada's Continents Margin and Offshore Petroleum Exploration, Can. Soc. Petroleum Geol. Mem. 4, pp. 103-121.

Grimm, E.C., 1987. CONISS: A FORTRAN 77 program for stratigraphically constrained cluster analysis by the method of incremental sum of squares. Computers and Geosciences, 13: 13-35.

Guex, J., 1977. Une nouvelle méthode d'analyse biochronologique, note préliminaire. Bull., Soc. Vaud. Sci. Nat., 73: 309-321.

Guex, J., 1980. Calcul, caractérisation et identification des associations unitaires en biochronologie. Bull. Soc. Vaud. Sci. Nat., 75: 111-126.

Guex, J., 1981. Associations virtuelles et discontinuités dans la distribution des espèces fossiles: un exemple intéressant. Bull., Soc. Vaud. Sci. Nat., 75: 179-197.

Guex, J., 1987. Corrélations biochronologiques et Associations unitaires. Presses Polytechniques Romandes, Lausanne, Switzerland, 264 pp.

Guex, J., 1988. Utilisation des horizons maximaux résiduels en biochronologie. Bull., Soc. Vaud. Sci. Nat., 79.2: 135-142.

Guex, J. and Davaud, E., 1984. Unitary associations method: Use of graph theory and computer algorithm. Computers and Geosciences, 10: 69-96.

Gyji, R.A. and McDowell, F.W., 1970. Potassium argon ages of glauconites from a biochronologically dated Upper Jurassic sequence of northern Switzerland. Eclogae Geol. Helvetiae, 63: 11-118.

Hald, A., 1957. Statistical Theory with Engineering Application. Wiley, New York, N.Y., 783 pp.

Hald, A., 1960. Statistical Tables and Formulas. Wiley, New York, 97 pp.

Hallam, A., 1975. Jurassic Environments. Cambridge Univ. Press, Cambridge, 269 pp.

Haq, D.U., Hardenbol, J. and Vail, T.R., 1987. Chronology of fluctuating sea levels since the Triassic. Science, 235: 1156-1166.

Hardenbol, J., Vail, P.R. and Ferrer, J., 1981. Interpreting paleoenvironments; subsidence history and sea-level changes of passive margins from seismics and biostratigraphy. Oceanologica Acta 1981, sp., pp. 33-44.

Harland, W.B., Cox, A.V., Llewellyn, Pickton, C.A.G., Smith, A.G. and Walters, R., 1982. A Geologic Time Scale. Cambridge Univ. Press, 131 pp.

Harper, C.W., Jr., 1981. Inferring succession of fossils in time: The need for a quantitative and statistical approach. J. Paleont., 55: 442-452.

Harper, C.W., Jr., 1984. A Fortran IV program for comparing ranking algorithms in quantitative biostratigraphy. Computers and Geosciences, 10: 3-29.

Hay, W.W., 1972. Probabilistic stratigraphy. Eclogae Geol. Helv., 65: 255-266.

Hay, W.W. and Southam, J.R., 1978. Quantifying biostratigraphic correlation. Annual Review of Earth and Planet Sc., 6: 353-375.

Hazel, J.E., 1977. Use of certain multivariate and other techniques in assemblage zonal biostratigraphy, examples utilizing Cambrian, Cretaceous, and Tertiary benthic invertebrates. In: E.G. Kauffman and J.E. Hazel (Editors), Concepts and Methods and Biostratigraphy, Dowden, Hutchinson and Ross, Stroudsburg, Pennsylvania, pp. 187-212.

Hedberg, H.D. (Editor), 1976. International Stratigraphic Guide. Wiley, New York, N.Y., 200 pp.

Heller, M., Gradstein, W.S., Gradstein, F.M. and Agterberg, F.P., 1983. RASC FORTRAN IV computer program for ranking and scaling of biostratigraphic events. Geological Survey of Canada Open File 922.

Heller, M., Gradstein, W.S., Gradstein, F.M., Agterberg, F.P. and Lew, S.N., 1985. RASC Fortran 77 computer program for ranking and scaling of biostratigraphic events. Geological Survey of Canada Open File 1203.

Hemelrijk, J., 1952. A theorem on the sign test when ties are present. Kon. Nederl. Akad. Wetensch., Proc., 55: 322.

Hibbert, P., 1990. Spline smoothing by means of an analogy to structural beams. In: F.P. Agterberg and G.F. Bonham-Carter (Editors), Statistical Applications in the Earth Sciences, Geol. Surv. Can. Paper 89-9.

Hill, M.O., 1979. DECORANA – a FORTRAN program for detrended correspondence analysis and reciprocal averaging: Ecology and Systematics. Cornell Univ. Ithaca, New York, 52 pp.

Hohn, M.E., 1978. Stratigraphic correlation by principal components: effects of missing data. J. Geol., 86: 524-532.

Hohn, M.E., 1985. SAS program for quantitative stratigraphic correlation by principal components. Computers and Geosciences, 11: 471-477.

Howell, J.A., 1983. A FORTRAN 77 Program for automatic stratigraphic correlation: Computers and Geosciences, 9: 311-327.
Hudson, C.B. and Agterberg, F.P., 1982. Paired comparison models in biostratigraphy. J. Math. Geol. 14: 141-159.
Jackson, A., Lew, S.N. and Agterberg, F.P., 1984. DISSPLA program for display of dendrograms from RASC output. Computers and Geosciences 10: 59-165.
Jasko, T., 1984. The first find; estimation of the precision of range zone boundaries. Computers and Geosc., 10: 133-136.
Jeletzky, J.A., 1965. Is it possible to quantify biochronological correlation? J. Paleont., 39: 135-140.
Jenkins, G.M. and Watts, D.G., 1968. Spectral Analysis and its Application. Holden-Day, San Francisco, 525 pp.
Johnson, N.I. and Kotz, S., 1969. Discrete Distributions. Houghton Mifflin Company, Boston, Massachusetts, 328 pp.
Jones, D.J., 1958. Displacement of microfossils. J. Sediment. Petrol., 28: 453-467.
Kemp, F., 1982. An algorithm for the stratigkraphic correlation of well logs. J. Math. Geol., 14: 271-285.
Kemple, W.G., Sadler, P.M. and Strauss, 1990. A prototype constrained optimization solution to the time correlation problem. In: F.P. Agterberg and G.F. Bonham-Carter (Editors), Statistical Applications in the Earth Sciences, Geol. Surv. Can. Paper 89-9.
Kendall, M.G., 1975a. Rank Correlation Methods. Griffin, London, 202 pp.
Kendall, M.G., 1975b. Multivariate Analysis. Hafner, New York, N.Y., 210 pp.
Kendall, M.G. and Stuart, A., 1961. The Advanced Theory of Statistics, Volume 2. Hafner, New York, 676 pp.
Kent, D.V. and Gradstein, F.M., 1985. A Jurassic and Cretaceous geochronology. Geol. Soc. America Bull., 96: 1419-1427.
Kent, D.V., and Gradstein, F.M., 1986. A Jurassic to Recent Chronology. In: P.R. Vogt and B.E. Tucholke (Editors), The western Atlantic region, Vol. M, The Geology of North America, Geol. Soc. Am., pp. 45-50.
King, C., 1983. Cainozoic micropaleontological biostratigraphy of the North Sea. Rept. Inst. Geol. Sciences No. 82/7, 40 pp.
Kwon, B.D. and Rudman, A.J., 1979. Correlation of geologic logs with spectral methods. Math. Geology, 11: 373-390.
Lapin, L.L., 1982. Statistics for Modern Business Decisions, 3rd Edition. Harcourt, Brace, and Jovanovich, Inc., New York, N.Y., 887 pp.
Lerche, I., 1990. Philosophies and strategies of model building. In: T.A. Cross (Editor), Quantitative Dynamic Stratigraphy, Prentice Hall, Englewood Cliffs, New Jersey, pp. 21-44.
McKenzie, R.M., 1981. The Hibernia - a classic structure. Oil and Gas J., September, 1981, pp. 243-247.
McKerrow, W.S., Lambert, R.St.J. and Chamberlain, V.E., 1980. The Ordovician, Silurian and Devonian time scales. Earth Plan. Sc. L., 51: 1-8.
McLaren, D.J., 1978. Dating and correlation, a review. In: G.V. Cohee, and others (Editors), Contributions to the geologic time scale. American Ass. Petroleum Geologists, Studies in Geology 6, pp. 1-7.
Macellari, C.E., 1986. Late Campanian-Maastrichtian ammonite fauna from Seymour Island (Antarctic Peninsula). J. Paleont., 60, II: 1-55.
Magara, K., 1976. Thickness of removed sedimentary rocks, paleopressure, and paleotemperature, southwestern part of western Canada Basin. Am. Assoc. Petroleum Geologists Bull., 60: 554-565.
Maher, L.J., 1972. Nomograms for computing 0.95 confidence limits of pollen data. Rev. Palaeobotany Palynology, 23: 85-93.
Maher, L.J., 1981. Statistics for microfossil concentration measurements employing samples spiked with marker grains. Rev. Palaeobotany Palynology, 32: 153-191.
Mann, C.J. and Dowell, T.P.L., Jr., 1979. Quantitative lithostratigraphic correlation of subsurface sequences. Computers and Geosciences, 4: 295-306.
Menning, M., 1989. A synopsis of numerical time scales, 1917-1986. Episodes, 12(1): 3-5.
Millendorf, S.A., Brower, J.C. and Dyman, T.S., 1978. A comparison of methods for the quantification of assemblage zones. Computers and Geosciences, 4: 229-242.
Miller, F.X., 1977. The graphic correlation method in biostratigraphy. In: E.G. Kauffman and J.E. Hazel (Editors), Concepts and methods of biostratigraphy, Dowden, Hutchison and Ross, Inc., Stroundsburg, USA, pp. 165-186.
Miller, K.G. and Fairbanks, R.G., 1985. Cainozoic $\delta^{18}O$ record of climate and sealevel. S. Afr. J. Sci., 81: 248-249.
Miller, R.G., 1974. The Jackknife – a review. Biometrika, 61: 1-17.

Mohan, M., 1985. Geohistory analysis of Bombay High region. Marine and Petroleum Geology, 2: 350-360.
Mosteller, F., 1951. Remarks on the method of paired comparisons, I, The least squares solution assuming equal standard deviations and equal correlations. Psychometrika, 16: 3-9.
Mouterde, R., Ruget, C. and Tintant, H., 1973. Le passage Oxfordien - Kimmeridgien au Portugal (régions de Torres-Vedras et du Montejunto). Com. Ren. Acad. Sc. Paris, 277 (Sér. D): 2645-2648.
Muller, C. and Willems, W., 1981. Nannoplankton en planktonische foraminiferen uit de Ieper-Formatie (Onder-Eoceen) in Vlaanderen (Belgie). Natuurw. Tijdschr., 62: 64-71.
Nazli, K., 1988. Geostatistical modelling of microfossil abundance data in upper Jurassic shale, Tojeira sections, central Portugal. Unpublished M.Sc. thesis, Univ. Ottawa, 369 pp.
Nowlan, G.S., 1986. Paleontology: ancient and modern. Geoscience Canada, 13 (2): 67-72.
Odin, G.S. (Editor), 1982. Numerical Dating in Stratigraphy, Parts I and II. Wiley- Interscience, Chichester, 1040 pp.
Olea, R.A., 1988. Correlator – an interactive computer system for lithostratigraphic correlation of wireline logs. Kansas Geol. Survey, Lawrence, Kansas, Petrophysical Ser. 4, 85 pp.
Oleynikov, N.A. and Rubel, M. (Editors), 1988. Quantitative Stratigraphy - Retrospective Evaluation and Future Development. Institute of Geology, Acad. Sc. Estonian SSR, Tallinn, U.S.S.R., 167 pp.
Palmer, A.R., 1954. The faunas of the Riley formation in central Texas. J. Paleont., 28: 709-786.
Postuma, J.A., 1971. Manual of Planktonic Foraminifera. Elsevier, Amsterdam, 420 pp.
Quenouille, M., 1949. Approximate tests of correlation in time series. J. Royal Statist. Soc. Ser. B., 11: 18-84.
Rao, C.R., 1973. Linear Statistical Inference and its Applications. Wiley, New York, N.Y., 625 p.
Reinsch, C.H., 1967. Smoothing by spline functions. Numerische Mathematik, 10: 177-183.
Reinsch, C.H., 1971. Smoothing by spline functions. II. Numerische Mathematik, 16: 451-454.
Reyment, R.A., 1980. Morphometrical Methods in Biostratigraphy Academic Press, London, 175 pp.
Reyment, R. and Sturesson, U., 1987. Correlation of chemical and physical environmental fluctuations in a late Cretaceous borehole sequence - A multivariate study. Sed. Geol. 53: 311-325.
Riedel, W.R., 1979. Recent and potential advances in DSDP biostratigraphy. Am. Ass. Petr. Geol. Bull., 63: 516.
Roberts, F., 1976. Discrete Mathematical Models. Prentice-Hall, Englewood Cliffs, N.J., 559 p.
Roberts, F., 1978. Graph Theory and its Applications to Problems of Society. Regional Conference Series in Applied Mathematics 29, SIAM, Philadelphia, Penn., 122 pp.
Royden, L., Sclater, J.G. and Von Herzen, R.P., 1980. Continental margin subsidence and heat flow: Important parameters in formation of petroleum hydrocarbons. Bull. Am. Assoc. Petr. Geol., 64: 173-187.
Russell, D.A., 1975. Reptilian diversity and the Cretaceous-Tertiary transition in North America. Geol. Ass. Can. Spec. Paper 13: 119-136.
Russell, D.A., 1977. The biotic crisis at the end of the Cretaceous period. National Museums of Canada, Syllogeus, no. 12, pp. 11-23.
Rubel, M., 1978. Principles of construction and use of biostratigraphical scales for correlation. Computers and Geosciences, 4: 243-246.
Rubel, M. and Pak, D.N., 1984. Theory of stratigraphic correlation by means of ordinal scales. Computers and Geosciences, 10: 97-105.
Salin, Yu. S., 1989. Computerized stratigraphic correlation by means of a geochronological scale. In: A. Oleynikov and M. Rubel (Editors), Quantitative Stratigraphy-Retrospective Evaluation and Future Development, Acad. Sciences Estonian S.S.R., Institute of Geology, Tallinn, pp. 73-80.
Sankoff, D. and Kruskal, J.B. (Editors), 1983. Time Warps, String Edits, and Macromolecules: The Theory and Practice of Sequence Comparison. Addison Wesley, London, 382 p.
Schindewolf, O.H., 1950. Grundlagen und Methoden der paläontologischen Chronologie, 3rd Ed. Borntraeger, Berlin, 152 pp.
Schlumberger, 1979. Log Interpretation Charts. Schlumberger Ltd., New York, 92 p.
Schoenberg, I.J., 1964. Spline functions and the problem of graduation. Proc. National Academy of Sciences of the U.S.A., 52: 947-950.
Schwarzacher, W., 1985a. Principles of quantitative lithostratigraphy – the treatment of single sections. In: Quantitative Stratigraphy, UNESCO, Paris and Reidel, Dordrecht, pp. 361-386.
Schwarzacher, W., 1985b. Lithostratigraphic correlation and sedimentation models. In: F.M. Gradstein et al., Quantitative Stratigraphy, UNESCO, Paris, and Reidel, Dordrecht, pp. 387-418.
Sclater, J.C., and Christie, P.A.F., 1980. Continental stretching: an explanation of the post mid-Cretaceous subsidence of the central North Sea basin. J. Geophys. Res., 85: 371-379.

Shaw, A.B., 1964. Time in Stratigraphy. McGraw-Hill, New York, 365 pp.
Shaw, B.R., 1978. Parametric interpolation of digitized log segments. Computers and Geosciences, 4: 277-283.
Signor, P.W. and Lipps, J.H., 1982. Sampling bias, gradual extinction patterns and catastrophes in the fossil record. In: L.T. Silver and P.H. Schulz (Editors), Geological Implications of Impacts of Large Asteroids and Comets on the Earth. Geol. Soc. Am., Special Pap. 190, pp. 291-296.
Silverman, B.W., 1984. A fast and efficient cross-validation method for smoothing parameter choice in spline regression. J. American Statistical Ass., 79: 584-589.
Smith, D.G. and Fewtrell, M.D., 1979. A use of network diagrams in depicting stratigraphic time correlation. Geol. Soc. London J., 136: 21-28.
Smith, T.F. and Waterman, M.S., 1980 New stratigraphic correlation techniques. J. Geol. 88: 451-457.
Southam, J.R., Hay, W.W. and Worsley, T.R., 1975. Quantitative formulation of reliability in stratigraphic correlation. Science, 188: 357-359.
Springer, M. and Lilje, A., 1988. Biostratigraphy and gap analysis: the expected sequence of biostratigraphic events. J. Geol., 96: 228-236.
Srivastava, S.P. (Editor), 1986. Geophysical maps and geological sections of the Labrador Sea. Geol. Survey Canada, Paper 85-16, 11 pp.
Stainforth, R.M., Lamb, J.L., Luterbacher, H., Beard, J.H. and Jeffords, R.M., 1975. Cenozoic planktonic foraminifera zonation and characteristics of index forms. Univ. Kansas Paleont. Contr., no. 62, pp. 1-162.
Stam, B., Gradstein, F.M., Lloyd, P. and Gillis, D., 1987. Algorithms for porosity and subsidence history. Computers and Geosciences, 13 (2).
Stam, B., 1987. Quantitative Analysis of Middle and Late Jurassic Foraminifera from Portugal and its Implications for the Grand Banks of Newfoundland. Utrecht Micropaleontological Bull. 34, 167 pp.
Strauss, D. and Sadler, P.M., 1989. Classical confidence intervals and Bayesian probability estimation for ends and local taxon ranges. Math. Geol., 21: 411-427.
Sullivan, F.R., 1965. Lower Tertiary nannoplankton from the California Coast Ranges; II. Eocene. Univ. Calif. Publ. Geol. Sc., 53: 1-52.
Thomas, F.C., Gradstein, F.M. and Griffiths, C.M., 1988. Bibliography and Index of Quantitative Biostratigraphy. Special Publ. No. 1, Comm. Quantitative Stratigraphy, Bedford Inst. Oceanogr., Dartmouth, N.S., Canada, 58 pp.
Tipper, J.C., 1988. Techniques for quantitative stratigraphic correlation: a review and annotated bibliography. Geol. Mag., 125 (5): 475-494.
Tjalsma, R.C. and Lohmann, G.P., 1983. Paleocene - Eocene bathyal and abyssal benthic Foraminifera from the Atlantic Ocean. Micropal., Spec. Publ., no. 4, 76 pp.
Tocher, K.D., 1950. Extension of the Newman-Pearson theory of tests to discontinuous variates. Biometrika, 37: 130.
Tukey, J., 1958. Bias and confidence in not quite large samples. Annals Math. Statist., 29: 614.
Tukey, J.W., 1977. Exploratory Data Analysis. Addison-Wesley, Reading, Massachusetts, 688 pp.
Utreras, F., 1981. Optimal smoothing of noisy data using spline functions. SIAM J. Stat. Comput., 2: 349-362.
Vail, P.R. and Mitchum, R.M., Jr., 1979. Global cycles of relative changes of sea-level from seismic stratigraphy. Am. Ass. Petr. Geol. Mem. 29: 469-472.
Vail, P.R., Mitchum, R.M., Jr. and Thompson, S., III, 1977. Seismic stratigraphy and global changes of sealevel. Part 4. Mem. Am. Assoc. Pet. Geol. 26: 83-97.
Van Hinte, J.E., 1978. Geohistory analysis, application of micropaleontology in exploration geology. Am. Assoc. Petrol. Geol. Bull., 62: 201-227.
Van Hinte, J.E., 1984. Synthetic seismic sections from biostratigraphy. Am. Ass. Petr. Geol. Mem. 34: 674-685.
Van Valen, L. and Sloan, R.E., 1977. Ecology and the extinction of the dinosaurs. Evolutionary Theory, 2: 37-64.
Vrbik, J., 1985. Statistical properties of the number of runs of matches between two random stratigraphic sections: Mathematical Geology, 17: 29-40.
Watts, A.B., and Steckler, M.S., 1981. Subsidence and tectonics of Atlantic-type continental margins. Oceanologica Acta, vol. 4, suppl. 1981, no. SP, pp. 143-153.
Wahba, G., 1975. Smoothing noisy data with spline functions. Numerische Mathematik, 24: 383-393.
Waterman, M.S. and Raymond, R., Jr., 1987. The match game: new stratigraphic correlation algorithms. Math. Geol. 19: 109-127.

Waterman, M.S., Smith, T.F. and Beyer, W.A., 1976. Some biological sequence metrics. Adv. Math., 20: 367-387.

Wegman, E.J. and Wright, I.W., 1983. Splines in statistics. J. American Statistical Ass., 78: 351-365.

White, J.M., 1990. Exploration of a practical technique to estimate the relative abundance of rare palynomorphs using an exotic spike. In: F.P. Agterberg and G.F. Bonham-Carter (Editors), Statistical Applications in the Earth Sciences, Geol. Surv. Can. Paper 89-9.

Whittaker, E.T., 1923. On a new method of graduation. Proc. Edinburg Math. Soc., 41: 63-75.

Wilkinson, E.M., 1974. Techniques of data analysis – seriation theory: Archaeo- Physika, 5: 1-142.

Williams, D.F., 1990. Selected approaches of chemical stratigraphy to time-scale resolution and quantitative dynamic stratigraphy. In: T. A. Cross (Editor), Quantitative Dynamic Stratigraphy, Prentice Hall, Englewood Cliffs, New Jersey, pp. 543-565.

Williams, D.F., Lerche, I. and Full, W.E., 1988. Isotope chronostratigraphy: Theory and Methods. Academic Press, San Diego, 352 pp.

Williamson, M.J., 1987. Quantitative biozonation of the Late Jurassic and Early Cretaceous of the East Newfoundland Basin. Micropaleontology, 33: 37-65.

Williamson, M.A. and Agterberg, F.P., 1990. A quantitative foraminiferal correlation of the late Jurassic and early Cretaceous offshore Newfoundland. In: F.P. Agterberg and G.F. Bonham-Carter (Editors), Statistical Applications in the Earth Sciences, Geol. Surv. Can. Paper 89-9.

Wilson, L.R., 1964. Recycling, stratigraphic leakage and faulty techniques in palynology. Grana Palynologica, 5: 427-436.

Wold, S., 1974. Spline functions in data analysis. Technometrics 16 (1): 1-11.

Wood, R.I., 1981. The subsidence history of the Conoco well 15/30-1, Central North Sea, Earth and Planetary Sci. Lett., 54: 306-312.

Worsley, T.R. and Jorgens, M.L., 1977. Automated biostratigraphy. In: A.T.S. Ramsay (Editor), Oceanic Micropaleontology, Academic Press, London, 2: 1201-1229.

Ziegler, P.A., 1981. Evolution of Sedimentary basins in Northwest Europe. In: L.V. Illing and G.D. Hobson (Editors), Petroleum Geology of the Continental Shelf of Northwest Europe. Inst. of Petroleum, London, pp. 3-39.

INDEX